INTRODUCTION TO BIOCHEMISTRY

INTRODUCTION TO BIOCHEMISTRY

SECOND EDITION

JOHN W. SUTTIE

University of Wisconsin

Holt Rinehart and Winston

New York Chicago San Francisco Atlanta Dallas Montreal Toronto London Sydney

cover design: Louis Scardino, based on work of
              M. Sundaralingam
Text design: Scott Chelius
Illustrations: Michael Zaidel

Library of Congress Cataloging in Publication Data

Suttie, John W   1934-
  Introduction to biochemistry.

  Includes bibliographies and index.
  1.   Biological chemistry.   I.   Title.   [DNLM:
1.   Biochemistry.   QU4 S967i]
QP514.2.S67 1977   574.1' 92   76-48885
ISBN 0-03-089713-0   (College Edition)
ISBN 0-03-048706-4 (International Edition)

International Edition is not for sale
in the United States of America, its
dependencies or Canada.

Printed in Great Britain by Butler & Tanner Ltd, Frome and London
Print number: 9 8 7 6 5 4 3 2 1

CONTENTS

Contents

PART TWO   DYNAMICS AND ENERGETICS OF BIOCHEMICAL SYSTEMS   150

chapter seven   structure and function of enzymes   152

chapter eight   biochemical energetics   183

Contents

Contents

PREFACE

The second edition of *Introduction to Biochemistry* has been prepared with two purposes in mind: first, to update and make changes in those areas where student feedback has indicated revision was warranted; and second, to retain the basic approach of bringing together only that amount of material which an instructor can reasonably expect to cover in a one-semester course while still giving the student an appreciation of the concepts and problems of modern biochemistry.

The one-semester, introductory biochemistry course at Wisconsin is taken by undergraduate chemistry, biology, and biochemistry majors, and by first-year biology graduate students who desire a concise one-semester course. For some, it is the only biochemistry they will be exposed to; for others, it serves as an introduction to an extensive two-semester biochemistry course sequence. The amount of material that a beginning biochemistry student is expected to assimilate and retain in such an introductory course is enormous, and my experience in teaching a course at this level has resulted in the firm conviction that the most useful text for most of these students is one that includes only that material which the student is expected to master in the course.

As in the first edition, the material is arranged around a rather classical order of presentation. There is, however, no need to cover the material in this order in a lecture series, and the text is written with the view that individual instructors will enlarge and expand on those areas that they feel are not adequately covered for their purposes. As before, the general emphasis is on mammalian metabolism as the central core of a general biochemistry course. There is a somewhat more extensive coverage of carbohydrate and lipid chemistry than is found in many elementary biochemistry texts because it has been my experience that students are not getting this material in modern organic courses. If not exposed to it at this level, they may miss this background area completely. Because many students taking a biochemistry course at this level have not taken physical chemistry, the concepts of pH and buffers which should have been covered in general chemistry are reviewed in Chapter 1, and only the most elementary enzyme kinetics and thermodynamics are covered. Most students taking a biochemistry course at this level have previously taken a modern biology or genetics course which has placed considerable emphasis on the molecular biological aspects of biochemistry. This material has therefore been covered in the text, but not as extensively as it would be in a text designed for courses which emphasize molecular biology rather than biochemistry.

The second edition includes an expanded coverage of membranes and active transport phenomena for which I express my appreciation to my colleague, David Nelson. There has been a general updating and expansion of a number of sections with an attempt to condense or delete other areas to keep the size of the book roughly the same as the first edition. There has been a complete change in format and style of illustrations which should make the text much more readable. Finally, an introductory chapter abstract, a number of problems, and a short list of suggested readings have been furnished for each chapter. The problems should provide the students with a self appraisal of their mastery of the material, and the references have been selected to give some examples of original papers as well as an access to more detailed reviews from various fields. Several reviewers have provided helpful suggestions for this edition, including Drs. John Anderson, University of Minnesota; David Ives, Ohio State University; James Thomas, Iowa State University; Wilbert Gamble, Oregon State University; Allan Bieber, Arizona State University; Edwin Liu, University of South Carolina; Jon Robertus, University of Texas, Austin; Neal Moir, California State Polytechnic University.

In a text of this size, difficult decisions of what material to include must be made, and undoubtedly others would have made somewhat different choices of material or would have emphasized different aspects. The material covered should be considered by most instructors to contain that biochemical knowledge which is essential for a student taking an introductory biochemistry course.

November, 1976   John W. Suttie
Madison, Wisconsin

INTRODUCTION
TO
BIOCHEMISTRY

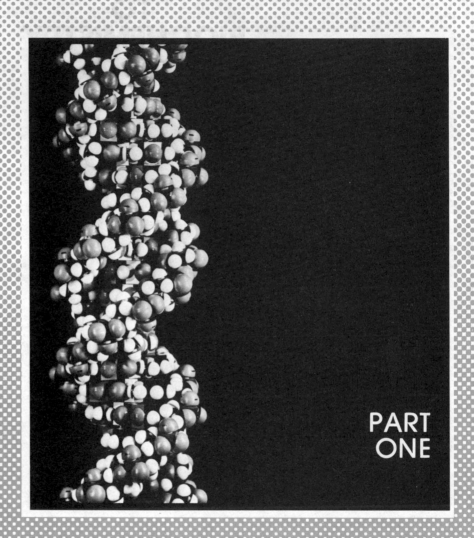

PART
ONE

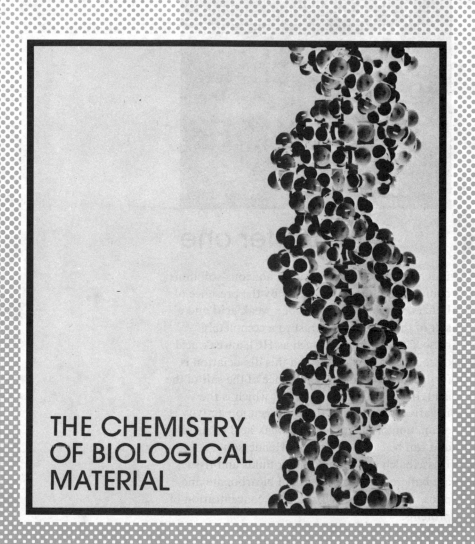

THE CHEMISTRY
OF BIOLOGICAL
MATERIAL

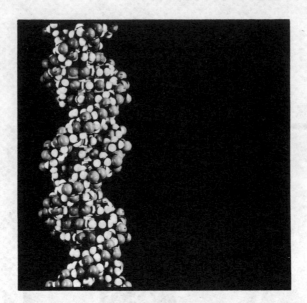

pH and buffers

chapter one

Biochemical reactions occur in aqueous solutions maintained close to neutrality by the presence of buffers, which are mixtures of a weak acid and a salt of that acid. In contrast to a completely dissociated strong acid such as HCl, a weak acid dissociates only slightly, and this dissociation is further depressed by the presence of the salt of the acid. In such a mixture, the pH, which is the negative logarithm of the hydrogen ion activity is a function of the ratio of the weak acid to its salt, and can be calculated by the Henderson-Hasselbalch equation. Cellular fluids and tissues are buffered by the presence of bicarbonate and phosphate salts as well as a high concentration of proteins.

Since biochemical reactions occur in an aqueous environment, and for the most part in an environment which is maintained close to neutrality, a thorough understanding of the properties of acids and bases and of a buffered system is essential. Because these subjects are adequately covered in general chemistry and quantitative analysis courses, only a brief review will be presented. For more help in this area, a number of paperbacks devoted exclusively to the solving of biochemical calculations are included in the references to this chapter.

IONIZATION OF ACIDS

The rather restricted definition of an *acid* as a substance that ionizes to furnish a proton, and of a *base* as a substance that ionizes to provide a hydroxyl ion, is not particularly useful in many biochemical situations. The Brönsted theory of acid dissociation is more generally applicable. According to the Brönsted definition, acids dissociate to a proton and to the conjugate base of the acid, and bases accept a proton to become an acid. In the generalized case,

$$HA \rightleftharpoons H^+ + A^-$$

The conjugate base of the acid HA is A^-. The base formed, A^-, need not carry only a single negative charge, and in some cases is a neutral molecule. The weak acid, HA, can also exist as a positively or negatively charged molecule as well as a neutral molecule (Table 1-1); it can be both an acid and a base. For example, the bicarbonate ion, HCO_3^-, is a weak acid, but it is also the conjugate base of carbonic acid, H_2CO_3.

Table 1-1 Ionization of Some Weak Acids

Weak Acid	$\rightleftharpoons$	Proton	+	Conjugate Base
CH_3COOH	$\rightleftharpoons$	H^+	+	CH_3COO^-
H_2CO_3	$\rightleftharpoons$	H^+	+	HCO_3^-
HCO_3	$\rightleftharpoons$	H^+	+	CO_3
NH_4^+	$\rightleftharpoons$	H^+	+	NH_3
$NH_3^+\text{-}CH_2COOH$	$\rightleftharpoons$	H^+	+	$NH_3^+\text{-}CH_2COO^-$
$NH_3^+\text{-}CH_2COO^-$	$\rightleftharpoons$	H^+	+	$NH_2\text{-}CH_2COO^-$

Strong acids, hydrochloric, for example, are almost completely dissociated in aqueous solution. Therefore the molar concentration of the proton formed is equal to the molar concentration of acid which was present.

$$HCl \longrightarrow H^+ + Cl^-$$

The acidity of aqueous solutions of strong acids is seldom expressed in terms of the hydrogen ion concentration, but as a logarithmic function of the concentration, that is, the pH of the solution. As originally defined by

Sörenson, the pH of a solution is equal to the negative logarithm of the

$$pH = \log \frac{1}{a_{H^+}} = -\log a_{H^+}$$

hydrogen ion activity. In dilute solutions, which are those encountered in most biochemical situations, the activity will approach the concentration. In this text no distinction will be made between the two terms and concentrations will be used in all calculations. Therefore, the expression becomes

$$pH = \log \frac{1}{[H^+]} = -\log[H^+]$$

where the use of brackets indicates molar concentrations. It must be stressed that the pH values are logarithmic functions of the hydrogen ion concentration and that a solution of pH 3 has a hydrogen ion concentration $(10^{-3}M)$ ten times that of a solution with pH 4 $(10^{-4}M)$.

MEASUREMENT OF PH

The pH of a solution can be determined in a number of ways. An estimation of the pH can be obtained by the use of indicators that change color at different pHs, but accurate measurements are made with the use of a pH meter and glass electrode. The pH meter measures the difference between the potential of the glass electrode, E_g, and the potential of a reference electrode (E_{ref}). The relationship between this difference at 25 °C is

$$pH = \frac{E_g - E_{ref}}{0.0591}$$

The value measured is that of the hydrogen ion activity, not of the hydrogen ion concentration. A discussion of the theory and use of the glass electrode is beyond the scope of this book, but a straightforward and comprehensive discussion of this subject can be found in the book by Bates which is included in the references to this chapter. It is very difficult to measure the potential difference, $E_g - E_{ref}$, accurately enough to standardize solutions. What is done in practice is to calibrate the pH meter with solutions of known pH.

IONIZATION OF WEAK ACIDS

Biochemists are more often concerned with the properties of weak acids than they are with strong acids, and when compared to the dissociation of strong acids, these compounds will show only a slight tendency to ionize. It is therefore possible to write an equilibrium constant which will quantitatively express this ionization:

$$HA \rightleftharpoons H^+ + A^- \qquad \text{therefore} \quad \frac{[H^+][A^-]}{[HA]} = K_{eq}$$

The term K_a, which describes the corresponding relationship between the activities of the various species, is a more useful expression, and the one that will be used in this text. The value for the K_a will be greater for those acids which are more highly ionized or "stronger." However, even for the relatively strong weak acids, such as acetic acid, the equilibrium is far in the direction of the undissociated acid. At 25 °C, the equilibrium expression for acetic acid is

$$\frac{[H^+][CH_3COO^-]}{[CH_3COOH]} = K_a = 1.86 \times 10^{-5}$$

If this equation is used to calculate for specific acid concentrations, it will be found that a one molar solution of acetic acid contains only about 0.5 percent of the dissociated form. As in the case of hydrogen ion concentrations, it is more convenient to express the dissociation constants as logarithmic functions, and comparisons of the strengths of various acids are usually made on the basis of their pK_a values, where

$$pK_a = -\log K_a$$

Acids with a low pK_a will therefore be stronger acids than those with a higher pK_a. If a weak acid has more than one acid dissociation, it will have a distinct pK_a associated with the loss of each proton. The best example of a multiple dissociation of importance in biological systems is probably the ionization of phosphoric acid (Table 1-2).

Table 1-2 Ionization of Phosphoric Acid

Dissociation	K_a	pK_a
$H_3PO_4 \rightleftharpoons H^+ + H_2PO_4^-$	7.5×10^{-3}	2.1
$H_2PO_4^- \rightleftharpoons H^+ + HPO_4^{2-}$	6.2×10^{-8}	7.2
$HPO_4^{2-} \rightleftharpoons H^+ + PO_4^{3-}$	2.2×10^{-13}	12.7

EFFECT OF SALTS ON WEAK ACID DISSOCIATION— BUFFERS

The addition of the soluble salt of a weak acid to an aqueous solution of that acid will increase the pH (lower the $[H^+]$) of the solution. The salt will be completely ionized and the increase in concentration of the conjugate base of the weak acid will, by the law of mass action, shift the equilibrium to the left and remove some hydrogen ions. Because of this, the pH of a solution of a weak acid and the salt of that weak acid is determined not so much by the concentration of the acid but by the ratio of the concentration of the salt (conjugate base) to the concentration of the undissociated acid. Such solutions are called buffers, and are able to resist a change in pH when a strong acid or base is added to them. The addition of a strong acid (free protons) to the excess of salt (anion) which is present will result in the formation of more of the weak acid. Because it dissociates only slightly, this increase in the concentration of the weak acid will have only a small effect on the pH. Conversely, when a strong base is added, it is neutralized by the weak acid and the major effect is an increase in salt concentration. The ionization of the remaining weak acid is suppressed by the increased salt concentration, and there is a slight increase in pH (Figure 1-1).

The effectiveness or *capacity* of such a buffer will depend on the total concentration of the salt and the acid, and also on the ratio of the two. If the molar concentrations of salt and acid are equal, a change in concentration of one or the other will have the least possible influence on the ratio.

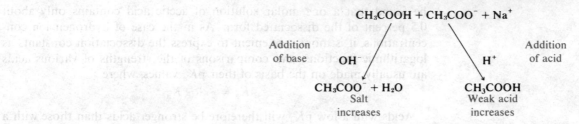

Figure 1-1   Addition of a strong acid or base to a buffer solution. The total concentration of the weak acid plus its salt does not change, but the ratio of the two can vary widely as the two forms are interconverted in acidic or basic solutions.

THE HENDERSON-HASSELBALCH EQUATION

The pH of buffers can be calculated by a rearranged form of the mass law called the Henderson-Hasselbalch equation.

In the buffer solution two dissociations are important

$$HA \rightleftharpoons H^+ + A^- \quad \text{(weakly dissociated)}$$
$$BA \rightleftharpoons B^+ + A^- \quad \text{(highly dissociated)}$$

the K_a was previously defined as

$$\frac{[H^+][A^-]}{[HA]} = K_a$$

therefore

$$[H^+] = K_a \frac{[HA]}{[A^-]}$$

If the logarithm of both sides of this equation is taken

$$\log[H^+] = \log K_a + \log \frac{[HA]}{[A^-]}$$

6

and the equation is multiplied by -1

$$-\log[H^+] = -\log K_a - \log\frac{[HA]}{[A^-]}$$

Therefore, by previous definitions

$$pH = pK_a + \log\frac{[A^-]}{[HA]}$$

In the use of this equation for routine calculations some further simplifications are usually made. Since the molar concentration of A^- contributed by the dissociation of HA is so small, it is usually ignored; A^- then equals the amount of the salt, BA, added. Also, because the amount of HA dissociated is small, its concentration is taken as the amount added. The expression then becomes

$$pH = pK_a + \log\frac{[BA]}{[HA]}$$

not

$$pH = pK_a + \log\frac{[BA + A^-]}{[HA - A^-]}$$

and is often simply written as

$$pH = pK_a + \log\frac{[salt]}{[acid]}$$

As an example, if a solution contains $0.05M$ sodium acetate and $0.1M$ acetic acid ($pK_a = 4.73$), the expression becomes

$$pH = 4.73 + \log\frac{0.05}{0.1} = 4.73 + \log 0.5$$
$$= 4.73 - 0.30 = 4.43$$

It should be clear that if the salt and acid concentrations are equal, the $\log[salt]/[acid]$ term becomes equal to 0, and the $pH = pK_a$. This situation is present, of course, if a weak acid is 50 percent neutralized with a strong base, and one useful definition of the pK_a is simply: that pH at which the acid is half neutralized. An examination of the Henderson-Hasselbalch equation will make it clear that if the pH of a solution is one pH unit above the pK_a of the weak acid involved, the ratio of salt to acid must be 10:1, and at two pH units above, the ratio will be 100:1. If it is not apparent that these high ratios of salt to acid will decrease the buffer capacity of the system, consult a text on biochemical calculations.

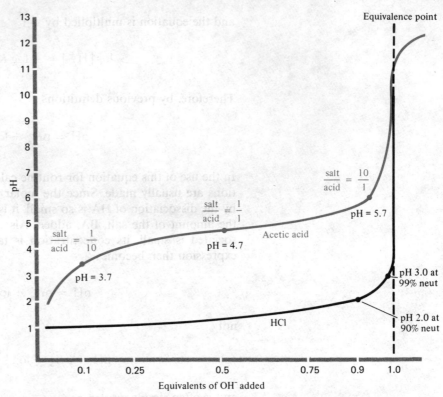

Figure 1-2   Titration curve of a 10-ml solution of 0.1 N HCl or 0.1 N acetic acid (pK_a 4.7) with a strong base. The curve is an idealized one which assumes that there has been no change in the volume of the solution being titrated, and the curve has been corrected for the amount of base needed to change the pH of the water alone.

TITRATION CURVES

A plot of the pH of a solution of an acid or a base against some measure of the degree of its neutralization is called a *titration curve*. Such a curve, comparing the titration of HCl and CH_3COOH by a strong base, is shown in Figure 1-2. The curve is an idealized one which ignores the fact that the addition of the base would cause some change in the volume of the solution and therefore in the concentration of the acid during the titration. It also assumes that a correction has been made for the amount of base needed to change the pH of the water alone. The change in pH as HCl is neutralized is simply due to the change in its concentration as some of it is converted to the corresponding salt. The curve goes up very steeply at the end because the amount of HCl left is small, and a small addition of base makes a relatively large change in concentration. The acetic acid curve shows a region of rapid increase in pH followed by a region, centered around the pK_a, where the system is rather well buffered (a large addition of base produces a small pH change). Beginning at about one pH unit above the pK_a, there is a loss of buffering action and a rapid increase in pH. The pH at the equivalence point, when all the acid is neutralized, will be above pH 7, because of the basic hydrolysis of the solution of sodium acetate which will be present at that point.

A compound like NH_3 in solution can be considered as a base (NH_4OII) and the titration curve obtained by the addition of acid to this base can be examined. It is more common, however, for biochemists to think of compounds containing amino groups in terms of their acid form, NH_4^+, than of the conjugate base, and to express their buffering ability as a titration curve of the acid form. These two possible titration curves for a compound such as ammonia are shown in Figure 1-3.

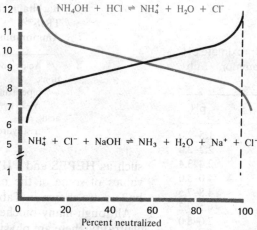

Figure 1-3   Titration of NH_3 with a strong acid and of NH_4^+ with a strong base. Both curves are correct representations of the buffering action of ammonium salts, but by convention, most biochemists tend to think in terms of the titration of the acid form, NH_4^+.

BUFFERS USED IN CHEMICAL REACTIONS

For a weak acid to be used as a buffer in an *in vitro* biochemical reaction, it should, of course, have a pK_a close to that of the desired pH to maximize its buffering capacity, and should meet a number of other criteria. It should be nontoxic to the biochemical reaction being studied, should be colorless, and should be free of ultraviolet absorption in the region where proteins and nucleic acids absorb. A good buffer is also one in which changes in temperature, concentration, or ionic strength have a minimal effect on dissociation. The second dissociation of phosphoric acid is in a good pH range, but phosphate solutions are easily precipitated by divalent cations, and phosphate is often a product or reactant in biochemical reactions. Tris [tris(hydroxymethyl)amino methane] is often used above pH 7, but its dissociation is rather temperature dependent, and other complex amines

Table 1-3 pK_a Values of Some Acids Important to Biochemists

Acid	pK_{a_1}	pK_{a_2}	pK_{a_3}
Phosphoric acid[a]	2.1	7.2	12.7
Citric acid	3.1	4.8	5.4
Formic acid	3.8		
Lactic acid	3.9		
Acetic acid	4.7		
Carbonic acid[a]	6.1	10.4	
PIPES[ab]	6.7		
Imidazole[a]	7.0		
HEPES[ac]	7.3		
Barbital (veronal)[a]	8.0		
TRIS[ad]	8.1		
Ammonium ion	9.3		

[a] Acids commonly used to buffer biochemical reactions.

[b] piperazine-N,N'-bis(2-ethanesulfonic acid).

[c] N-2-hydroxyethylpiperazine-N'-2-ethanesulfonic acid.

[d] tris(hydroxymethyl)amino methane.

Table 1-4 Approximate pH Values of Common Substances

Substance	pH
Gastric juice	1.5–2.0
Lemons	2.2–2.4
Vinegar	2.4–3.4
Beer	4.0–5.0
Urine	4.8–7.5
Milk	6.5–7.2
Intestinal juice	7.0–8.0
Blood plasma	7.3–7.5
Eggs	7.6–8.0

such as HEPES and PIPES (see Table 1-3) are now often used. The pK_a values of some of the more commonly used weak acids and biological buffers are shown in Table 1-3.

Although many biochemical reactions do normally occur at a pH near neutrality, there are physiological solutions that have pH values drastically removed from pH 7. This is illustrated by the range of pH values for some of the more common biological fluids shown in Table 1-4.

BUFFER SYSTEMS IN THE ANIMAL BODY

The pH of the blood of higher animals is maintained in a very narrow range around pH 7.4 in spite of the continual production of CO_2 by respiration. Production of CO_2 may amount to 10–20 moles per day in an adult human and is accompanied by the production of up to 0.1 mole per day of sulfuric, lactic, and β-hydroxybutyric acid. The CO_2 formed in cells by metabolic reactions is enzymatically hydrated to form H_2CO_3, which then dissociates to a proton and bicarbonate anion. This increase in $[H^+]$ is buffered chiefly by ionizable groups of proteins, mainly erythrocyte hemoglobin. After it is transported to the lungs, hemoglobin is converted to oxyhemoglobin. Oxyhemoglobin is a stronger acid than hemoglobin, so that protons which were picked up in the cells are dissociated from the protein. The increase in $[H^+]$ then shifts the bicarbonate equilibrium back to H_2CO_3 and eventually to CO_2, which is expired. The small amount of nonvolatile acids produced are effectively buffered by bicarbonate and phosphate salts and to some extent by the plasma proteins. Even though the pK_a of the carbonic acid-bicarbonate dissociation is only 6.1, its high concentration in the plasma makes it a rather effective buffer at pH 7.4. This is also a nontypical buffer system, because the ratio of salt to acid in plasma is regulated not only by the pH of the system but also by the CO_2 tension in the lungs.

PROBLEMS

1. What is the pH of a solution with a hydrogen ion concentration (activity) of (a) $5 \times 10^{-6}M$, (b) $6.5 \times 10^{-10}M$, (c) $1.5 \times 10^{-5}M$?

2. What is the hydrogen ion concentration of solutions with pH of (a) 3.82, (b) 6.53, (c) 11.11?

3. Calculate the pH of a dilute solution which contains the following molar ratios of K acetate: acetic acid ($pK = 4.70$): (a) $2:1$, (b) $1:3$, (c) $1:1$, (d) $1:10$.

4. How much K_2HPO_4 must be added to a solution containing 0.02 moles of KH_2PO_4 ($pK_a = 7.3$) to bring the pH to (a) 6.0, (b) 7.0?

5. What would be the pH of a solution obtained by mixing equal amounts of a solution containing 0.24 eq/l of NH_4Cl ($pK = 9.30$) and a solution containing 0.06 eq/l of NaOH?

6. How much 0.1 N NaOH would be required to bring the pH of 100 ml of 0.01 N acetic acid ($pK = 4.7$) to 4.1?

7. How much 0.02 N NaOH would be required to bring the pH of 500 ml of $0.01M$ H_3PO_4 ($pK_{a_1} = 2.1$, $pK_{a_2} = 7.3$, $pK_{a_3} = 12.6$) to 7.3?

8. How would you prepare a 30 mM, pH 7.8, Tris $\cdot$ Cl buffer?

Suggested Readings

Books

R. G. Bates, *Determination of pH Theory and Practice.* Wiley, New York (1964).

Finlayson, J. S. *Basic Biochemical Calculations.* Addison-Wesley, Reading, Mass. (1969).

Montgomery, R., and C. A. Swenson. *Quantitative Problems in Biochemical Sciences,* 2nd ed. Freeman, San Francisco (1976).

Segel, H. *Biochemical Calculations,* 2nd ed. Wiley, New York (1976).

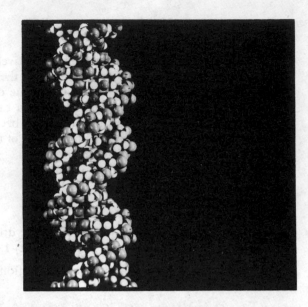

cellular composition and morphology

chapter two

All living material is composed largely of water with varying amounts of protein, lipid, carbohydrate, and ash. A metabolically active animal cell such as a liver cell contains about 20 percent protein, 5 percent lipid, 1 percent ash, and only a trace of carbohydrate. In general, plant cells contain much more carbohydrate than animal cells. Eucaryotic (nucleated) cells contain a number of subcellular organelles: a nucleus, mitochondria, lysosomes, and a series of specialized membranes called the endoplasmic reticulum. In addition to these basic organelles, cells which have specialized functions contain additional organelles.

The subsequent chapters of Part 1 will deal in some detail with the chemistry of various classes of biological compounds. Before beginning these discussions of the chemistry of the compounds involved, however, it is necessary to have a general understanding of the gross chemical composition of living material.

CHEMICAL COMPOSITION OF LIVING MATERIAL

Data on the chemical composition of biological material have been available for many years based on a concept called a proximate analysis. Such an analysis indicates nothing about the content of individual compounds, but expresses the composition of the material in terms of some very crude chemical classifications:

Water: what is removed by drying at 100–105°
Ash: what remains after high-temperature oxidation
Crude protein: Kjeldahl nitrogen times a conversion factor of 6.25
Fat: what can be extracted from the dry matter by ether
Carbohydrate: what is left

The Kjeldahl procedure for nitrogen is one in which almost all chemical forms of nitrogen in a sample are converted to $(NH_4)_2SO_4$ by H_2SO_4 digestion, and in which the amount of NH_3 is subsequently determined. Because most proteins are about 16 percent nitrogen, a conversion factor of 6.25 converts the amount of nitrogen which was determined to a protein basis. In this procedure most of the nucleic acid nitrogen in a sample would be digested and determined as protein. The amount of carbohydrate is determined by subtracting everything else from 100 percent, and it is sometimes further divided into a crude fiber and a soluble carbohydrate portion.

When the results of such an analysis on a number of organisms are compared in Table 2-1, some generalizations are apparent. It is clear that water is the major constituent of all living tissue, and that plants can, in general, be distinguished from animals by their low-protein and high carbohydrate content. The amount of mineral matter is higher in those organisms with a calcified skeleton. The mineral content of plant material is very low, because it depends on a rigid carbohydrate-rich cell wall for structural support.

Further relationships between chemical composition and function can be seen in the data in Table 2-2. Except for skeletal tissue itself, animal cells have no more mineral matter than any other type of cell. The amount of lipid in a tissue is to some extent inversely related to its metabolic activity, and a high lipid content is usually achieved at the expense of water in the cell. It is interesting that blood, when considered as a tissue, does not have a great deal more water than other tissues and has as much protein as other tissues. The majority of the protein content of whole blood is due to the presence of one protein, hemoglobin, which is the oxygen-carrying pigment in the red blood cells. None of the organs or tissues in Table 2-2 contains

Table 2-1 Gross Composition of Organisms

			Proximate Analysis (average values in percent)			
Type	Example	H₂O	Protein (N × 6.25)	Carbohydrate	Lipid	Ash (Mineral)
Plants	Cabbage	92	1.4	6.3	0.2	0.8
	Corn	76	2.0	20	0.7	1.3
	Spinach	93	2.3	3.8	0.3	0.6
Micro-organisms	*E. coli*	78	18.0	1.0	1.0	2.0
	Yeast	72	12.0	13.0	1.0	2.0
Invertebrates	Fly	73	20.0	3.0	3.0	1.0
	Scallop	80	15.0	3.4	0.1	1.4
Vertebrates	Halibut	75	18.0	Trace	5.2	1.3
	Hen	56	21.0	Trace	19.0	3.2
Mammals	Hog	58	15.0	Trace	24.0	2.8
	Man	59	18.0	Trace	18.0	4.0

more than a small amount of carbohydrate, with the highest concentration probably being present in skeletal tissue and liver. As mentioned previously, in this crude type of analysis nucleic acids would be included in the crude protein fraction. The total nucleic acid content of cells of higher organisms, other than sperm cells, would usually be considerably less than 1 percent. In rapidly growing microorganisms, however, the amount of nucleic acid might be as much as 20 percent of the amount of protein.

Table 2-2 Chemical Composition of the Adult Human Body [a]

		Percent of Wet Weight			
Organ or Tissue	Percent of Body	H₂O	Lipid	Crude Protein	Ash
Muscle	40	70	7	22	1.0
Skeleton	18	28	25	20	26.0
Adipose tissue	11	23	72	6	0.2
Blood	8	79	<1	20	0.2
Skin	6	57	14	27	0.6
Nervous tissue	3	75	12	12	0.1
Liver	2.5	71	3	22	1.4
Heart	0.5	63	16	17	0.6

[a] Only a trace of carbohydrate would be found in these tissues.

ELEMENTAL COMPOSITION

Other than those nonmineral elements found in the organic constituents of the cell, only a few elements are found in high concentrations in biological material. The data in Table 2-3 show that the majority of the ash fraction

Table 2-3 Approximate Elemental Composition of the Human Body (wet weight basis)

Class	Element	Percentage
Nonmineral elements	Oxygen	65
	Carbon	18
	Hydrogen	10
	Nitrogen	3.0
Major mineral nutrients	Calcium	1.5
	Phosphorus	1.0
	Sulfur	0.25
	Potassium	0.2
	Sodium	0.15
	Chlorine	0.15
	Magnesium	0.05
Trace mineral nutrients[a]	Iron	0.006
	Zinc	0.003
	Copper	0.0001

[a] These elements in order of their abundance are also considered trace essentials in human nutrition, and are present at less than 0.0001 percent of the wet weight: I, Mn, Cr, Mo, Co, and Se.

Table 2-4 Approximate Ionic Composition of Body Fluids

Ion	meq/Liter of Fluid	
	Plasma	Muscle
Cations		
Na^+	142	10
K^+	5	148
Ca^{2+}	5	2
Mg^{2+}	3	40
Total	155	200
Anions		
Cl^-	103	Trace
HCO_3^-	27	8
HPO_4^{2-} and organic acids	8	116
SO_4^{2-}	1	20
Protein	16	56
Total	155	200

of an animal body is made up of calcium and phosphorus. Other than these elements, sulfur, which is a constituent of proteins, and the elements that furnish the main electrolytes for the cellular and extracellular fluid compartments are present in the highest concentration. The major cations in body fluids are Na^+ in the extracellular fluid, and K^+ and to some extent Mg^{2+} in the cells (Table 2-4). In the extracellular fluids, Cl^- is the major anion with some contribution from bicarbonate and negatively charged proteins. In the intracellular water the major anion is phosphate with a small contribution by organic acids and a more substantial contribution from charged protein molecules.

The chemical analysis of an organism or cell says nothing about the morphology or function of that cell. Since many biochemical events are localized in certain morphologically distinct areas of the cell, a brief description of cellular structure is necessary for an understanding of metabolism.

CELL MORPHOLOGY

There is no typical or universal cell. The diagram in Figure 2-1 indicates those features which are commonly found in nucleated (eucaryotic) cells, and the electron micrograph accompanying it is of a rat liver parenchymal cell. The majority of the metabolic reactions to be discussed in this text are those which are common to a large number of cells of higher organisms, and the rat liver cell might well be taken as a general example.

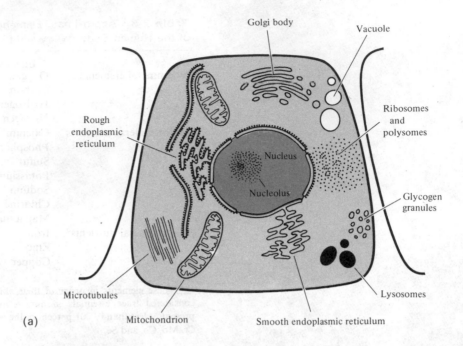

(a)

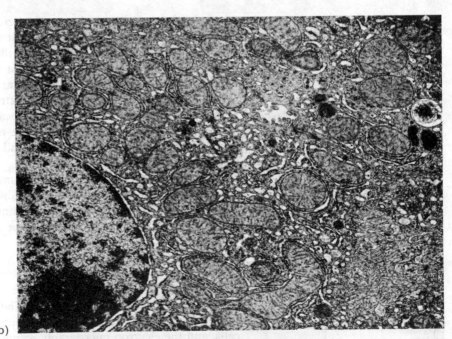

(b)

Figure 2-1 (*a*) Diagrammatic illustration of a generalized animal cell; (*b*) Electron micrograph of a liver cell from a rat (×18,800). (Courtesy of M. Barnhart, Wayne State University)

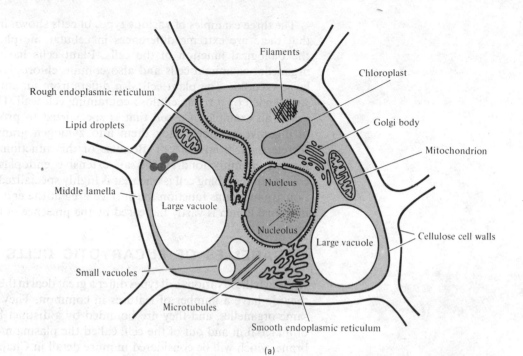

Filaments

Chloroplast

Rough endoplasmic reticulum

Golgi body

Lipid droplets

Mitochondrion

Nucleus

Middle lamella

Large vacuole

Nucleolus

Large vacuole

Cellulose cell walls

Small vacuoles

Microtubules

Smooth endoplasmic reticulum

(a)

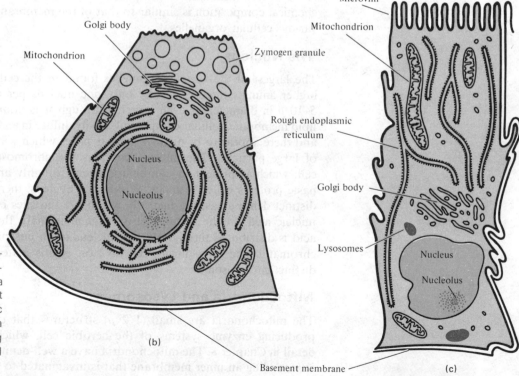

Microvilli

Golgi body

Mitochondrion

Mitochondrion

Zymogen granule

Rough endoplasmic reticulum

Nucleus

Nucleolus

Golgi body

Lysosomes

Nucleus

Figure 2-2 Diagrammatic illustrations of: (*a*) a generalized plant cell; (*b*) a pancreatic acinar cell; (*c*) a columnar absorbing cell of the intestinal mucosa.

Nucleolus

(b)

Basement membrane

(c)

17

The three examples of various types of cells shown in Figure 2-2 indicate that there are extreme differences in cellular morphology depending on the biological function of the cells. Plant cells have many of the same organelles as animal cells and also contain chloroplasts and a number of large vacuoles. The plant cell also differs from an animal cell in that it is surrounded by a rigid cellulose-containing cell wall. The pancreatic acinar cell is an example of a cell that is specialized to produce large amounts of digestive enzymes, pack them into zymogen granules, and eventually secrete them from the cell. Because of this function, it contains a large number of mitochondria and an extensive endoplasmic reticulum. The columnar absorbing cell is one that is highly specialized for efficient absorption. To meet this function the surface area at the end of the cell facing the intestinal lumen is vastly increased by the presence of the microvilli.

ORGANELLES OF EUCARYOTIC CELLS

Although these various cell types differ a great deal in their basic morphology, they do have a number of features in common. They contain most of the same organelles, and they are bounded by a distinct barrier to movement of material in and out of the cell, called the plasma membrane. This membrane, which will be considered in more detail in Chapter 5, is on the order of 80–100 Å thick and is composed of protein and lipid complexes. Its chemical composition is similar to that of the membranes surrounding the various cellular organelles.

The Nucleus

The largest of the various organelles found in the cell is the nucleus. In higher animals there is usually only one nucleus per cell and it may be 5–10 μ in diameter or even larger. Although it is bounded by a protein-lipid membrane, cellular components easily diffuse in and out of the nucleus, and there appear to be distinct nuclear pores which will allow the passage of large particles. The nucleus contains the chromosomal DNA of the cell, which is present in combination with roughly an equal amount of basic proteins called *histones*. Within the nucleus, there are one or more distinct dense areas, the nucleoli, which are the sites of the nuclear ribonucleic acid. In the resting cell of higher animals, the deoxyribonucleic acid is distributed throughout the nucleus in an unstructured form called chromatin, and only during cell division is this material organized into distinct chromosomes.

Mitochondria and Lysosomes

The mitochondria are small (1–2 μ) structures that contain the energy-producing enzyme systems of the aerobic cell, which are described in detail in Chapter 8. The mitochondria have a well-defined outer membrane surrounding an inner membrane that is invaginated to form vesicles called *cristae.* The material inside the inner membrane, the matrix, contains many of the enzymes needed to degrade metabolizable substrates to CO_2. To a large extent, the distribution of mitochondria within cells is a function of the cellular requirement for energy. Cells producing a large amount of protein for export from the cell or having increased energy requirements will contain a large number of mitochondria with closely packed cristae.

Lysosomes are membrane-enclosed cellular organelles similar in size to mitochondria, but without the characteristic involuted inner membranes of the mitochondria. These particles are packed with hydrolytic enzymes, and their function appears to be that of digesting cellular material. The enclosure of these enzymes within the lysosomal membrane will prevent the random destruction of cellular contents.

Other Membranous Structures

The cytoplasm of most cells is rather completely penetrated by a complex series of vesicular elements called the endoplasmic reticulum. The membranes that make up these vesicles either may be of the "smooth" or agranular form, or they may be studded with ribonucleoprotein particles called ribosomes to form "rough" or granular endoplasmic reticulum. The membranes that make up the endoplasmic reticulum of the cell are continuous with both the plasma membrane of the cell and the nuclear membrane.

Many of the key enzymes of the metabolic pathways for triglyceride, phospholipid, sterol, and glycogen synthesis, as well as the enzymes active in drug detoxification, are located in the smooth, endoplasmic reticulum. The endoplasmic reticulum appears to be the major adaptive organelle of the cell. The rough endoplasmic reticulum, with its attached ribosomes, is the portion of the cell actively engaged in the synthesis of protein for export from the cell. Another system of membrane-bound vesicles, which is particularly prominent in secretory cells, is the Golgi complex. These closely packed membranes are the site of concentration and accumulation of proteins or other secretory material in secretory cells. In addition to these basic organelles, which are common to almost all cells, specialized cells contain various granules, filaments, microbodies, and microtubules.

STRUCTURE OF PROCARYOTIC CELLS

In contrast to the complex assortment of membrane-bound organelles in eucaryotic cells, the procaryotes lack a membrane-defined nucleus and the other subcellular organelles. Procaryotic cells, bacteria, and blue-green algae are bounded by an outer cell wall that contains a large amount of complex carbohydrate, and inside of this a plasma membrane much simpler in composition than that found in eucaryotic cells. The genetic information in these cells appears to be on a single chromosome consisting of one large deoxyribonucleic acid molecule. There is no membrane to separate it from the rest of the cellular interior, and it is not associated with basic proteins. No internal membranes, such as the endoplasmic reticulum, are present in these cells, and the majority of the ribosomes appear to be free within the cell. Although procaryotic cells lack defined organelles, it is likely that some areas of the plasma membrane have specialized functions, and it has been shown that a number of the cellular enzymes are present as large multienzyme complexes. The electron micrograph of an *A. vinelandii* cell in Figure 2-3 indicates that the distinguishable intracellular features of animal and plant cells are not found in bacteria.

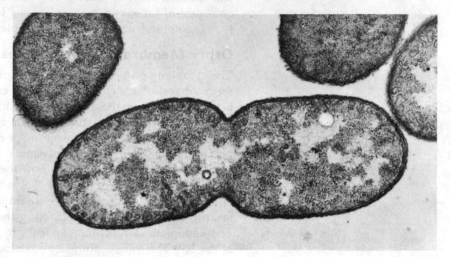

Figure 2-3 Electron micrograph of an *A. vinlandii* cell (× 35,000). (Courtesy of J. Pate, University of Wisconsin)

PROBLEMS

1. On the basis of the data in Table 2-1, would you expect a rat to contain more or less of the following constituents than a potato plant? (a) lipid, (b) ash, (c) carbohydrate, (d) water, (e) crude protein.

2. Why would the use of the Kjeldahl procedure to determine the amount of protein in cells obtained from a rapidly growing bacterial culture be less accurate than the use of the same procedure to determine the amount of protein in a rat liver preparation?

3. What four elements make up greater than 95 percent of the mass of living organisms?

4. If someone gave you two test tubes containing the ash from one ml of blood or from one gram of muscle, what are two simple chemical analyses you could use to decide which was which?

Suggested Readings

Books

Fawcett, D. W. *The Cell: An Atlas of Fine Structure.* Saunders, Philadelphia (1966).

Lehninger, A. L. *The Mitochondrion.* Benjamin, New York (1965).

Loewy, A., and P. Siekevitz. *Cell Structure and Function,* 2nd ed. Holt, Rinehart and Winston, New York (1970).

Novikoff, A. B., and E. Holtzman. *Cells and Organelles,* 2nd ed. Holt, Rinehart and Winston, New York (1976).

Articles and Reviews

Allison, A. Lysosomes and Disease. *Sci. Amer.* **217**:62–72 (1967).

Neutra, M., and C. P. Leblond. The Golgi Apparatus. *Sci. Amer.* **220**:100–107 (1969).

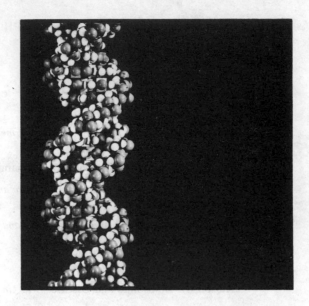

carbohydrate chemistry

chapter three

Carbohydrates are optically active polyalcohols possessing an aldehyde or a ketone function. They are found in a free sugar or monosaccharide form, or a number of monosaccharides are linked by glycoside bonds to form oligosaccharides. As carbohydrates are polyfunctional they exhibit many of the chemical properties of both alcohols and carbonyl compounds. Most of the carbohydrate in natural products is in the form of polysaccharides such as starch, glycogen, or cellulose, but monosaccharides such as glucose and fructose, and disaccharides such as sucrose and lactose are biochemically important compounds. The structures of complex, high molecular weight polysaccharides can be determined by chemical and enzymatic analysis.

Carbohydrates, as implied by their name, are those compounds which are composed of carbon, hydrogen, and oxygen in the ratio of two hydrogens and one oxygen for each carbon. In the simplest chemical sense, they can therefore be considered to be composed of a number of hydrated carbon atoms. Carbohydrates are polyalcohols which also possess, in either a free or a combined form, an aldehyde or a ketone function. More complex carbohydrates may contain nitrogen, phosphorus, or sulfur in addition to carbon, hydrogen, and oxygen, and are often oxidized or reduced so that their formula no longer appears to be that of a simple carbon hydrate.

The classical scheme for the grouping of carbohydrates into various classes divides them into monosaccharides, or simple sugars; oligosaccharides, which contain a number of monosaccharide units; and polysaccharides, which are high molecular weight polymers of monosaccharides. The latter two classes of compounds can be broken down to monosaccharides by acid hydrolysis. The conditions usually used are 1–2N acid at 100 °C for from one to several hours. Some examples of these different classes of carbohydrates are indicated in Table 3-1.

Table 3-1 Examples of Classes of Carbohydrates

General Class	Formula	Example of a Naturally Occuring Compound
I *Monosaccharides*		
Trioses	$C_3H_6O_3$	Glyceraldehyde
Pentoses	$C_5H_{10}O_5$	Ribose
Hexoses	$C_6H_{12}O_6$	Glucose (an aldohexose) and fructose (a ketohexose)
II *Oligosaccharides*		
Disaccharides	$C_{12}H_{22}O_{11}$	Maltose (two glucose residues) and sucrose (glucose and fructose)
Trisaccharides	$C_{18}H_{32}O_{16}$	Raffinose (fructose, glucose, and galactose)
III *Polysaccharides*		
Pentose polymers	$(C_5H_8O_4)_x$	Xylan (polyxylose)
Hexose polymers	$(C_6H_{10}O_5)_x$	Starch (polyglucose)

STEREOISOMERS

Although almost all compounds of biological origin contain asymmetric centers, the situation is particularly apparent in the case of the carbohydrates. Compounds having four different substituents on a single carbon atom, as well as a few other compounds, are said to be optically active; that is, they have the property of rotating a beam of plane-polarized light. If a solution of such an optically active compound is placed in a polarimeter (an instrument that generates a beam of polarized light from ordinary light), some rotation of the plane of polarization of the light passing through the solution can be measured. If the beam of light is rotated counterclockwise as the observer looks toward the light source, the compound is said to be levorotatory (levo, *l*, or −) and if clockwise, dextrorotatory (dextro, *d*, or +). A polarimeter can also be used to measure concentration if the specific optical rotation is known. The specific optical rotation, which is dependent on temperature and the wavelength of the light used, is defined as

$$[\alpha]_\lambda^t = 100 \frac{\alpha_{obs}}{cl}$$

where *c* is the concentration in g/100 ml; α_{obs} the observed rotation; and *l*, the length of the polarimeter tube in decimeters. These measurements are most often made at 589 nanometers, which is the D line of sodium and at 25 C, so the values commonly seen are $[\alpha]_D^{25}$.

Structural Formulas for Stereoisomers

There are a number of ways to represent the arrangement in space of the four different groups around an asymmetric carbon atom. Those illustrated in Figure 3-1 range from a ball-and-stick model, which clearly shows the the three-dimensional orientation, through the tetrahedral and perspective formula representations, to the simplest form, which is called a *Fischer projection*. The pair of compounds shown in Figure 3-1 are said to be *enantiomers* (*antipodes*), or *mirror images* of one another. If one of the pair has a negative specific optical rotation, the other will have a specific optical rotation equal in magnitude but positive in sign. An equimolar mixture of the two enantiomers would have no observable rotation and would be called a *racemic* mixture. Because of the convenience they give in writing formulas, the Fischer projections are the type of structures most often used for writing compounds with asymmetric centers. When the Fischer projections are used, it must be remembered that they are produced by essentially flattening a perspective formula against the plane of the paper. Two compounds are identical if the groups around their asymmetric center can be superimposed on one another. This can be done with the projection formulas only as long as they are rotated in the plane of the paper; they cannot be picked up from the paper and laid down again. As shown in Figure 3-2, this can be done with a perspective formula but not with a projection formula.

In addition to the *d* and *l*, and the equimolar solution of the two, the racemic mixture, there is the possibility of a third, or *meso*, form if the compound has more than one asymmetric center. This is illustrated for the case of tartaric acid in Figure 3-3. The meso form is optically inactive because it has an internal plane of symmetry, as is indicated by the dashed line, and is said to have internal compensation. A concept of particular importance to biochemists is that of compounds which contain a meso carbon atom. These are compounds which are not optically active themselves, but which can be metabolized to an optically active compound. One of the best examples of this type of compound is citric acid, which will be considered in some detail in the metabolism section.

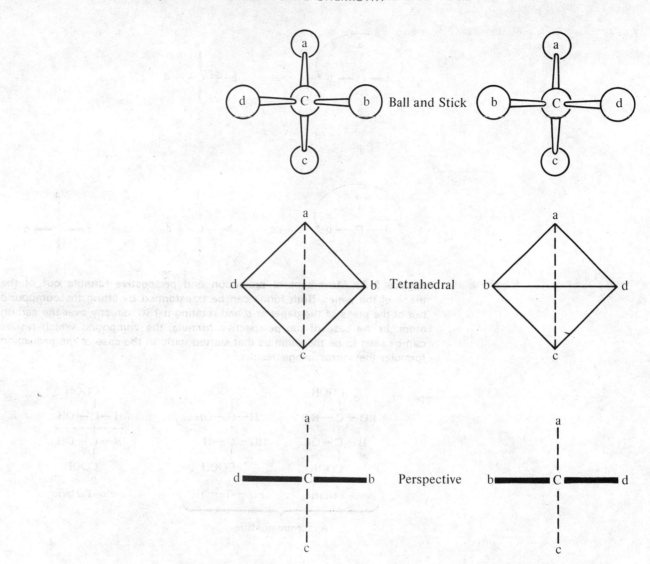

Ball and Stick

Tetrahedral

Perspective

Projection

Figure 3-1 Four ways of representing a pair of enantiomers. Each of the compounds in the left-hand column is a mirror image of the corresponding compound in the right-hand column. In the perspective formula the groups on the thickened lines are assumed to be in front of the plane of the paper, and those on the broken lines, behind the plane of the paper.

25

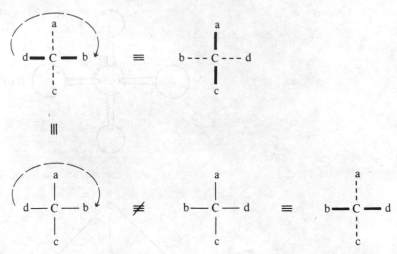

Figure 3-2 Movement of projection and perspective formula out of the plane of the paper. Both forms can be transformed by lifting the compound out of the plane of the paper at *d* and rotating it 180° directly over the carbon atom. In the case of the perspective formula, the compound which results can be seen to be the same as that started with. In the case of the projection formula, the mirror image results.

COOH	COOH	COOH
HO — C — H	H — C — OH	H — C — OH
H — C — OH	HO — C — H	H — C — OH
COOH	COOH	COOH
levo-Tartaric	*dextro*-Tartaric	*meso*-Tartaric

A racemic mixture

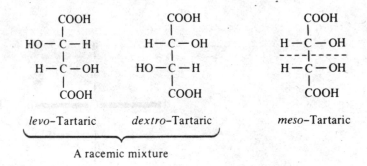

Citric acid

Figure 3-3 The three stereochemical forms of tartaric acid. The dotted line indicates the internal plane of symmetry through the meso form, which can be written with both hydroxyl groups to either the right or the left. Compounds, such as citric acid, with two like and two unlike groups on a carbon atom, are said to contain meso carbons.

Optical Activity and Configuration

The designation of a compound as ($+$) or ($-$) [or *d* or *l*] can be used to indicate which of a pair of optical isomers is being considered, but tells nothing about the configuration of the compound. The configuration of a carbohydrate is designated by relating the structure of the compound to that of the simple monosaccharide, glyceraldehyde. The glyceraldehyde with a positive optical rotation has been called D-glyceraldehyde; the one with a negative rotation, L-glyceraldehyde. The rest of the simple sugars have then been divided into a D and L series according to the configuration about the highest numbered asymmetric carbon. If, when the sugar is written with the aldehyde group at the top, the hydroxyl at that center points to the right, as in D-glyceraldehyde, it is a D sugar. If this hydroxyl points to the left, it is an L sugar. Sugars with more than four asymmetric carbon atoms in the straight-chain form have special rules for determining their configuration and they will not be considered here. Placing a sugar in the D or the L series indicates nothing about its optical activity; some D sugars have a ($+$), and some a ($-$), rotation. In general, to avoid confusion, the designation *d* or *l* should not be used to indicate optical activity, but rather D or L if the configuration is known, and if not, ($+$) or ($-$).

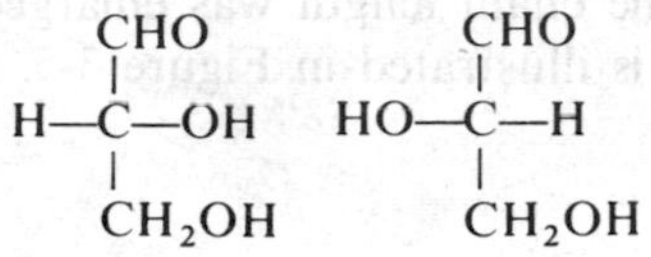

Types and Numbers of Stereoisomers

Compounds which are stereoisomers of one another, but are not enantiomers, are called *diastereoisomers*. These, in contrast to mirror images, are distinctly different chemical compounds and have different chemical and physical properties. Diastereoisomers which differ at only one asymmetric center are a special class called *epimers*. These relationships, using some of the common aldohexoses as examples, are illustrated in Figure 3-4.

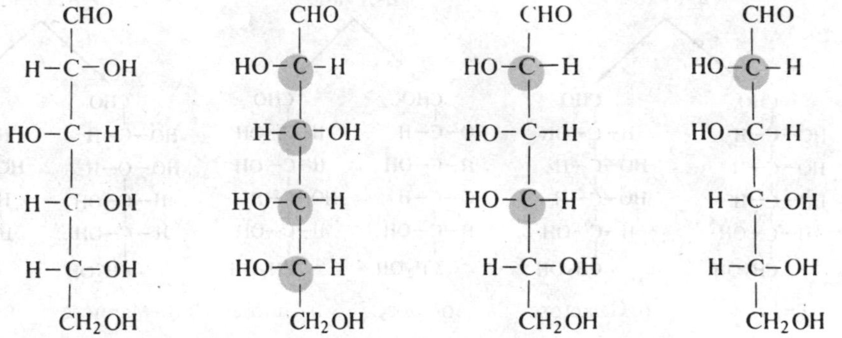

CHO H—C—OH CH₂OH — D-Glyceraldehyde

CHO HO—C—H CH₂OH — L-Glyceraldehyde

Figure 3-4 D-Glucose, an aldo-hexose, and three of its fifteen possible stereoisomers. The shaded carbons are those about which the configuration differs from that at the same carbon in D-glucose. Note that L-glucose, the enantiomer, is the compound with the configuration about all asymmetric centers reversed.

D-Glucose

L-Glucose — Its enantiomer

D-Talose — A diasterioisomer

D-Mannose — A diasterioisomer and also an epimer

27

In both the D and the L series of sugars, there are two aldotetroses, four aldopentoses, eight aldohexoses, and so on. Therefore many simple sugars contain the same number and kinds of functional groups and yet are distinctly different chemical compounds. The Vaht Hoff rule, which states that the maximum possible number of optical isomers of a compound is equal to 2^n, where n is the number of asymmetric centers, would predict that there would be sixteen different aldohexoses. The relationship between the various simple sugars can be looked at as if a new CHOH group were inserted just under the carbonyl group as the chain length was enlarged. This relationship for the common D sugars is illustrated in Figure 3-5. It

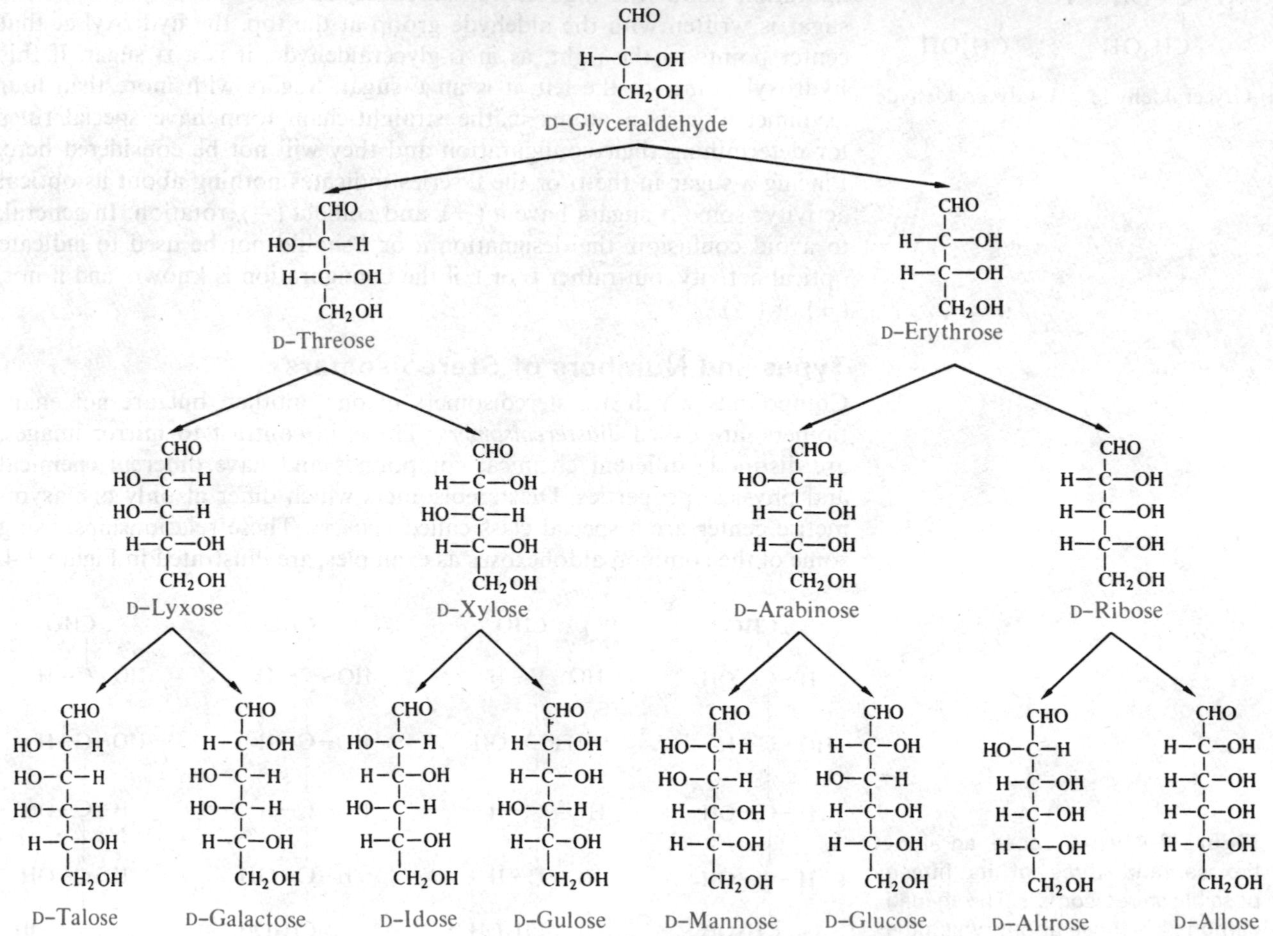

Figure 3-5 Stereoisomeric aldoses of the D-series. There will also be an equal number of isomers in the L-series, which are the enantiomers of each of these compounds.

should be clear that for each of the aldoses in Figure 3-5, there is a corresponding mirror image or L form. The ketoses are a similar series of compounds based on the addition of a CHOH group to dihydroxy acetone.

These sugars are often named in relationship to the corresponding sugar in the aldose series. Some of the biologically important sugars of the ketose series are shown in Figure 3-6.

$$
\begin{array}{cccc}
\text{CH}_2\text{OH} & \text{CH}_2\text{OH} & \text{CH}_2\text{OH} & \text{CH}_2\text{OH} \\
| & | & | & | \\
\text{C}{=}\text{O} & \text{C}{=}\text{O} & \text{C}{=}\text{O} & \text{C}{=}\text{O} \\
| & | & | & | \\
\text{H}{-}\text{C}{-}\text{OH} & \text{HO}{-}\text{C}{-}\text{H} & \text{H}{-}\text{C}{-}\text{OH} & \text{HO}{-}\text{C}{-}\text{H} \\
| & | & | & | \\
\text{H}{-}\text{C}{-}\text{OH} & \text{H}{-}\text{C}{-}\text{OH} & \text{HO}{-}\text{C}{-}\text{H} & \text{H}{-}\text{C}{-}\text{OH} \\
| & | & | & | \\
\text{CH}_2\text{OH} & \text{CH}_2\text{OH} & \text{H}{-}\text{C}{-}\text{OH} & \text{H}{-}\text{C}{-}\text{OH} \\
& & | & | \\
& & \text{CH}_2\text{OH} & \text{CH}_2\text{OH} \\
\text{D-Ribulose} & \text{D-Xylulose} & \text{D-Sorbose} & \text{D-Fructose}
\end{array}
$$

Figure 3-6 Some of the more common keto sugars. Note that they have one less asymmetric center than the corresponding aldoses.

STRUCTURE OF THE MONOSACCHARIDES

For convenience, the Fischer projection formulas are seldom written out completely, but are further abbreviated as stick models, where a short horizontal bar sticking out from a vertical line represents a hydroxyl group.

$$
\begin{array}{ccc}
\text{CHO} & & \text{CH}_2\text{OH} \\
| & & | \\
\text{HO}{-}\text{C}{-}\text{H} & \text{CHO} & \text{C}{=}\text{O} \quad\quad \text{CH}_2\text{OH}\\
| & | & | \\
\text{H}{-}\text{C}{-}\text{OH} & | & \text{HO}{-}\text{C}{-}\text{H} \quad\quad {=}\text{O}\\
| & \text{or} & | \quad\quad \text{or}\\
\text{HO}{-}\text{C}{-}\text{H} & | & \text{H}{-}\text{C}{-}\text{OH}\\
| & | & | \\
\text{HO}{-}\text{C}{-}\text{H} & \text{CH}_2\text{OH} & \text{H}{-}\text{C}{-}\text{OH} \quad\quad \text{CH}_2\text{OH}\\
| & & | \\
\text{CH}_2\text{OH} & & \text{CH}_2\text{OH} \\
\text{L-Glucose} & & \text{D-Fructose}
\end{array}
$$

Although the simple sugars are clearly aldehydes or ketones, there was evidence which suggested to early carbohydrate chemists that these sugars were not behaving as such in solution. A simple sugar like glucose did not give the normal reactions of aldehydes as would be predicted, and when dissolved, its optical rotation initially changed with time but eventually approached an equilibrium value. The chemical basis for these changes is explained by the observation that sugars can undergo the typical hemiacetal–acetal reactions of an aldehyde and an alcohol (Figure 3-7) and that, in the case of the sugars, the hemiacetal formed is an internal one. The closing of the hemiacetal ring at carbon 1 of an aldose will produce a new asymmetric center, and will therefore have an influence on the optical rotation. The only stable ring systems that can be formed when carbohydrates are placed in solution are five- or six-membered, and because they contain an oxygen atom, they are named as derivatives of the pyran or furan ring system. An aqueous solution of D-glucose is an

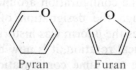

Pyran   Furan

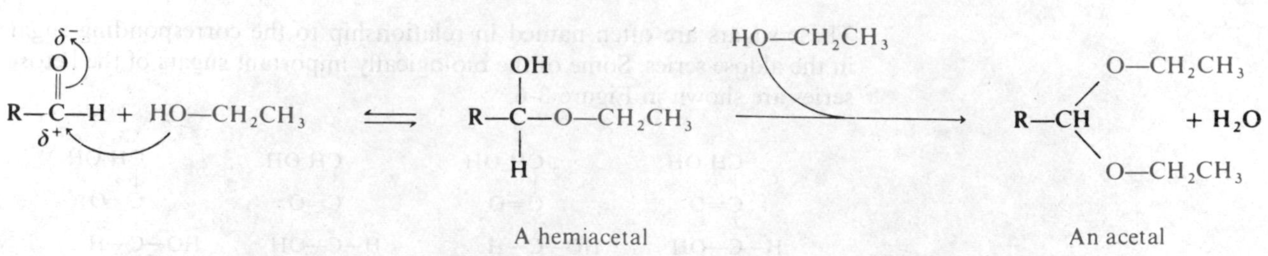

$$R-\underset{\delta^+}{\overset{\delta^-}{\underset{}{\overset{O}{C}}}}-H + HO-CH_2CH_3 \rightleftharpoons R-\underset{H}{\overset{OH}{\underset{|}{\overset{|}{C}}}}-O-CH_2CH_3 \longrightarrow R-CH\overset{O-CH_2CH_3}{\underset{O-CH_2CH_3}{}} + H_2O$$

A hemiacetal   An acetal

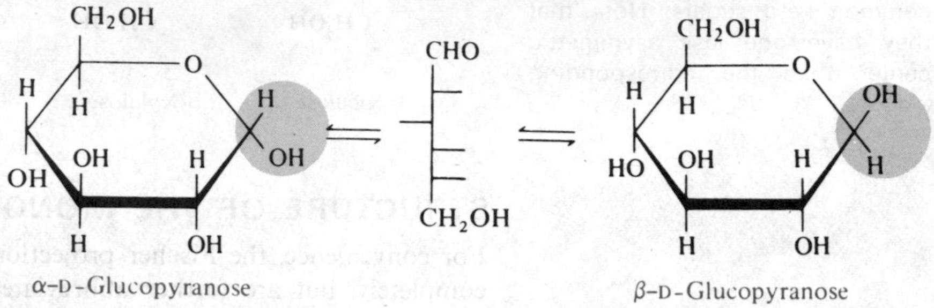

Figure 3-7 The type reaction of an aldehyde and an alcohol to form a hemiacetal and an acetal, and an example showing the two different internal hemiacetals that can be formed from D-glucose. Note that the only difference in the two compounds is the placement of the hydroxyl group at carbon-1.

α-D-Glucopyranose   β-D-Glucopyranose

equilibrium mixture of roughly one-third α-D-glucopyranose, $[\alpha]_D = +112°$, and two-thirds β-D-glucopyranose, $[\alpha]_D = +19°$, and has an $[\alpha]_D$ of $+52°$. If a pure solution of either of these two isomers is formed, the optical rotation will change with time as this equilibrium value is approached. This change in optical activity is the phenomenon called mutarotation. The five-member, or furanose ring, form of glucose is a less stable form and only a very small amount of this form is present in the equilibrium mixture. However, appreciable amounts of the furanose form are present in the equilibrium solution of galactose and some of the rarer hexoses. The furanose forms of some of the simple sugars are also found in a combined form in some of the common oligo- and polysaccharides.

Haworth Formulas

Depictions of the ring forms of the sugars such as those shown for α- and β-D-glucopyranose in Figure 3-7 are called *Haworth perspective formulas* and the substituents on the carbon atoms are represented as extending above or below the plane of a ring. The lower half of the hexagon is thickened to indicate that it is the portion of the ring that is directed out of the plane of a printed page toward the reader. In practice, the formulas are often simplified even more by indicating the hydroxyl groups of the ring by a short vertical line (see Figure 3-8) and assuming that the lower half of the hexagon is the part which would normally be drawn with a thickened line.

Sugars differing from one another only in the configuration around the reducing carbon are called *anomers*. Although the designation of one anomer of a sugar as the α form and the other as the β form was historically made on the basis of the magnitude of its optical rotation, it is now based on configuration. Those anomers which have the same configuration at both the anomeric carbon and the asymmetric carbon most removed from the aldehyde group, the D or L reference carbon, are the α anomers.

A few simple rules are all that are needed to correctly change the structures of the monosaccharides from the open-chain "stick," or Fischer projection, form to the correct Haworth perspective formula. The Fischer projection is essentially rotated 90°, so that the carbonyl group is to the right, and then a five- or six-member ring is closed by bringing the two ends together. The correct configuration of the groups on the ring can be determined by remembering three simple rules. First, if the ring closes on a hydroxyl which is on the right in the Fischer projection, the hydroxymethyl group (tail) points up; if it closes on a hydroxyl which was on the left in the Fischer projection, the tail points down. Second, the ring hydroxyls point down if they are on the right in the Fischer projection, and up if they are on the left in the Fischer projection. Note that this is also the position they are in after the Fischer projection has been rotated 90° to lie on its side. Third, the hydroxyl on the anomeric carbon points down in the D series if it is α and up if β. In the L series, α is up and β is down. The application of these simple rules is illustrated in Figure 3-8.

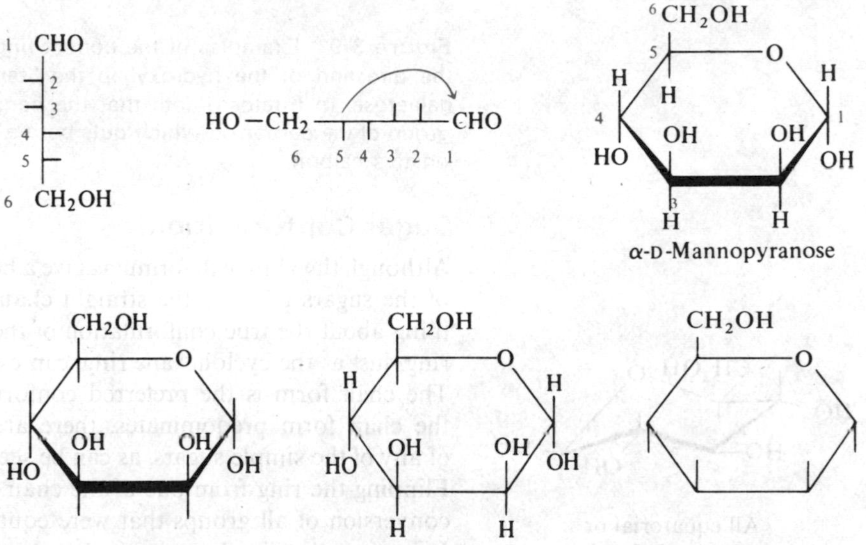

α-D-Mannopyranose

Figure 3-8 Formation of the pyranose ring form of D-mannose. The α-anomer is represented. Carbon-5, which reacts with the aldehyde group to form a hemiacetal, is a D-hydroxyl; therefore, the hydroxymethyl group points up out of the ring. The three structures at the bottom are alternative ways of writing this structure.

Some complications arise when the furanose form of the aldohexoses are written. The "tail" contains an asymmetric carbon as the ring closes on carbon 4. If this points up, the hydroxyl will be on what appears to be the wrong side until it is realized that this part of the molecule is really written upside down. The structures of the ketoses can be formed by simply forming the correct structure of the corresponding aldose with one carbon atom less, and then replacing the H on the anomeric (carbonyl) carbon with a -CH₂OH group. The examples in Figure 3-9 along with a study of structures in other figures should sufficiently illustrate the correct conversions.

31

β-D-Glucofuranose

α-D-Galactofuranose

α-L-Mannopyranose

β-D-Fructofuranose

Figure 3-9 Examples of the correct ring forms of some of the sugars. Note the direction of the hydroxyl in the "tail" of the examples of glucose and galactose. In fructose, note that the hemiacetal ring is closed with the keto group of the carbon-2, which puts both a hydroxyl and a hydroxymethyl group on this carbon.

Sugar Conformation

Although the Haworth formulas give a better indication of the true structure of the sugars than do the straight-chain forms, they do not indicate anything about the true conformation of the sugars. The six-member pyranose ring, just as the cyclohexane ring, can exist in either a chair or a boat form. The chair form is the preferred conformation. Even if it is assumed that the chair form predominates, there are still two possible conformations of any of the simple sugars, as can be seen in the case of β-D-glucopyranose. Flipping the ring from one of the chair forms to the other will lead to the conversion of all groups that were equatorial to axial in the flipped form. It is clear that in the case of β-D-glucopyranose, the form which has all the hydroxyls equatorial, the "normal form," will be favored over the form which puts all these rather bulky groups axial. Any change in the configuration of aldohexoses from the structure of β-D-glucose, such as α-D-glucose or β-D-galactose, results in the introduction of an axial hydroxyl into the ring. As a general rule, the most stable form of any of the sugars will be the one with the fewest axial constituents. An understanding and a knowledge of the conformation of the sugars are particularly important to chemists seeking to postulate the rate of reaction of various sugars in any chemical reaction. Whether or not a group is protected by the ring will greatly influence its reactivity.

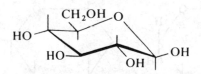

All equatorial or "normal" form

All axial or "alternate" form

β-D-Glucopyranose

CHEMICAL REACTIONS OF SUGARS

The carbohydrates, because of the various reactive groups present in the molecule, undergo a large number of specialized chemical reactions. In weak alkaline solutions, such as lime water, an isomerization of the carbonyl group to the 1,2-enediol form is promoted, and a quasi-equilibrium mixture of sugars is formed (Figure 3-10). This reaction, which is a general one for all monosaccharides, is called the *Lobry de Bruyn transformation*. When no other methods were available, it was used as a method of monosaccharide synthesis.

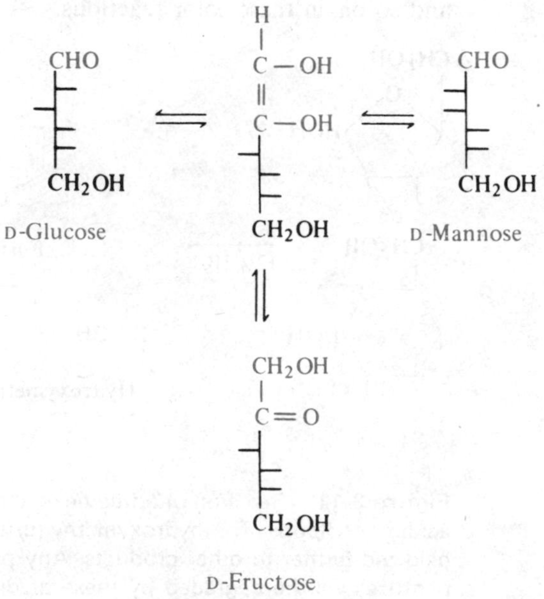

Figure 3-10 Equilibrium of different sugars which would result if either D-glucose, D-mannose, or D-fructose were placed in a weak alkaline solution. The amount of the enediol form in the equilbrium will be very small.

The 1,2-enediol form of the sugar is also the reactive species in what is called a *reducing sugar test*. When carbohydrates are heated with cupric ion in a weakly alkaline solution, the carbonyl group is oxidized and a number of products, including the corresponding sugar acid, are formed. This reduction of Cu^{2+} to Cu^+ and the subsequent precipitation of insoluble cuprous oxide, Cu_2O, is a qualitative test for sugars with a free aldehyde or ketone function. *Fehlings solution*, a mixture of $NaOH$, $CuSO_4$, and Na-K tartrate, has most often been used for this purpose. This reaction can be used as a quantitative assay for carbohydrate by allowing the reduced copper to react with arsenomolybdic acid, iodine, or other oxidizing agents. Those sugars which do not have a potentially free aldehyde or ketone function cannot react under these conditions and are called nonreducing sugars.

A second general qualitative test for carbohydrates, which can also be used as a quantitative assay, is the reaction of carbohydrates with strong nonoxidizing acids (Figure 3-11). Heating a solution of pentoses under these conditions leads to the formation of the unsaturated aldehyde, furfural. Under the same conditions hydroxymethylfurfural and further oxidation products are formed from hexoses. The condensation between these activated aldehydes and phenolic compounds such as resorcinol, α-naphthol, anthrone, or various activated amines results in the formation of colored compounds which serve as the basis for a number of qualitative carbohydrate tests. One of the most common quantitative but nonspecific assays for carbohydrate is based on the reaction of anthrone in concentrated sulfuric acid with sugars. By varying the conditions and reagents, it is possible to obtain a certain degree of specificity for pentoses, ketoses, and so on, in these color reactions.

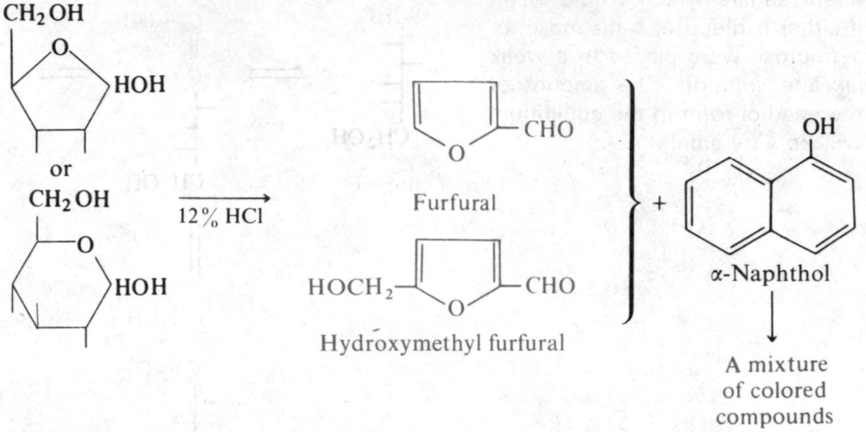

Figure 3-11 Reaction of a free hexose or pentose under strong nonoxidizing acidic conditions. The hydroxymethylfurfural formed from a hexose is usually oxidized further to other products. Any polysaccharide containing hexoses or pentoses will be degraded by these acidic conditions and will give the same reactions.

Reactions of the Carbonyl Group

One of the most typical reactions of the carbohydrates is the addition of substituted hydrazines to the carbonyl group. In the presence of excess hydrazine the carbon adjacent to the carbonyl group is oxidized so that the bis-hydrazones or osazones are formed rather than the hydrazones. Because of this, sugars which differ in configuration only at carbons 1 or 2, such as glucose, mannose, and fructose, will give the same osazone. The classical reaction is carried out with phenylhydrazine to give the bright yellow phenylosazones (Figure 3-12). These compounds are easily crystallized, have high melting points, unique crystal structures, and form at characteristic rates under standardized conditions. Because of these properties they have often been used as derivatives for the characterization of sugars. At the present time, most of the characterization of carbohydrates is carried out by physical methods, chiefly NMR, and gas-liquid chroma-

34

H
C = N—OH
|
|
|
CH$_2$OH

D-Glucose-
oxime

↑

NH$_2$—OH
Hydroxylamine

CH$_2$OH
O
HOH

D-Glucose

H H
HN—N

Phenylhydrazine

↓

H
C = N — N
H
C = N — N
|
|
CH$_2$OH

D-Arabino–hexose phenylosazone
(D-Glucose phenylosazone)

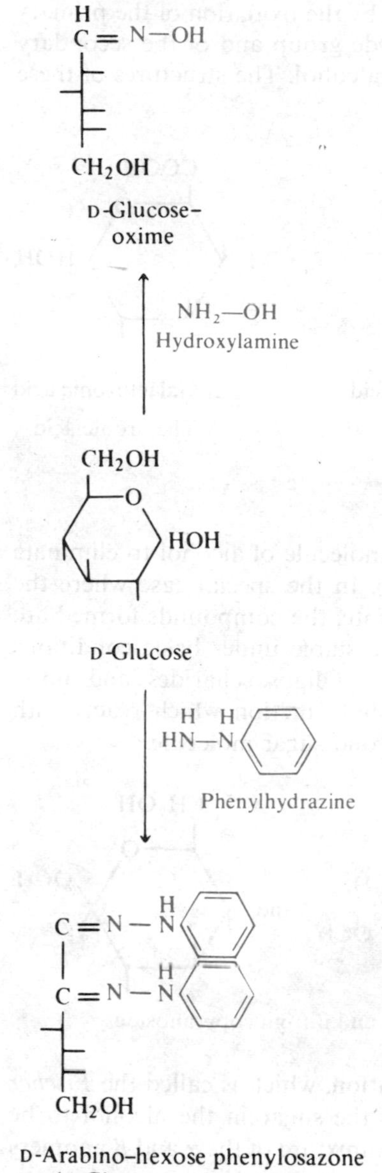

Figure 3-12 Two common addition reactions obtained with carbohydrates that have a free aldehyde group.

tography of the silyl derivatives. These addition reactions are therefore of more historical than practical interest. In a similar addition reaction, the reducing sugars will react with hydroxylamine to form oximes.

The addition of HCN to the carbonyl group is the reaction which forms the starting point for the *Kiliani synthesis* of carbohydrates. If an aldose of known configuration is treated with HCN, two cyanohydrins, which differ in configuration only at the α carbon, are formed. If these compounds are converted to aldehydes, two new sugars are formed which, as they are diastereoisomers of one another, can usually be derivatized and separated by fractional crystallization (Figure 3-13). Under appropriate conditions, usually NaBH$_4$ treatment, the simple aldoses can also be reduced to the corresponding sugar alcohols or glycitols.

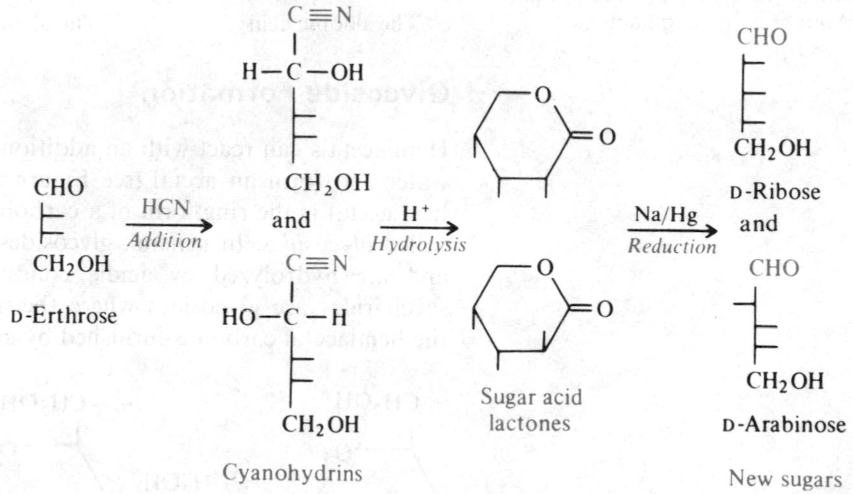

C≡N
|
H—C—OH
|
CH$_2$OH

CHO
|
CH$_2$OH

D-Erthrose

HCN
Addition

and

C≡N
|
HO—C—H
|
CH$_2$OH

Cyanohydrins

H$^+$
Hydrolysis

O
O

O
O

Sugar acid
lactones

Na/Hg
Reduction

CHO
|
CH$_2$OH

D-Ribose

and

CHO
|
CH$_2$OH

D-Arabinose

New sugars

Figure 3-13 Reactions involved in the Kiliani synthesis of two new aldopentoses from an aldotetrose.

Formation of Sugar Acids

Three different sugar acids can be derived from each of the simple sugars: *aldonic acids*, in which the aldehyde group has been oxidized to a carboxylic acid; *uronic acids*, in which the primary alcohol group has been oxidized to a carboxylic acid; and *aldaric acids*, in which both the aldehyde and primary alcohol carbons have been oxidized to form the dibasic acid. The simplest acids to form are the aldonic acids, which can be obtained in good yield usually as the δ-lactone, by the oxidation of the parent sugar with bromine water. Strong oxidizing agents, such as nitric acid, will oxidize not only the hemiacetal carbon but also the primary alcohol carbon and will result in the formation of the dibasic aldaric acids. The oxidation of galactose results in the formation of galactaric (mucic) acid, which is extremely insoluble, and this reaction has been used as a qualitative test for galactose or galactose-containing carbohydrates. Because of the ease of oxidation of aldehydes, the formation of uronic acids by the oxidation of the primary hydroxyl requires protection of the aldehyde group and of the secondary hydroxyls before oxidation of the primary alcohol. The structures of these sugar acids are illustrated in Figure 3-14.

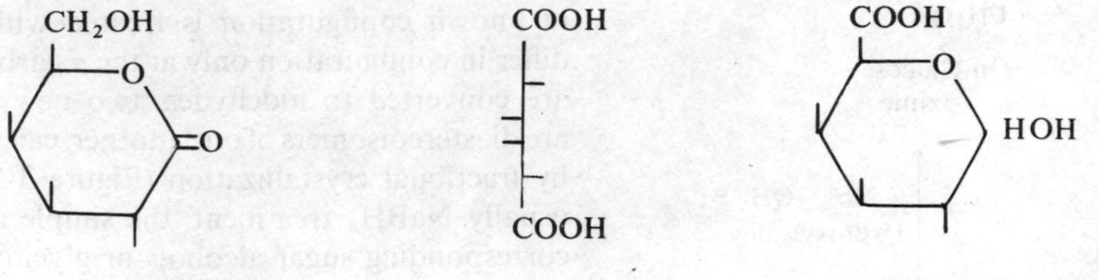

Figure 3-14 Examples of the three different types of sugar acids formed by oxidation of galactose. Note that the same aldaric acid will be formed from the oxidation of either D- or L-galactose.

D-Galactono-δ-lactone
The aldonic acid

Galactaric acid
The aldaric acid

D-Galacturonic acid
The uronic acid

Glycoside Formation

Hemiacetals can react with an additional molecule of alcohol to eliminate water and form an acetal (see Figure 3-7). In the special case where the hemiacetal is the ring form of a carbohydrate, the compounds formed are called *glycosides*. In general, glycosides are stable under basic conditions and are hydrolyzed by acidic conditions. Oligosaccharides and polysaccharides are glycosides where the alcohol function which reacts with the hemiacetal carbon is furnished by a second sugar molecule.

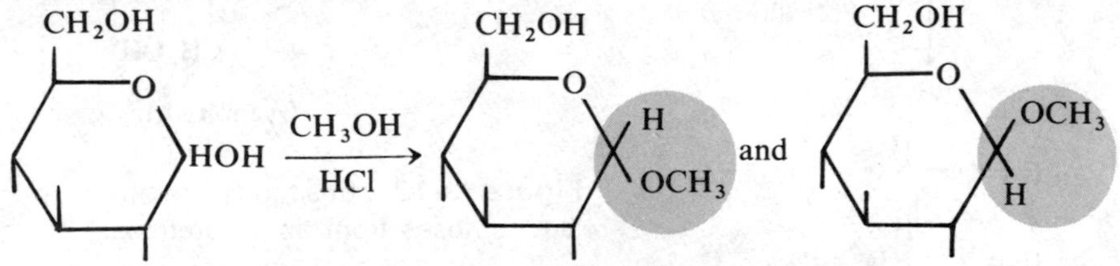

D-Glucose

Methyl α- and β-D-glucopyranosides

The simplest method of glycoside formation, which is called the *Fischer synthesis*, involves refluxing a solution of the sugar in the alcohol to be added in the presence of an acid catalyst. A mixture of the α and β anomers is formed, but depending on which particular sugar is being reacted it may be possible to adjust conditions to give predominantly one form. In addition to the two isomers shown in the preceding example, a small amount of the furanosides would be formed. For the preparation of specific isomers, the glycosyl halides, protected by previous acylation, are condensed with the desired alcohol in the presence of silver or mercury salts. The conformation of the acylated sugar will then determine whether the α or the β anomer will be formed. This procedure has the advantage that almost any alcohol can be used, whereas the Fischer synthesis has a more limited application.

Reactions of the Alcohol Groups

Two important general reactions of the sugars depend on derivatizing the free alcohol groups in the molecule. Sugars can be acylated by reaction with an acyl chloride or acid anhydride and a catalyst (Figure 3-15). The treatment results in complete acylation of the molecule, and variations in the conditions can determine whether the α or β anomer is formed. The resulting esters are easily hydrolyzed by alkali but are relatively stable under acidic conditions.

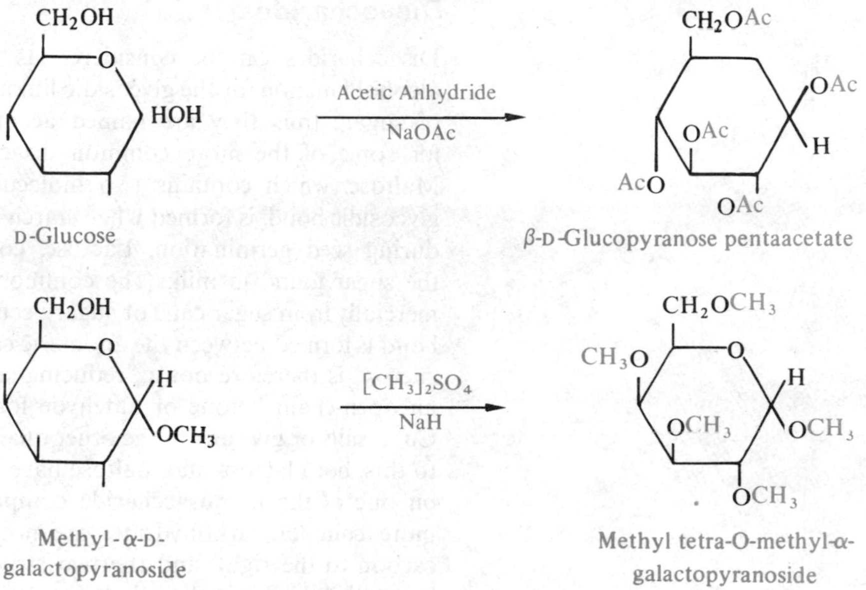

Figure 3-15 Acetylation and methylation of the free alcohol groups of sugars.

One derivative that has been of the most value in the characterization of carbohydrates is the completely methylated sugar. The formation of ether derivatives on the free hydroxyl is a difficult chemical reaction and must sometimes be repeated a number of times to force the reaction to completion. The glycoside is treated with methyl iodide in dimethyl sulfoxide with sodium hydride as a catalyst (Figure 3-15). The products are isolated; if the reaction has not gone to completion, it is repeated. The entire process is called an *exhaustive methylation*. The value of these derivatives in characterization is that the ethers formed are completely resistant to acidic conditions that will hydrolyze glycosidic bonds (see Figure 3-20). It is important to realize that although all the methoxy groups on the molecule (Figure 3-15) look the same, the one on the anomeric carbon is an acetal (a glycoside) and therefore has chemical properties much different from the true ether functions on the rest of the carbons of the sugar.

37

BIOLOGICALLY IMPORTANT SUGARS

Biochemically, the most important sugar is undoubtedly glucose. Large amounts of glucose are present in glycosidic linkages in starch and cellulose, and a small amount of the free sugar is present in fruits and in blood plasma. Fructose, the important ketohexose (see Figure 3-6), is found in small quantities in fruits and also as a component of some polysaccharides. Galactose is another common hexose found as a component of lactose, and it, as well as mannose, is a component of a number of polysaccharides.

Disaccharides

Disaccharides can be considered as if they were glycosides where the alcohol function for the glycosidic linkage is furnished by a second molecule of sugar; thus they are named accordingly. The correct nomenclature for some of the more common disaccharides is shown in Figure 3-16. Maltose, which contains two molecules of glucose joined by an α-1,4-glycosidic bond, is formed when starch is degraded by the action of amylase during seed germination. Lactose, containing glucose and galactose, is the sugar found in milk. The common table sugar that is obtained commercially from sugar cane or sugar beets is sucrose. In sucrose the glycosidic bond is formed between the anomeric carbons of both glucose and fructose; sucrose is therefore not a "reducing sugar." As it cannot equilibrate with an open-chain ketone or aldehyde form of the sugars, it will not reduce Cu^{2+} salts or give any of the other qualitative tests of aldehydes. In contrast to this, both lactose and maltose have a potentially free aldehyde available on one of the monosaccharide components. Note that the structures of more complex carbohydrates are not always written with the anomeric carbon to the right, and the two structures presented for sucrose should be studied until it is clear that they are really equivalent.

Polysaccharides

The majority of the carbohydrate found in living tissue is in the form of various polysaccharides. These high molecular weight polyglycosides may be composed of only one type of sugar residue, or they may contain more than one in an alternating, repeating sequence. Both linear and branched-chain polysaccharides are abundant natural products. They function mainly as cellular structural components or as reserve food sources which, when needed for energy, can easily be degraded. The composition of some of the more common polysaccharides is shown in Table 3-2.

Cellulose is the most abundant organic compound found in nature; as much as 50 percent of all the carbon in vegetable material may be present as this one polysaccharide. It is a linear compound with D-glucose residues linked β-(1,4). There are probably at least 3000 molecules of glucose in the linear cellulose chain and it forms a fibrous, partially crystalline, structure which is extremely insoluble. The large amount of glucose present in plants in the form of cellulose is not available as an energy source for man or other monogastric animals, because they lack an enzyme that can cleave the β-(1,4)-glycosidic bond. Only ruminants, such as cattle, sheep, and goats, which have a bacterial population in the rumen to metabolize cellulose, can make effective use of it.

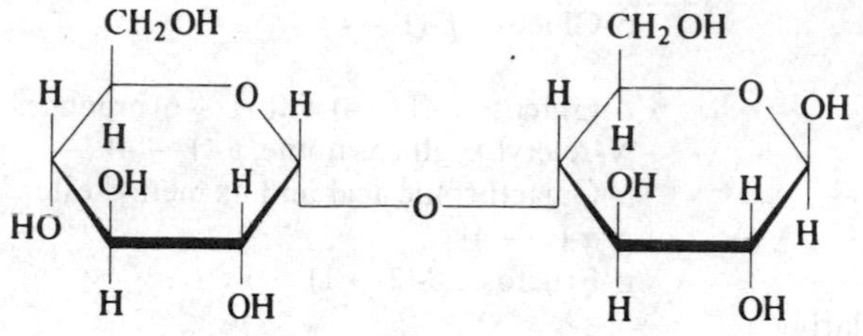

(4-O-α-D-Glucopyranosyl-β-D-glucopyranose)
MALTOSE (β-form)

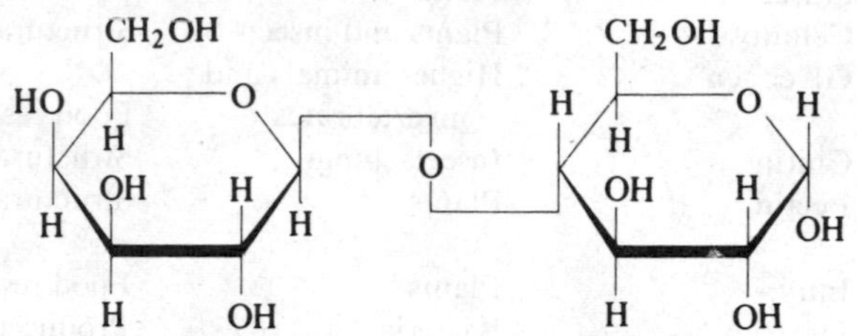

(4-O-β-D-Galactopyranosyl-α-D-glucopyranose)
LACTOSE (α-form)

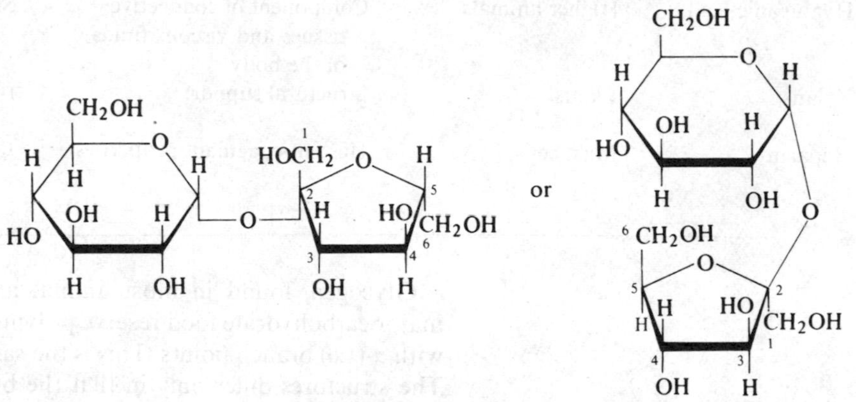

(α-D-Glucopyranosyl-β-D-fructofuranoside)
SUCROSE

Figure 3-16 Structures of three common disaccharides. Note that the fructose ring in the left-hand structure for sucrose has been flipped so that carbon-2 is to the left. This is the way the structure is usually written.

Starch, found in high concentrations in the common cereal grains and tubers, consists of a mixture of two molecules—amylose and amylopectin. Amylose is a linear polymer analogous to cellulose with an α-(1,4), rather than a β-(1,4), link, whereas amylopectin is an α-(1,4) polymer with α-(1,6) branch points. It is the amylose fraction of starch which, because of its helically coiled structure, can interact with iodine to give the typical deep blue color of the starch-iodine test. Purified amylopectin will give a red color with iodine. The molecular weight of starch molecules varies greatly, and a wide range of sizes is found in any preparation.

Table 3-2 Some Important Polysaccharides

Polysaccharide	Major Sources	Function or Source	Sugar and Linkage
Starch	Plants	Food reserve	D-Glucose, α-$(1 \rightarrow 4)$ with $(1 \rightarrow 6)$ branches
Cellulose	Plants and insects	Structural support	D-Glucose, β-$(1 \rightarrow 4)$
Glycogen	Higher animals and invertebrates	Food reserve	D-Glucose, α-$(1 \rightarrow 4)$ with $(1 \rightarrow 6)$ branches
Chitin	Insects, fungi	Structural support	N-Acetyl-D-glucosamine, β-$(1 \rightarrow 4)$
Pectin	Plants	Structural support	D-Galacturonic acid and its methyl ester, α-$(1 \rightarrow 4)$
Inulin	Plants	Food reserve	D-Fructose, β-$(2 \rightarrow 1)$
Dextran	Bacteria	Produced extracellularly by bacteria	D-Glucose, α-$(1 \rightarrow 6)$ some branching
Chondroitin-SO$_4$	Higher animals	Component of cartilage, tendons, and skin	D-Glucuronic acid, N-acetyl-D-galactosamine β-$(1 \rightarrow 3)$, β-$(1 \rightarrow 4)$ with sulfate esters at 4 or 6 of the galactosamine
Hyaluronic acid	Higher animals	Component of connective tissues and viscous fluids of the body	Structure similar to chondroitin-SO$_4$ without the sulfate esters, glucosamine rather than galactosamine
Xylans	Plants	Structural support	D-Xylose, β-$(1 \rightarrow 4)$ usually with some substitution by other pentoses
Heparin	Mast cells	Has anticoagulant properties	L-Iduronic and D-glucuronic acids, N-Sulfo-D-glucosamine, α-$(1 \rightarrow 4)$ with several O-sulfate groups

Glycogen, found in most animal and in some microbial cells as the major carbohydrate food reserve polymer, is also an α-(1,4)-glucose polymer with α-(1,6) branch points. This is the same structure as that of amylopectin. The structures differ only in that the branch points in glycogen are about twice as frequent as they are in amylopectin. The basic structure of these important polyglucans is illustrated in Figure 3-17. It should be noted that these polymers will have a sugar residue at one end which still has a free aldehyde or ketone group. This is termed the *reducing end* of the polysaccharide, and the molecule will have one or more *nonreducing ends* depending on whether it is a linear or a branched polymer.

Modified Sugars

In addition to simple aldoses and ketoses, a number of other commonly occurring compounds are classified as carbohydrates. Amino sugars are found in a number of polysaccharides, and although the 2-amino sugars are most common, others have been identified. Deoxysugars of a number of types are known, and some branched-chain carbohydrates are found in plant and bacterial sources. The N-acetyl derivative of neuraminic acid, which can be considered a condensation product of D-mannosamine and pyruvic acid, is contained in a large number of complex lipids and

β (1 → 4) linked glucose residues in cellulose

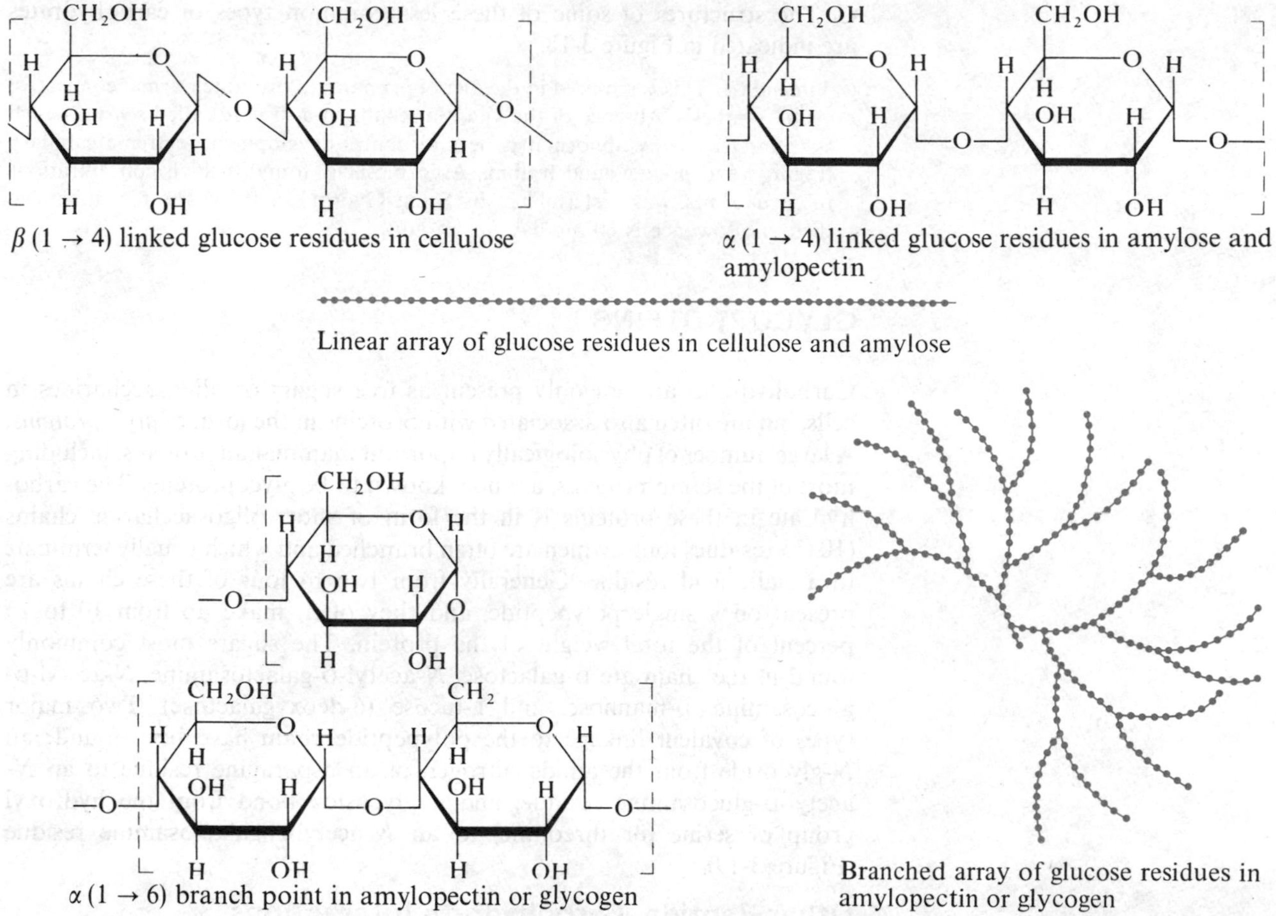

α (1 → 4) linked glucose residues in amylose and amylopectin

Linear array of glucose residues in cellulose and amylose

α (1 → 6) branch point in amylopectin or glycogen

Branched array of glucose residues in amylopectin or glycogen

Figure 3-17 Basic structures of the common polyglucans. Note that a branched structure like glycogen or amylopectin will have a large number of nonreducing end groups.

glycoproteins. These neuraminic acid derivatives are called *sialic acids*.

Some of the important polysaccharides (Table 3-2) contain sugar residues that have been structurally altered or derivatized. Acid polysaccharides, such as the chondroitins, contain galactasamine residues that have been esterified with sulfate at the 4 or 6 position, and a number of polysaccharides contain uronic acids (Figure 3-14), which may be present as the methyl ester. Glucuronic acid is also involved in detoxification reactions.

Although the cyclitols, polyhydroxy derivatives of cyclohexane, lack a carbonyl group and are therefore not true carbohydrates, they are usually discussed with this class. The biochemically important cyclitols are the inositols which have a hydroxyl group on each of their six carbons. There are nine possible isomers, but only one, *myo*-inositol is found in large quantities in natural products. Small amounts of the free inositol are found in many tissues, and large amounts are often found in plants as the hexaphosphate derivative, phytic acid. Another physiologically important compound which is a carbohydrate derivative is L-ascorbic acid, or vitamin

C. The structures of some of these less common types of carbohydrates are indicated in Figure 3-18.

Vitamin C   This is required in the diet of primates to promote normal connective tissue synthesis. Absence of the vitamin results in a disease called *scurvy*, which is characterized by abnormal bone and dental development, extreme capillary fragility, and poor wound healing. Ascorbic acid, found in high concentrations in citrus fruits, was first isolated by Szent-Gyorgyi in 1928. The recommended dietary allowance is 60 mg/day for an adult.

GLYCOPROTEINS

Carbohydrates are not only present as free sugars or oligosaccharides in cells, but are often also associated with proteins in the form of *glycoproteins*. A large number of physiologically important mammalian proteins, including most of the serum proteins, are now known to be glycoproteins. The carbohydrate in these proteins is in the form of short oligosaccharide chains (10–15 residues long), which are often branched and which usually terminate in a sialic acid residue. Generally from two to four of these chains are present on a single polypeptide, and they often make up from 10 to 15 percent of the total weight of the protein. The sugars most commonly found in the chain are D-galactose, N-acetyl-D-galactosamine, N-acetyl-D-glucosamine, D-mannose, and L-fucose (6-deoxygalactose). Two major types of covalent linkage to the polypeptide chain have been found; an N-glycoside from the amide nitrogen of an asparagine residue to an N-acetyl-D-glucosamine residue, and a glycoside bond from the hydroxyl group of serine (or threonine) to an N-acetyl-D-galactosamine residue (Figure 3-19).

Other Peptide–Carbohydrate Interactions

There are also proteins which are covalently linked to complex polysaccharides, such as hyaluronic acid, chondroitin sulfate, and heparin. These substances contain a much higher percentage of carbohydrate than protein, and they are commonly called *mucopolysaccharides* rather than glycoproteins. The covalent bond in these compounds is usually a glycoside from the serine residue of the protein to a xylose residue at the reducing end of the polysaccharide chain.

A third type of peptide–carbohydrate association is found in bacterial cell walls. In these organisms a complex cross-linked structure called a *peptidoglycan* covers the entire cell surface. This substance varies considerably in its composition in different bacteria, but there are some common features. The basic polysaccharide chain is composed of alternating residues of N-acetyl-D-glucosamine and N-acetyl muramic acid (similar to neuraminic acid), which are joined by a β-(1,4) linkage. These polysaccharide chains are linked by short polypeptide cross links. The peptides are held

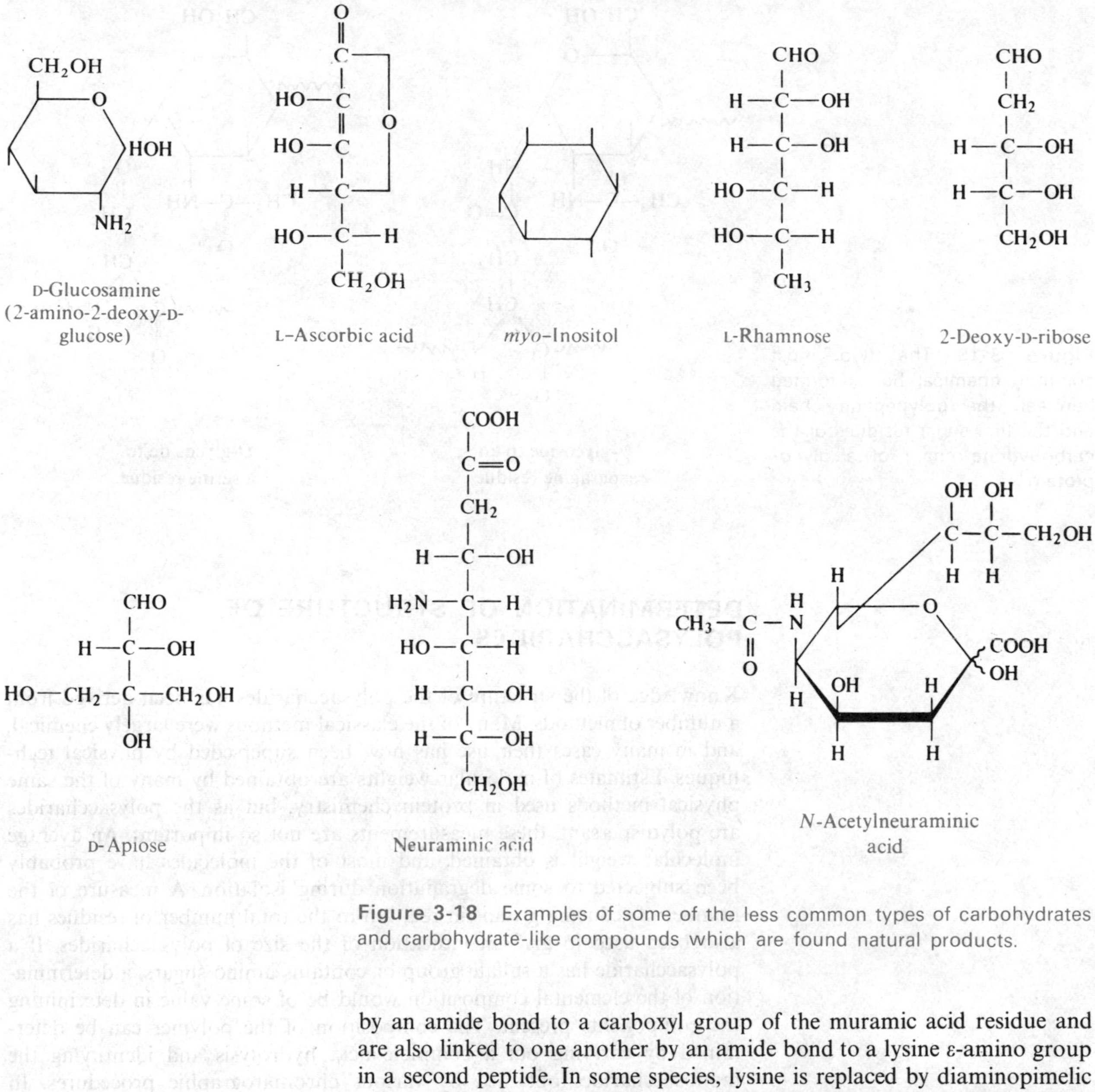

D-Glucosamine
(2-amino-2-deoxy-D-
glucose)

L–Ascorbic acid

myo–Inositol

L-Rhamnose

2-Deoxy-D-ribose

D-Apiose

Neuraminic acid

N-Acetylneuraminic
acid

Figure 3-18 Examples of some of the less common types of carbohydrates and carbohydrate-like compounds which are found natural products.

by an amide bond to a carboxyl group of the muramic acid residue and are also linked to one another by an amide bond to a lysine ε-amino group in a second peptide. In some species, lysine is replaced by diaminopimelic acid which has an additional carboxyl group on the ε carbon. These peptides usually begin with the sequence L-ala-γ-D-glu-L-lys-D-ala, whereas the number and composition of additional residues vary with the organism. The net effect of this extensive cross linking is that the entire surface of the bacterial is covered with what is essentially a single molecule.

43

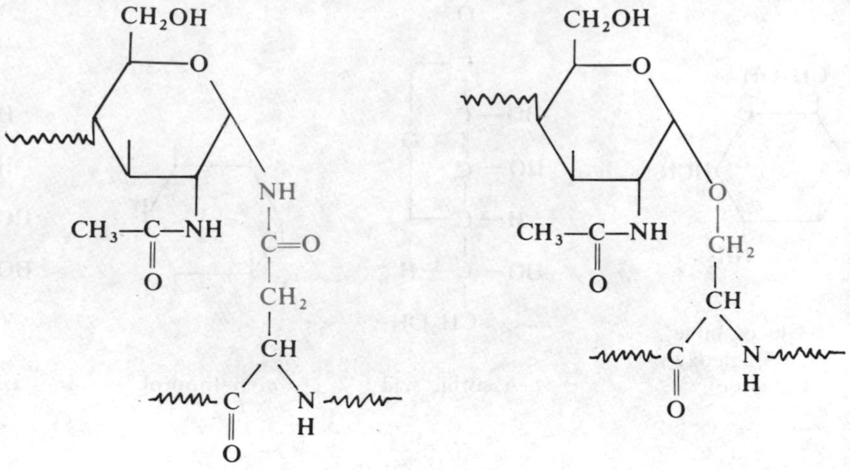

Figure 3-19 The two most common chemical bonds formed between the polypeptide chain and the first sugar residues of the carbohydrate chain of a glycoprotein.

N-glycoside to an
asparagine residue

O-glycoside to
a serine residue

DETERMINATION OF STRUCTURE OF POLYSACCHARIDES

Knowledge of the structure of the polysaccharides has been derived from a number of methods. Many of the classical methods were largely chemical, and in many cases their use has now been superseded by physical techniques. Estimates of molecular weights are obtained by many of the same physical methods used in protein chemistry, but as the polysaccharides are polydispersant, these measurements are not so important. An average molecular weight is obtained, and most of the molecules have probably been subjected to some degradation during isolation. A measure of the number of reducing groups in relation to the total number of residues has also been used to give an indication of the size of polysaccharides. If a polysaccharide has a sulfate group or contains amino sugars, a determination of the elemental composition would be of some value in determining the components present. The composition of the polymer can be determined by carrying out a complete acid hydrolysis and identifying the monosaccharides obtained by various chromatographic procedures. In some cases a partial acid hydrolysis will result in the formation of di- or tri-saccharides of known sequences which can be identified. This information can be used to establish some of the linkages in the molecule. There are hydrolytic enzymes or glycosidases which have also been used for characterization purposes. The specificity of these enzymes varies considerably. Some are specific for either α- or β-glycosides; the use of these enzymes

44

has been of value in establishing the configuration of the residues in a polypeptide.

Exhaustive Methylation

The method classically used to determine the position of the linkages between sugar residues in polysaccharides was exhaustive methylation followed by hydrolysis. The various hydrolyzed products were then isolated and characterized, and from a knowledge of which hydroxyl groups had been free and available for methylation in the native compound, the position of the glycosidic bonds could be determined. The use of this method is illustrated in Figure 3-20. None of the residues isolated has a methylated hydroxyl at the carbon-5 position, indicating that the glucose must have been in the pyranose form. The detection of residues with methoxy groups on the 2, 3, and 6 position would establish that the basic linear chain was linked $(1 \rightarrow 4)$, and the identification of the 2,3-dimethyl-glucose would

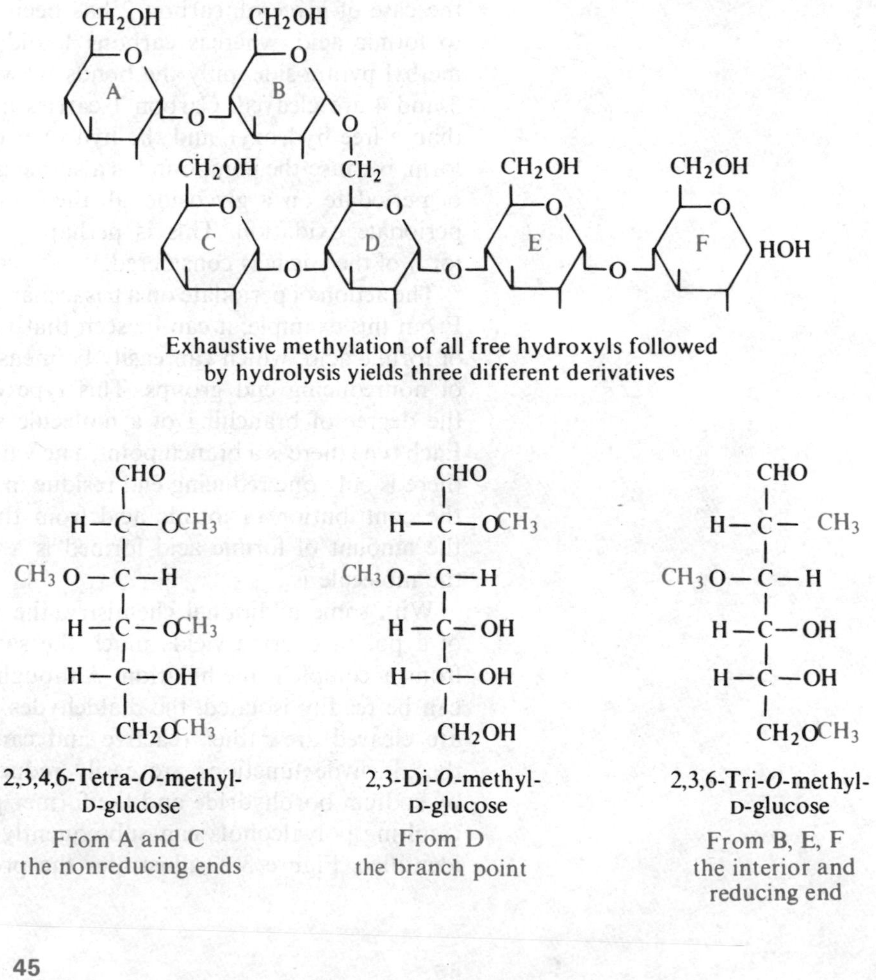

Figure 3-20 A hypothetical oligosaccharide and the products that would be obtained from each of its residues following exhaustive methylation and hydrolysis.

Exhaustive methylation of all free hydroxyls followed by hydrolysis yields three different derivatives

2,3,4,6-Tetra-*O*-methyl-D-glucose

From A and C
the nonreducing ends

2,3-Di-*O*-methyl-D-glucose

From D
the branch point

2,3,6-Tri-*O*-methyl-D-glucose

From B, E, F
the interior and
reducing end

prove that there are branch points where both $(1 \rightarrow 4)$ and $(1 \rightarrow 6)$ bonds are present. Those sugars which have neither the O-5 or O-4 position methylated could have presumably been in the form of $(1 \rightarrow 5)$-linked furanose rings, but these are rather rare. It should be noted that even though the anomeric carbon of residue F might well have been derivatized during the methylation, this would be a glycoside, not an ether methyl, group and would be removed when the rest of the molecule was hydrolyzed.

Periodate Oxidation

Exhaustive methylation, which depends on complete methylation, isolation, and characterization of a large number of variously methylated sugars is time-consuming and difficult. Some of the same information can be gained through periodate oxidation of the carbohydrate. Periodic acid will oxidize and cleave carbon-carbon bonds which contain adjacent free hydroxyl groups or a hydroxyl group and an aldehyde or ketone. The oxidation state of the carbon atom will be raised one level each time it is oxidized. This action is illustrated for some simple compounds in Figure 3-21. In the case of glycerol, carbon 2 has been oxidized twice, so it is converted to formic acid, whereas carbons 1 and 3 are left as formaldehyde. In a methyl pyranoside, only the bonds between carbons 2 and 3 and carbons 3 and 4 are cleaved. Carbon 1 carries a glycosidic methoxy group rather than a free hydroxyl, and the hydroxyl on carbon 5 remains in the acetal form, because the compound is a stable glycoside. In contrast to the action of periodate on a glycoside, all the bonds in a free sugar are broken by periodate oxidation. This is perhaps easiest to see if the straight-chain form of the sugar is considered.

The action of periodate on a trisaccharide is also illustrated in Figure 3-21. From this example, it can be seen that in a polysaccharide the production of formic acid, which can easily be measured, is a measure of the number of nonreducing end groups. This type of assay can be used to measure the degree of branching of a molecule such as glycogen or amylopectin. Each time there is a branch point, a new nonreducing end is formed. Because there is only one reducing end residue in a high molecular weight polymer, the contribution of formic acid from the reducing end is negligible, and the amount of formic acid formed is a measure of how highly branched the molecule is.

With some additional chemistry, the results of the periodate oxidation of a polysaccharide yields much the same information as can be gained from a complete methylation. Although the periodate oxidized polymer can be readily isolated, the dialdehydes formed when the acetal functions are cleaved are rather reactive and cannot be easily isolated. However, the aldehyde functions are easily reduced to the corresponding alcohol by sodium borohydride and the former glycosidic bonds remaining in the resulting polyalcohols can subsequently be hydrolyzed and the products identified. Figure 3-22 illustrates the products that would be obtained if

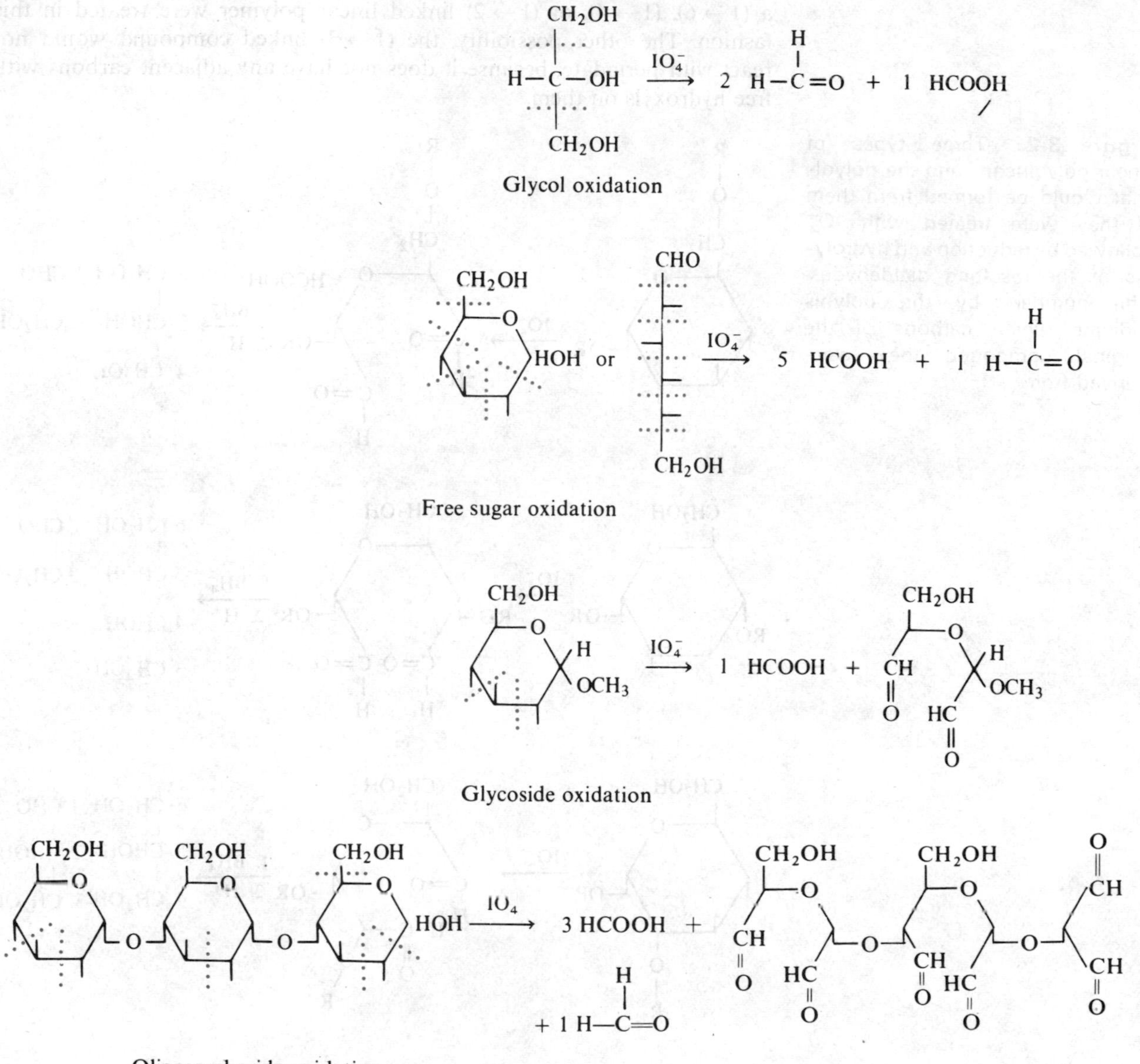

Glycol oxidation

Free sugar oxidation

Glycoside oxidation

Oligosaccharide oxidation

Figure 3-21 Action of HIO_4 on glycols, free sugars, glycosides, and oligosaccharides. The dotted lines indicate the carbon-carbon bonds that would be broken during the oxidation. Free sugars behave as if they were in their open-chain forms, and it is easier to see why the bond between carbon-5 and carbon-6 of an aldohexose is cleaved if the compound is written as a free aldose.

47

a $(1 \to 6)$, $(1 \to 4)$, or $(1 \to 2)$ linked linear polymer were treated in this fashion. The other possibility, the $(1 \to 3)$ linked compound would not react with periodate, because it does not have any adjacent carbons with free hydroxyls on them.

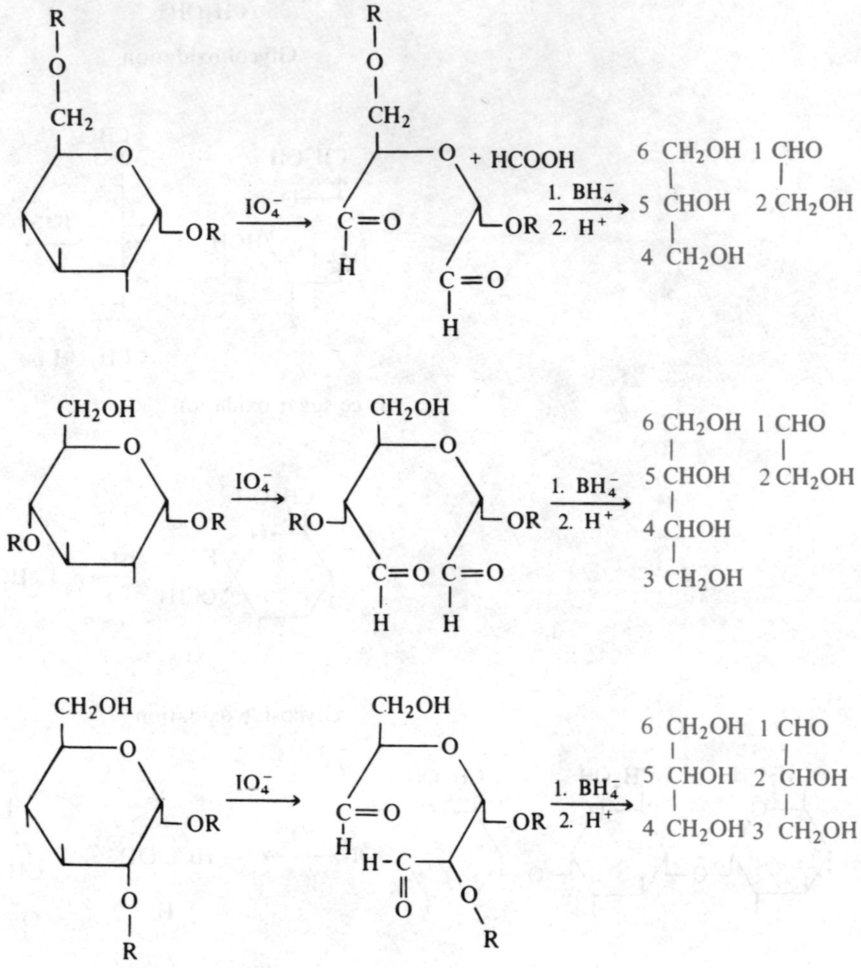

Figure 3-22 Three types of linear polyglucans and the polyols that would be formed from them if they were treated with IO_4^- followed by reduction and hydrolysis of the resulting dialdehydes. The numbers by the polyols indicate which carbons of the original compound they were derived from.

Enzymatic Degradation

It is now possible to obtain a large number of specific glycosidases, which have been useful in structural analysis. These enzymes will cleave specific residues from the nonreducing end of a polysaccharide chain. Enzymes such as galactosidase, mannosidase, *N*-acetyl-glucosaminidase and neuraminidase have been successfully used to determine the sequence of the short heteropolysaccharide chains found in glycoproteins. The polysaccharide is incubated with one, or a combination, of these enzymes, and the sequence of residues can be determined by following the hydrolysis of specific sugars. As many of these enzymes are specific for either an α- or a β-glycoside bond, this additional information about the structure can be obtained by the same procedure.

PROBLEMS

1. Write the correct Haworth perspective formulas for the following sugars or sugar derivatives. (See Figure 3-5 for the correct stereochemistry of the monosaccharides.) (a) β-methyl-D-talofuranoside, (b) α-L-mannofuranose, (c) 4-α-D-glucopyranosyl-α-methyl-D-galactopyranoside, (d) 2,6-dimethyl-α-methyl-D-allpyranoside.

2. Which two aldohexoses in Figure 3-5 would you expect to give the same phenylosazone as D-sorbose (Figure 3-6)?

3. The specific optical rotation of maltose in water is +138°. What would be the concentration (g/100 ml) of a maltose solution that had an observed optical rotation of +23° when measured in a 10-cm polarimeter tube?

4. Assume that a bacterial polysaccharide was isolated, subjected to exhaustive methylation, and hydrolyzed. Following this, it was possible to identify a large amount of 2,3,4-tri-O-methyl-D-glucose and much smaller, but relatively equal, amounts of 2,4-di-O-methyl-D-glucose and 2,3,4,6-tetra-O-methyl-D-glucose. What would this tell you about the basic structure of the polysaccharide?

5. Assume that you have been given a mono-O-methyl-α-methyl-glycoside of D-glucose, but do not know what position the hydroxymethyl group is on. The sugar is oxidized with periodate, and the products of the oxidation are reduced with borohydride and hydrolyzed. The only products that can be identified are glycerol and glyceraldehyde. What was the structure of the original sugar?

Suggested Readings

Books

Davidson, E. A. *Carbohydrate Chemistry*. Holt, Rinehart and Winston, New York (1967).

Florkin, M., and E. H. Stotz (eds.). *Comprehensive Biochemistry*, vols. 5 and 26. Elsevier, Amsterdam-New York (1963, 1968).

Pigman, W. W., and D. Horton (eds.). *The Carbohydrates*, vols. 1A and 1B. Academic Press, New York (1972).

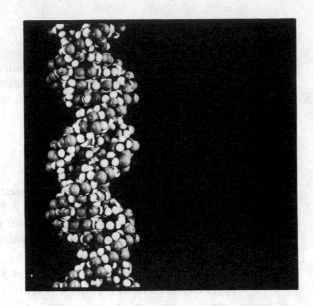

protein chemistry

chapter four

Proteins are macromolecular polymers composed of L-α-amino acids linked by peptide bonds. Some of the twenty commonly occurring amino acids found in proteins have ionizable functional groups which give proteins their charged properties in solution. Protein structure is usually discussed in terms of *primary structure*, the linear sequence of amino acids in the peptide chain; *secondary structure*, the hydrogen-bond stabilized interactions between peptide carbonyl oxygens and amide hydrogens along this linear polymer; *tertiary structure*, the association between different portions of the protein to give it a unique three-dimensional structure; and *quaternary structure*, the noncovalent association between more than one polypeptide chain to form a multisubunit protein. Proteins function in cells both as structural elements and in providing the cells enzymatic capability.

Proteins—macromolecules that impart both structural and catalytic properties to all living cells—could well be considered the basic material of the biochemist. Early knowledge of the properties of proteins was obtained from the study of a few readily available natural materials which are largely protein in nature. It was found that common substances such as blood serum, egg white, and milk would coagulate when heated or when treated with strong acids. When these substances were subjected to chemical analysis, they were found to contain a large amount of nitrogen. On the basis of the similar chemical composition of these substances, chemists initially thought that all proteinaceous material was made up of different combinations of some uncharacterized basic subunit. As methods for the separation of proteins and techniques of chemical analyses improved, it was apparent that this was not true. Although many proteins had very similar properties, they were drastically different compounds.

PROTEIN CLASSIFICATION

All proteins are mixed polymers of amino acids, and as such, all contain carbon, hydrogen, oxygen, and nitrogen. Most contain sulfur, a large number contain phosphorus, and some contain various other mineral elements. Proteins can be subdivided and classified into a number of general groups, but most of these classification schemes are of less value today than they were years ago when they were developed. At that time it was more difficult to purify and study single proteins, and general classes of proteins were studied in much the same manner as a single pure protein would be today. Proteins have often been divided into globulins, albumins, and other classes on the basis of their solubility in water or salt solutions. One general classification which is still useful is based on the nonprotein (prosthetic group) portion of conjugated proteins.

In that sense, a protein that was tightly associated with a nucleic acid would be called a *nucleoprotein*; one associated with lipid, usually phospholipid, would be called a *lipoprotein*. *Phosphoproteins* contain covalently bound phosphate groups, and *glycoproteins* contain varying amounts of carbohydrate bound to the polypeptide chain. There are a large number of *metalloproteins* which have ions such as Cu^{2+} or Zn^{2+} associated with them, and *chromoproteins* which contain colored heme or flavin prosthetic groups.

AMINO ACIDS

When subjected to hydrolysis, proteins are broken down into a small number of individual α-amino acids. Although a few additional amino acids are found in low concentration in specialized proteins and others are found in nonprotein natural products, only about twenty amino acids are generally found in most proteins. The commonly occurring amino acids can be conveniently divided into a number of groups based on the constituent on the α carbon, and these are illustrated in Figure 4-1. Proline and hydroxyproline are usually called amino acids, even though the presence of a secondary, rather than a primary, amino group would place them into the chemical class of imino acids.

NEUTRAL AMINO ACIDS

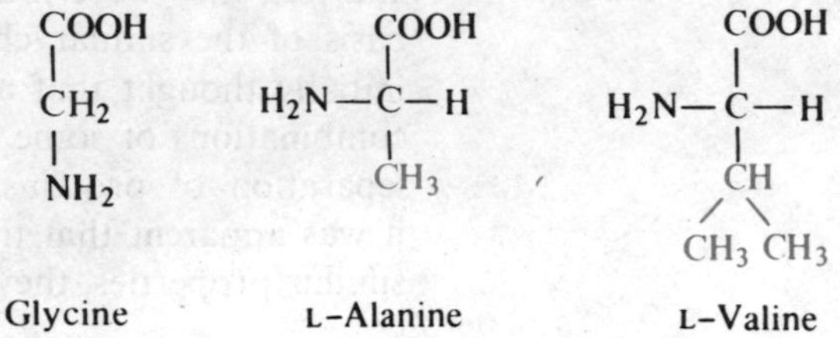

Glycine
(Gly)

L-Alanine
(Ala)

L-Valine
(Val)

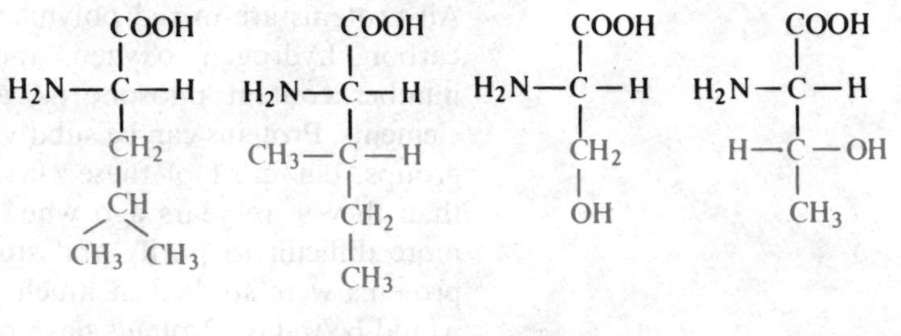

L-Leucine
(Leu)

L-Isoleucine
(Ile)

L-Serine
(Ser)

L-Threonine
(Thr)

AROMATIC AMINO ACIDS

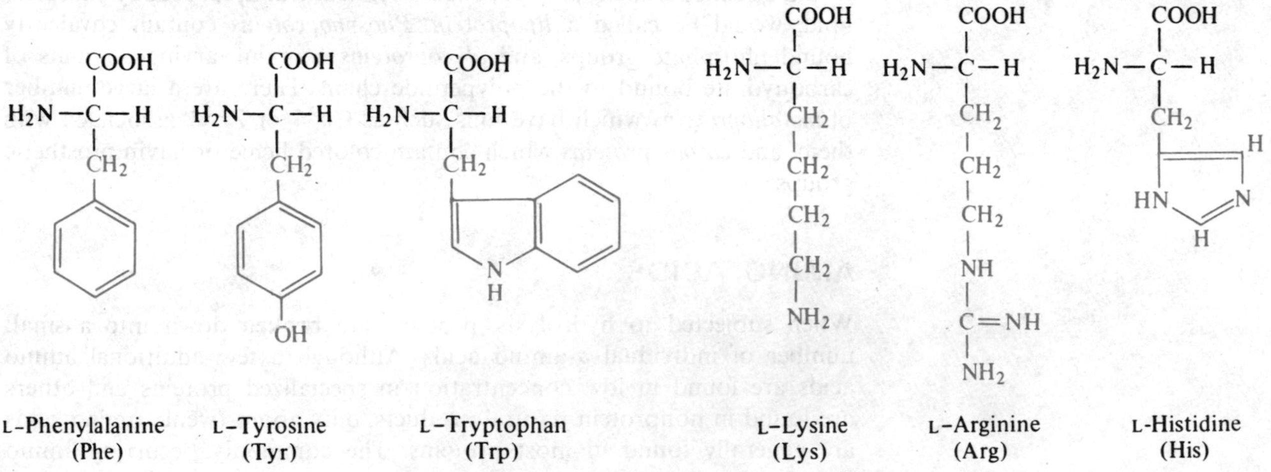

L-Phenylalanine
(Phe)

L-Tyrosine
(Tyr)

L-Tryptophan
(Trp)

BASIC AMINO ACIDS

L-Lysine
(Lys)

L-Arginine
(Arg)

L-Histidine
(His)

SULFUR–CONTAINING AMINO ACIDS

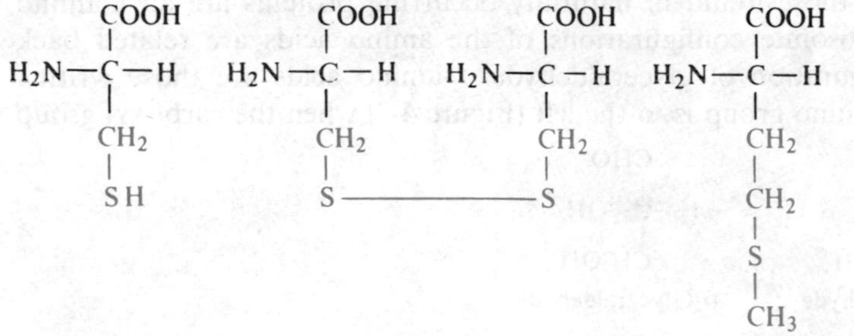

L–Cysteine (CySH)
L–Cystine (Cys)
L–Methionine (Met)

IMINO ACIDS

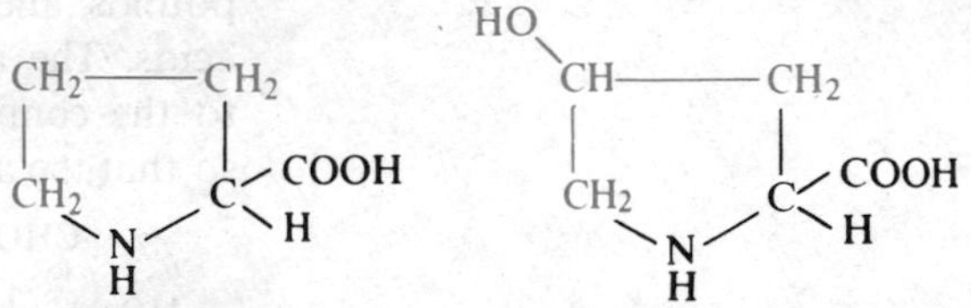

L–Proline (Pro)
L–Hydroxyproline (Pro–OH)

Figure 4-1 Structures of the amino acids commonly found in proteins.

ACIDIC AMINO ACIDS AND THEIR AMIDES

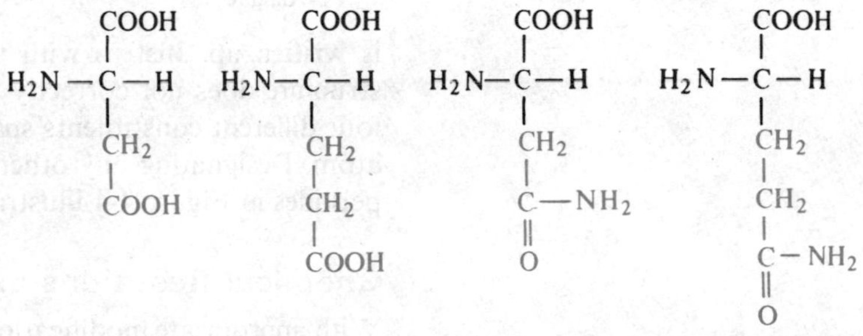

L–Aspartic Acid (Asp)
L–Glutamic Acid (Glu)
L–Asparagine (Asn)
L–Glutamine (Gln)

The grouping of the amino acids in Figure 4-1 makes it easy to remember the structures, but does not have much functional significance. The contribution of each amino acid to the properties of a protein depends to a large extent on the ionic properties of the residue at physiological pH values, and the amino acids can also be grouped on the basis of the properties of the group attached to the α carbon, as shown in Table 4-1.

Table 4-1 Classification of Amino Acids Based on Properties of the α-Carbon Substituent

Nonpolar	Polar, But Neutral	Charged Polar
Gly	Ser	Asp
Ala	Thr	Glu
Val	Asn	His
Leu	Gln	Lys
Ile	Tyr	Arg
Pro	Trp	
Phe	Cys	
Met		

53

With the exception of glycine, the amino acids are optically active compounds, and those found in naturally occurring proteins are all L-amino acids. The absolute configurations of the amino acids are related back to the configuration of glyceraldehyde. L-amino acids are those written so that the amino group is to the left (Figure 4-2) when the carboxyl group

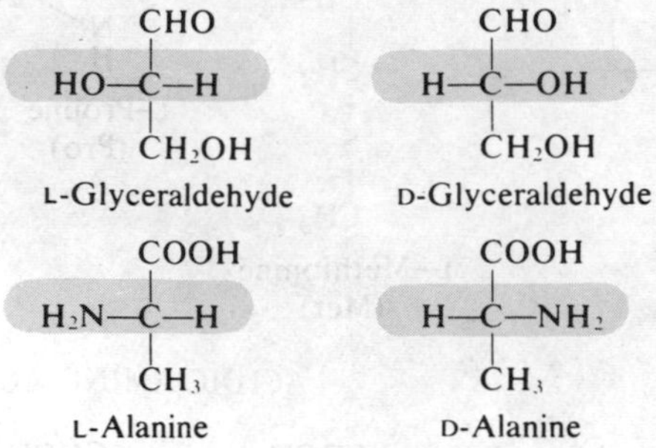

Figure 4-2 Stereochemistry of the amino acids. The D and L configurations of amino acids are related back to the structure of D- and L-glyceraldehydes.

L-Glyceraldehyde   D-Glyceraldehyde

L-Alanine   D-Alanine

is written up. Just as with the simple sugars, an amino acid or peptide structure does not correctly indicate configuration unless it clearly shows four different constituents spaced 90° apart around the asymmetric carbon atom. Designating any other type of structure as D or L is incorrect. The peptides in Figure 4-4 illustrate the correct form.

Chemical Reactions of Amino Acids

With appropriate modifications of procedures, the amino acids will undergo many of the common reactions of simple carboxylic acids and also many

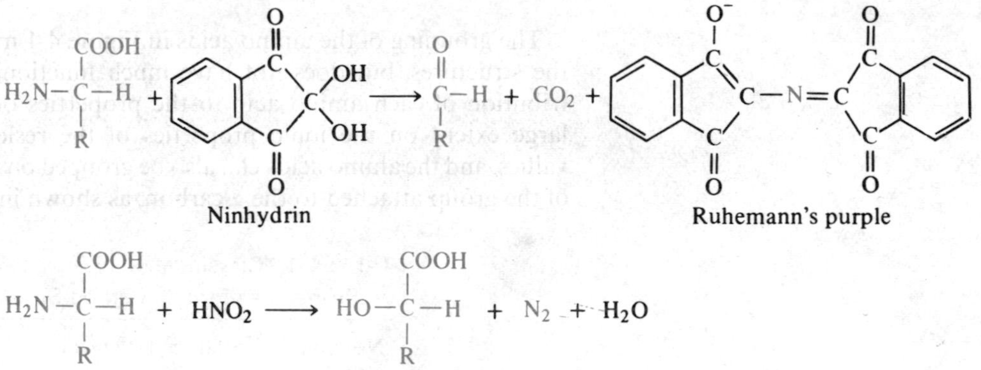

Ninhydrin   Ruhemann's purple

Figure 4-3 Ninhydrin and Van Slyke reactions. In the ninhydrin reaction, the colored compound which is produced forms the basis for a colorimetric quantitative assay of the amino acid. The color produced depends somewhat on the amino acid used. It is also possible to measure the CO_2 released. The identification of the aldehyde produced may be used for characterization purposes. Other types of amino compounds can cause interference with the ninhydrin assay for amino acids. The liberation of nitrogen from amino acids by HNO_2 is called the Van Slyke reaction, and if the amount of N_2 formed is measured manometrically, it can be used as a quantitative assay. Amines other than α-amino groups will also give this reaction, but at a slower rate.

reactions that would be expected from an amino group. Two of the chemical reactions which have been adapted as quantitative methods for determining the concentration of amino acids in solution, the *ninhydrin* reaction and the *Van Slyke* reaction, are illustrated in Figure 4-3. In alkaline solution, the peptide bond of proteins reacts with Cu^{2+} to form a blue color, and this *biuret* reaction is also used as a quantitative assay for proteins. The amino acids which contain a chemically reactive group such as tyrosine, tryptophan, or arginine will undergo reactions that form colored derivatives. These reactions, such as the *Millons* reaction for phenolic groups, the *xanthoproteic* test for aromatic rings, the *Sakaguchi* reaction for arginine, or the *Hopkins-Cole* test for tryptophan, have been used as the basis for various qualitative tests for proteins. Other chemical reactions of the carboxyl and amino groups have also been extremely important in the determination of protein structure. These will be considered later in the chapter.

PEPTIDE BOND THEORY

The amino acids that are present in peptides and proteins are joined together by peptide bonds, which are chemically similar to a substituted amide (Figure 4-4). The hypothesis that peptide bonds determine the basic structure of proteins was independently proposed by Fischer and Hofmeister in 1902. The evidence available to them at that time included the following observations: first, there was a large increase in both free amino groups and free carboxyl groups during hydrolysis of proteins, and the magnitude

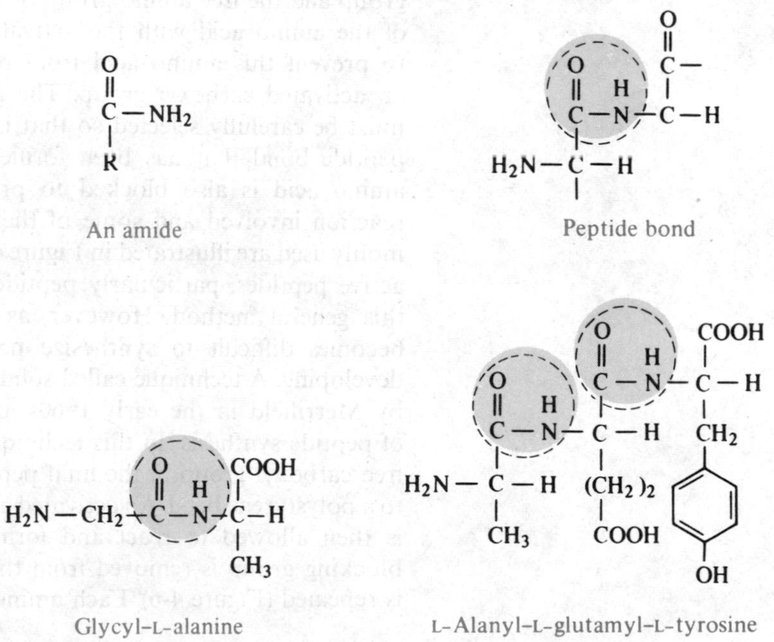

Figure 4-4 The peptide bond. Peptides are named following the convention that the amino acid residue with the free α-amino group is the first one named. Structures of peptides are usually written with this free α-amino group to the left. If the configurations of the amino acids are indicated in the name of the peptide, the structure should be written so that the configuration of each amino acid can be seen.

An amide

Peptide bond

Glycyl–L–alanine

L–Alanyl–L–glutamyl–L–tyrosine

of this increase was roughly the same for both; second, enzymes which could hydrolyze proteins could also hydrolyze synthetic di- and tripeptides; third, peptides identical to known synthetic peptides could be isolated from partial hydrolysates of proteins; and fourth, proteins were known to give a biuret reaction with alkaline $CuSO_4$.

By convention, the structures of peptides are written with the free amino group, the "amino terminal" residue, to the left. They are then named as if they were derivatives of the amino acid with the free carboxyl group, the "carboxyl terminal" residue (Figure 4-4). For any peptide larger than a few amino acids, structural formulas are seldom written; the accepted abbreviations used to designate these structures are given in Figure 4-1. The convention of writing the residue with the free amino group to the left must be followed if the structural formula is to be unambiguous.

All proteins are, of course, composed of polypeptide chains, but not all polypeptides that have physiological activity are called proteins. In general, an amino acid polymer with a molecular weight of over 5000, about forty amino acid residues, would be referred to as a protein. Molecules smaller than this, such as many of the peptide hormones, would more commonly be called a polypeptide.

CHEMICAL SYNTHESIS OF PEPTIDES

Protein chemists have always been interested in synthesizing small peptides of known sequence. The general chemical reaction used to form the peptide bond must take place between an amino acid having an activated carboxyl group and the free amino group of a second amino acid. The amino group of the amino acid with the activated carboxyl group must be protected to prevent this amino acid from reacting with another molecule having an activated carboxyl group. The group used to block the amino group must be carefully selected so that it can be removed without cleaving the peptide bond that has been formed. The carboxyl group of the second amino acid is also blocked to prevent side reactions. The generalized reaction involved and some of the blocking and activating groups commonly used are illustrated in Figure 4-5. A number of biologically important active peptides, particularly peptide hormones, have been synthesized by this general method. However, as the size of the molecule increases, it becomes difficult to synthesize peptides without serious side reactions developing. A technique called solid-state peptide synthesis was introduced by Merrifield in the early 1960s and promises to revolutionize the field of peptide synthesis. In this technique, the amino acid which will have the free carboxyl group in the final peptide is attached by the carboxyl group to a polystyrene bead. An activated amino acid with its amino group blocked is then allowed to react and form a polystyrene-bound dipeptide. The blocking group is removed from the dipeptide, and the reaction sequence is repeated (Figure 4-6). Each amino acid is added stepwise to the growing chain, and the peptide is not removed from the solid support until the amino terminal residue has been added. The major advantage of the procedure is that each intermediate need not be purified by crystallization before proceeding to the next step. Because the growing chain is bound to the insoluble support, it can be washed free of contaminating reagents and will not be lost. The entire procedure has been automated so that the amount of labor involved is greatly reduced.

The technique was first applied to the synthesis of the bradykinins, a series of peptide hormones containing nine amino acids, and in 1965, to the synthesis of insulin, a hormone with fifty-one amino acids in two different peptide chains. More recently, Merrifield's group has synthesized ribonuclease, an enzyme with 124 amino acids in a single peptide chain, and has shown that the synthetic enzyme has most of the same properties

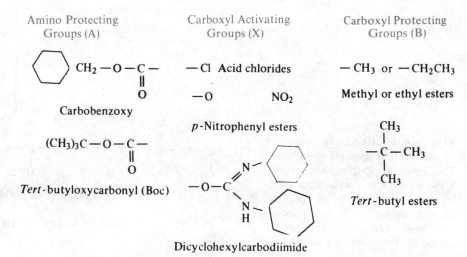

Figure 4-5 Chemical synthesis of peptides. The general reaction is between an amino acid with its carboxyl group activated and the free amino group of a second amino acid to form a peptide bond. The carboxyl group of the second amino acid and the amino group of the first must be protected. Some of the common blocking groups are illustrated. The *tert*-butyloxycarbonyl, Boc, group is the one most commonly used in solid-state peptide synthesis (Figure 4-6).

as the natural protein although it does not have the full activity. Although much smaller amounts of both insulin and another form of ribonuclease have also been synthesized by the more classical methods, there is no doubt that the automated solid-state peptide synthesis will be the method used in the future for the synthesis of most large peptides. One interesting application will be the synthesis of proteins which differ in only a few residues from the native protein. This experiment is an attempt to delineate the contribution of individual amino acids to protein structure or function. The recent synthesis of growth hormone by this general method also suggests that these synthetic products may have possible clinical applications.

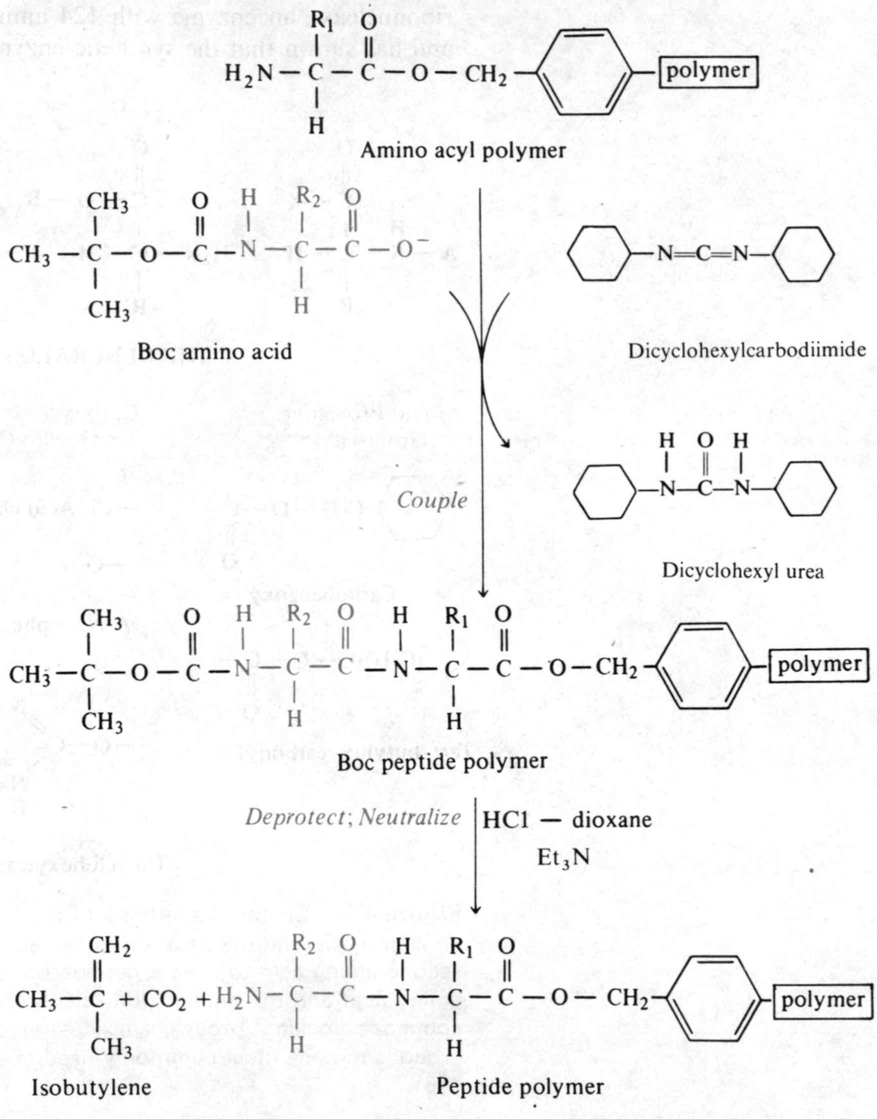

Figure 4-6 Reactions involved in solid-state peptide synthesis. This series of coupling and deprotecting reactions can be continued to form any length of peptide desired. When the desired peptide has been produced, it can be removed by treatment with HBr in trifluoroacetic acid. This will remove the final blocking group and also remove the peptide from the polymer. The first amino acid is put onto the polymer by the reaction of a Boc amino acid with the chloromethyl derivative of the polymer.

IONIC PROPERTIES OF AMINO ACIDS AND PROTEINS

The structures of the amino acid in the previous figures have been written as if they were neutral, uncharged molecules. If any of the simple amino acids are dissolved in an aqueous solution, they will exist in the doubly charged or zwitterion form, in which the amino group carries a positive charge and the carboxyl group is ionized. As each of these groups has a distinct pK_a associated with its ionization, the actual ionic configuration of any of the amino acids depends on the pH of the solution. The ionic forms of glycine which predominate at various pHs are illustrated in Figure 4-7. Because a simple neutral amino acid has two ionizable groups,

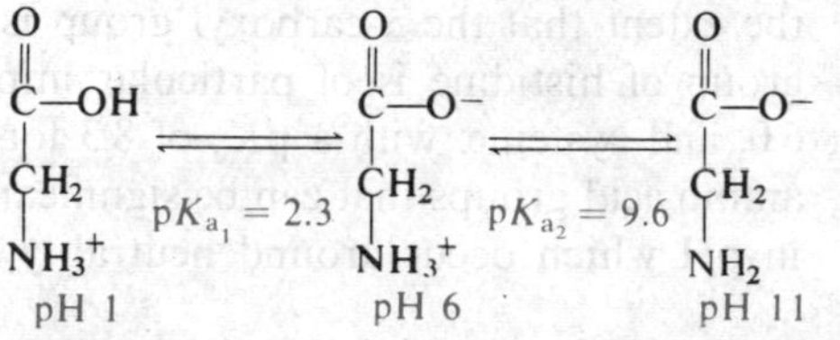

Figure 4-7 Ionic forms of glycine which predominate at different pHs.

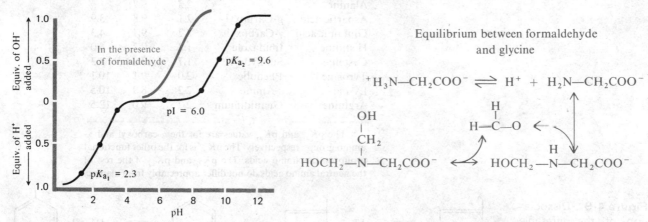

it will act as a buffer over two distinctly different pH regions. If glycine is placed in water, the pH will be around 6 and titration of the solution with either acid or base will involve a buffering region. The titration curve in Figure 4-8 illustrates the biphasic nature of the titration curve of any

Equilibrium between formaldehyde and glycine

$$^+H_3N{-}CH_2COO^- \rightleftharpoons H^+ + H_2N{-}CH_2COO^-$$

Figure 4-8 Titration curve of glycine. If glycine is placed in water, the solution will have a pH of about 6.0 and it can be titrated with acid or base to demonstrate two buffering regions. If formaldehyde is added to the solution at a pH near neutrality, the equilibrium shown will be achieved. As formaldehyde reacts with the $-NH_2$ form of the amino acid, it will pull the equilibrium in that direction and reduce the amount of the $-NH_3^+$ form. The net effect is that the pH at which the removal of the proton from the $-NH_3^+$ form is essentially complete is markedly reduced. The entire curve is therefore shifted to a more acidic pH, and the end point is about pH 8.

of the simple amino acids such as glycine, and also the behavior of the alkaline portion of this curve when the titration is carried out in the presence of formaldehyde. This modification, which is called a *formol titration*, brings the end point of the titration into the region where the color change of phenolphthalein can serve as an indicator. Because of this, it was once used as a simple method for quantitatively determining amino acid concentrations.

In addition to the α-amino and the carboxyl groups of the amino acids, the other charged groups will also have characteristic pK_as (Table 4-2). The pK_a of the ω carboxyl of glutamic and aspartic acid is about 4, which is much less acidic than the α carboxyl, and near the pK_a for other simple carboxylic acids. Its pK_a is higher because its dissociation is not influenced by the presence of the neighboring positively charged amino groups to the extent that the α-carboxyl group is. The dissociation of the imidazole group of histidine is of particular importance because it, with a pK_a of 6.0, and cysteine, with a pK_a of 8.3 for the sulfhydryl group, are the only amino acid groups that can be significantly influenced by the small changes in pH which occur around neutrality (Figure 4-9). If an amino acid has

Table 4-2 pK_a Values for Some Amino Acids[a]

Amino acid	Functional Group	pK_{a_1}	pK_{a_2}	pK_{a_3}
Alanine	—	2.3	9.7	
Aspartic acid	β-Carboxyl	2.1	9.8	3.9
Glutamic acid	γ-Carboxyl	2.2	9.7	4.3
Histidine	Imidazole	1.8	9.2	6.0
Cysteine	Sulfhydryl	1.7	10.8	8.3
Tyrosine	Phenolic	2.0	9.1	10.1
Lysine	ε-Amino	2.2	8.9	10.5
Arginine	Guanidinium	2.2	9.0	12.5

[a] The pK_{a_1} and pK_{a_2} values are for the α-carboxyl and α-amino groups, respectively. The pK_{a_3} is for the other functional group on the amino acids. The pK_{a_1} and pK_{a_2} of the rest of the neutral amino acids do not differ appreciably from alanine.

Figure 4-9 Dissociation of the titratable groups in amino acids. The lysine, glutamic acid, or second amino or carboxyl group on aspartic acid will have the same type of dissociation as that seen for α-amino and α-carboxyl groups of amino acids.

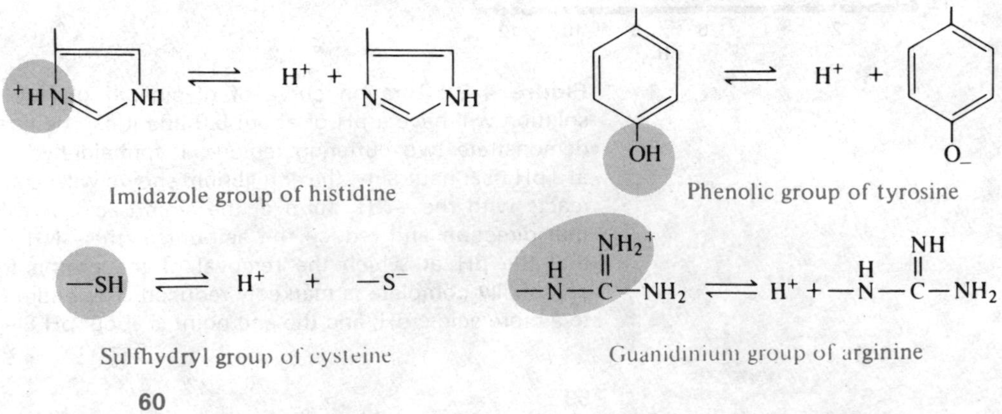

Imidazole group of histidine

Phenolic group of tyrosine

Sulfhydryl group of cysteine

Guanidinium group of arginine

more than one charged group, this will be reflected in a more complex titration curve than that shown in Figure 4-8. Each ionizable group will have a buffering region associated with it, and it will require as many equivalents of base to titrate the amino acid as there are charged groups. Two examples of such titration curves are shown in Figure 4-10.

Figure 4-10 Titration curve of glutamic acid and histidine. These curves are drawn assuming that the amino acid was in acid solution and all the groups were titrated with base. Note that in the case of glutamic acid, the pK_as of the two carboxylic acid groups are so close that what is seen is essentially one extended buffering region. The dotted lines indicate the shape of the individual curves.

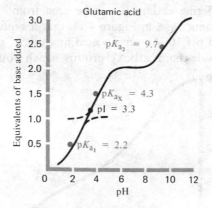

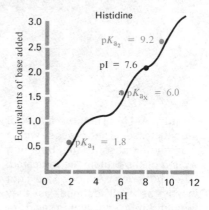

Isoelectric Points

Because of the presence of the charged groups, an amino acid in solution will have a positive charge at a low pH. As the pH is raised, it will pass through a region where the positive and the negative charges are equal, and will eventually have a net negative charge. The pH at which the net charge on an amino acid is 0 is defined as the pI, or the *isoelectric* point. The pI can be experimentally determined as being that pH at which the amino acid will show no migration in an electric field. The pI of a simple amino acid will be halfway between the two pK_a values, that is,

$$pI = \frac{pK_{a_1} + pK_{a_2}}{2}$$

With glycine, for example,

$$pI = \frac{2.4 + 9.6}{2} = 6.0$$

It should be clear that the pI is not the pH where there is no charge on the amino acid, but rather the point where the positive and negative charges are equal. For an amino acid with three ionizable groups, the pI can be determined by a careful consideration of the pK_a values of the ionizable groups involved. For example, the pI of lysine is 9.7. A consideration of the Henderson-Hasselbalch equation will make it clear that at this point the total contribution of positive charge from the two amino groups will be +1, which will be balanced by the complete negative charge on the carboxyl group.

Titration of Proteins

The titration curve of a protein will reflect the total of all the ionizable groups on the amino acid residues which it contains. Because the pK_as of the individual groups are to some extent influenced by neighboring groups, the buffering regions tend to overlap even more than would be apparent from simply considering the groups involved. Therefore, the titration curves of proteins do not usually contain distinct sharp breaks. These curves will, however, give some indication of the composition of the protein in terms of which amino acid residues furnish most of the buffering action. It can be seen from the titration curve of the protein, ribonuclease, in Figure 4-11, that it buffers rather strongly around pH 3–4 and pH 10. An amino acid analysis of this protein shows that it contains 10 side-chain carboxyl groups, which would contribute the low pH buffering,

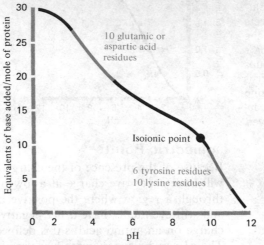

Figure 4-11   Titration curve of ribonuclease. The groups which contribute most to the buffering action of the protein are indicated. The buffering regions are not nearly as well defined as they would be in an amino acid or small peptide. In addition to the charged groups indicated, ribonuclease contains six histidine and four arginine residues and the amino and carboxyl terminal groups, which would contribute to its buffering action.

and 16 side-chain amino and phenolic groups, which would buffer around pH 10. The calculated isoelectric point, based on the known amino acid composition of the protein may differ somewhat from the experimentally determined value because of the influence of buffer salts used when the isoelectric point is determined.

DETERMINATION OF MOLECULAR WEIGHTS

Because proteins have high molecular weights and are temperature sensitive, most of the thermodynamic methods of molecular weight determination commonly used for small molecules cannot be applied. Simple calculations will show that it is impossible to get sufficient amounts of most proteins in solution to cause an appreciable freezing point depression or boiling point elevation, and the lability of proteins to high temperatures would also rule out the latter method.

If the amino acid composition of a protein is known, a minimal molecular weight can be obtained by assuming that the amino acid present in the least amount is contained only once in the protein. On the basis of this assumption, the number of each of the other amino acids present can be calculated, and the molecular weight obtained by adding them together.

If the protein contains a prosthetic group such as a flavin, or is a metalloprotein, a minimum molecular weight can also be calculated on the assumption that the protein contains only one mole of this constituent. The molecular weight calculated by this means will be some fraction of the true molecular weight. A second molecular weight value can be obtained by some physical measurement, and this will indicate how many times the minimal molecular weight must be repeated to give the true value.

In general, physical measurements are able to furnish information about both molecular weight and homogeneity of protein solutions. Several techniques are available to determine the molecular weight of protein solutions by physical measurements, but if the solution available is not pure, the molecular weight obtained will be an average of the molecular weight of the various protein species present. If a protein solution does consist of a mixture of components, they can usually be detected by the different rates of movement they will exhibit in a centrifugal or an electrical field. These types of physical methods are therefore used to give an indication of the molecular homogeneity of protein preparations.

Osmotic Pressure and Light Scattering

An indication of the molecular weight of a protein can be obtained by osmotic pressure measurements. If a protein solution and its solvent are divided by a membrane which is permeable to solvent but not to protein, there will be a net migration of solvent molecules from the solvent side to the protein solution. The resulting increase in hydrostatic pressure in the protein solution can be measured by an instrument called an osmometer. For any given weight of protein there will be more particles in solution from a lower molecular weight protein; therefore, the osmotic pressure is inversely related to the molecular weight.

A second method that can be used to yield molecular weight data is light scattering. Proteins in solution behave as any particle; they will scatter light from a beam passing through a protein solution. Large particles will scatter more light than small ones. It is therefore possible to calculate the molecular weight of a protein by measuring the percent of light scattered at a particular angle when passed through a protein solution.

Ultracentrifugation

Although light scattering and osmotic pressure measurements have been of some value to protein chemists, the classical method used to obtain an indication of molecular weight of a protein and also an indication of purity is ultracentrifugation. If very high centrifugal forces are employed (up to $500,000 \times g$), it is possible to sediment protein molecules at a reasonable rate. Other factors being held equal, large molecules will sediment faster than small ones. In the technique of sedimentation velocity, the rate at which protein particles sediment at a given centrifugal force is determined. From this information and a knowledge of either the diffusion coefficient or frictional coefficient, a molecular weight can be calculated. Sedimentation velocity is also widely used to determine purity of a protein. Proteins of different molecular weights will sediment at different rates, and the presence of two components will be detected by the optical system used to follow the movement of the proteins (Figure 4-12).

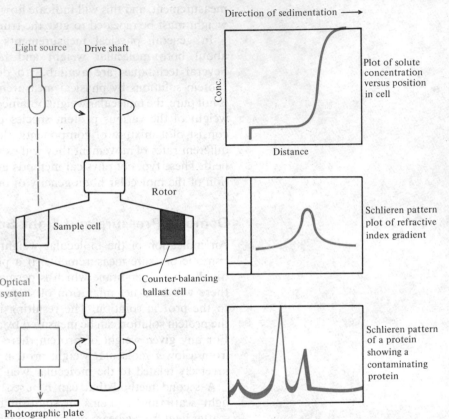

Figure 4-12 Analytical ultracentrifuge. At the left is a diagram of the rotor showing the optical system which allows optical measurements to be made while the sample is being centrifuged. At the right are shown idealized plots of protein concentration, and an actual Schlieren pattern of a protein solution.

In the technique of sedimentation equilibrium, the speed of the centrifuge rotor (and therefore the strength of the gravitational field) is adjusted so that the force tending to sediment the protein is balanced by diffusion of the protein molecules. The molecular weight can then be calculated from values of the concentration distribution observed in a given gravitational field.

Density Gradient Centrifugation

In the technique of ultracentrifugation, a complicated optical system must be used to follow the changes of the protein solution during the centrifugation. In a more recently developed technique, density gradient sedimentation, this is not needed. Two general techniques have been used. The first involves sedimentation velocity through a gradient.

A concentration gradient of sucrose, which increases from the top to the bottom of the tube, is prepared, and the macromolecular solution is layered on top of this gradient. The system is centrifuged for several hours, and after any given time the higher molecular weight molecules will have migrated further and will be in the higher density area of the gradient. The solution is then removed carefully from the tube (Figure 4-13) and the macromolecular concentration in each part of the tube is determined.

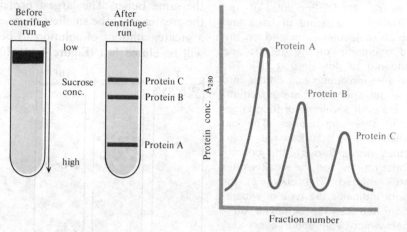

Figure 4-13 Density gradient centrifugation. A gradient of increasing sucrose concentration is prepared from the top to the bottom of the tube, and the protein layered on top. After centrifugation a hole is punched in the bottom of the centrifuge tube, and drops are collected in separate fractions. Both total protein (absorbence at 280 nm) and enzyme activity in each fraction can be measured.

If particles of known molecular weight are also used as markers, it is possible to make rough calculations of macromolecular weights by this technique. This method depends on the velocity with which the macromolecule moves for separation. The gradient tends to stabilize the moving bands of different molecular weight and to prevent the more rapidly moving components from reaching the bottom of the tube before the lighter components are resolved.

The second method involves density gradient sedimentation equilibrium, or isopycnic equilibrium. This is usually carried out in high concentrations of a dense salt such as cesium chloride, and the concentration gradient of the salt is generated by an extended centrifugation. If the salt concentration is chosen so that the gradient is at some point equal in density to the particles being separated, they will move down the gradient until they reach the point, and will be held there. Depending on their density, different particles will reach different equilibrium positions in the gradient. The gradient can then be removed from the tubes and the particles recovered.

Both types of density gradient techniques have been extremely valuable in investigating the properties of the nucleic acids (see Chapter 6) and large multiunit protein particles, and sedimentation velocity through gradients has been useful for studies of particles as small as proteins.

Gel Filtration

One of the most commonly used methods to obtain some indication of molecular weight is to use the technique of molecular sieve chromatography or gel filtration. Molecular sieves are high molecular weight polymeric carbohydrate or polyacrylamide resins with varying degrees of cross-linking between chains which have been formed into beads. In solution, these materials form gels which, because of the cross-linking, act as if they were sieves with different pore sizes. A low molecular weight protein will be able to diffuse into the beads, whereas one with a high molecular weight will be excluded. When used to separate proteins, a column of this material is prepared, and a mixture of proteins is added to the top and eluted with the same buffer. The largest proteins will move through the column at the greater rate; the smallest ones, as they can diffuse into the gel and have a greater amount of solution to be distributed in, will move slower and will be eluted last (Figure 4-14). If the column has been calibrated with

Figure 4-14 Principle of gel filtration. A mixture of large and small molecules is placed on top of a column of the gel (I) and allowed to flow into it (II). The smaller molecules can diffuse into the gel and will therefore equilibrate with a volume of fluid larger than the bigger ones. The net effect will be that the larger molecules move through the column most rapidly and the smaller ones are retarded. The curve at the right indicates the type of separation that can be achieved on a gel which will allow even very large particles to diffuse into the gel. The numbers are a rough indication of molecular weight.

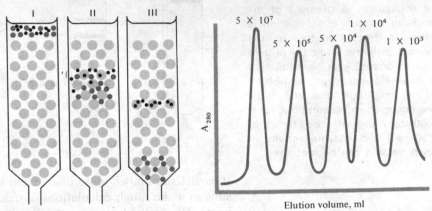

proteins of known molecular weight, an estimation of the molecular weight of an unknown protein can be obtained. Although the method is not as accurate as some others, the simplicity of the technique and the low cost of the equipment needed have made it very popular.

Electrophoresis

Because of the large number of ionizable groups present on the amino acid side chains, proteins in solution behave as large charged polyelectrolytes, and the presence of the charge will cause the molecules to migrate in an electrical field. Because their amino acid composition varies, different proteins will possess different charges at any given pH and the rate of their migration in a charged field will differ. In the original techniques which were developed, a specially designed electrophoresis cell was used. By the use of suitable optical methods the presence of more than one species of protein migrating through the solution in the cell can be followed (Figure 4-15). This basic technique is relatively complicated, not readily adapted to routine assay of large numbers of samples, and is seldom used by contemporary biochemists today.

Modifications of the electrophoretic technique where the protein solution is held on an inert support such as starch, paper, or an inert gel are, however, frequently used. In these cases the proteins are applied at a single point and are allowed to migrate for sufficient time to separate different components. Then the supporting medium is treated with a protein-sensitive stain to detect the number of components in the mixture (Figure 4-15).

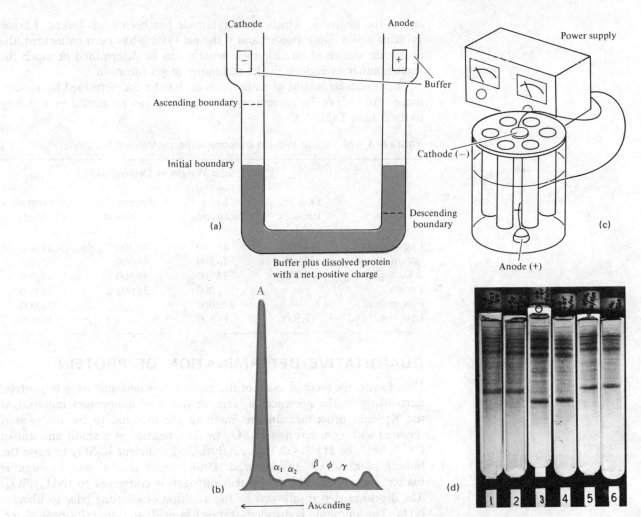

Figure 4-15 (*a*) Schematic diagram of a moving boundary electrophoresis apparatus. (*b*) Electrophoretic pattern of human blood plasma proteins (pH 8.6). A = serum albumin; ϕ = fibrinogen; α_1, α_2, β, and γ are various globulins. (*c*) Diagram of the apparatus used for acrylamide gel electrophoresis. The protein solution is put on top of a small tube full of the gel and is allowed to move down through the gel when the electrical field is applied. After some time the gel can be removed and stained to visualize the proteins. (*d*) Photograph of an extract of intestinal tissue that has been treated in this way. (Courtesy of H. F. DeLuca)

Modification of this technique, in which the proteins are allowed to migrate into a tube of acrylamide gel, has been particularly useful. It is possible to carry out these separations in the presence of a high concentration of an anionic detergent such as sodium dodecylsulfate which will effectively neutralize the charge differences of various proteins. Under these conditions the rate of migration of a protein is a function of its size, and of the degree to which the acrylamide has been cross-linked. Larger proteins will migrate slower, and if the gel system has been calibrated, the molecular weight of an unknown protein can be determined in much the same manner as is done in the technique of gel filtration.

The molecular weight of many proteins has been determined by various means. An idea of the agreement between them can be gained by referring to the data in Table 4-3.

Table 4-3 Molecular Weight Determination by Various Methods

Protein	Molecular Weight as Determined by			
	Osmotic Pressure	Light Scattering	Sedimentation Equilibrium	Sedimentation Velocity
Egg albumin	44,000	45,700	40,500	44,000
Serum albumin	73,000	76,600	68,000	70,000
β-Lactoglobulin	38,000	35,700	38,500	41,500
Pepsin	—	37,000	39,000	36,000
Fibrinogen	—	340,000	—	330,000
Lysozyme	17,500	14,800	—	16,000

QUANTITATIVE DETERMINATION OF PROTEIN

Historically, the method used for the quantitative determination of protein, particularly in the presence of large amounts of nonprotein material, is the Kjeldahl procedure. In this method, the material to be analyzed is digested with concentrated H_2SO_4 in the presence of a small amount of Cu^{2+}, Se^{2+}, or Hg^{2+} salt as a catalyst, and sufficient K_2SO_4 to raise the boiling point of the sulfuric acid. Under these conditions, the organic matter is oxidized, and the protein nitrogen is converted to $(NH_4)_2SO_4$. The digestion step is followed by the addition of a strong base to liberate NH_3. The ammonia is distilled, trapped in acid, and quantitatively determined by titration or colorimetric assay. Almost all organic forms of nitrogen will be converted to NH_3 by the conditions of the digestion. If the procedure is to be made specific for protein nitrogen, a preliminary separation of proteins from other nitrogenous material is needed. The results of Kjeldahl analyses are usually expressed as crude protein by multiplying the percentage of nitrogen in the sample by 6.25. This conversion is based on an average content of 16 percent nitrogen in many proteins. Although this average may not be valid for specific purified proteins, it is sufficiently accurate for most purposes.

The Kjeldahl procedure is very useful for the determination of the amount of protein in crude mixtures, and if a micro modification is used, it can be used in the range of 1–10 mg of protein. It is, however, not a rapid or convenient assay for large numbers of samples of relatively pure soluble proteins. The two assays most commonly used for routine assay are the biuret reaction, which is a measure of the number of peptide bonds present, and various modifications of the Folin-Ciocalteu procedure, which depends on the tyrosine and tryptophan residues in the protein. Both these methods are simple colorimetric assays sensitive to small amounts of protein and which can be conveniently applied to a large number of samples. The biuret reaction is often used when protein concentrations are in the range of 0.5–5 mg/ml. The Folin-Ciocalteu method is ten to one hundred times more sensitive, and the Lowry modification of it is the most commonly used method of protein assay.

If solutions contain no other ultraviolet-absorbing material, an estimation of the amount of protein can be obtained from the ultraviolet absorption of the solution. This absorption is due to the presence of the tryptophan, tyrosine, and phenylalanine residues present in the protein (Figure 4-16).

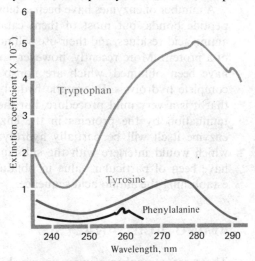

Figure 4-16 Ultraviolet absorption spectra of tryptophan, tyrosine, and phenylalanine.

The measurement is usually made at 280 nm, and if there is any indication that the sample is contaminated with nucleic acids, a correction which is based on the 260 nm absorption is made. The sensitivity of this method depends on the amino acid composition of various proteins, but for many proteins a solution of 1 mg of protein/ml will have an absorbance of about 1.0 at 280 nm.

PROTEIN HYDROLYSIS

An assessment of the amount of each amino acid present in a protein is, of course, the first step to a real understanding of its structure, and would require that all the peptide bonds be hydrolyzed and the free amino acids liberated. This hydrolysis can be accomplished by an acid, base, or an enzyme-catalyzed reaction.

The method most commonly used is to treat the protein with 6 *N* HCl at 100 °C for twelve to forty-eight hours in a sealed vial. Under these conditions there is a complete hydrolysis of the peptide bonds, but the amino acid tryptophan is almost completely destroyed. There is also some destruction of serine and threonine. A correction for the loss of these two amino acids can be made if samples are analyzed after different periods of hydrolysis and the concentration of serine and threonine extrapolated back to the start of the hydrolysis.

Base-catalyzed hydrolysis will liberate tryptophan without destruction, but it does cause degradation of the sulfur and hydroxy amino acids and some deamination of other amino acids. The conditions used, 2–4 *N* NaOH for four to eight hours at 100 °C, also cause amino acid racemization. The major use of this procedure has been to obtain a hydrolyzate for the estimation of tryptophan, although the tryptophan concentration of proteins is most commonly calculated from careful measurements of its ultraviolet absorption spectra.

A number of enzymes have been extensively used in the past to hydrolyze peptide bonds, but most of them catalyze protein cleavages at specific amino acid residues, and their use will not result in the complete hydrolysis of a protein. More recently, however, some bacterial and fungal proteases have been obtained, which are nonspecific and which will give a very complete hydrolysis. Enzymatic hydrolysis of proteins has the advantage that it is a very mild procedure, but there is the problem that some contamination by the proteins in the enzyme preparation or some of the enzyme itself will be partially hydrolyzed and will furnish amino acids which would interfere with the subsequent analysis. Enzymatic methods have been of particular value in obtaining partial digests of proteins for establishing the amino acid sequence.

AMINO ACID ANALYSIS

There are a number of ways in which the content of the various amino acids in the hydrolyzed mixture can be determined. It has been possible to obtain microbial strains that have a growth requirement for each of the amino acids. These have been used to develop a microbiological assay to determine the concentration of a single amino acid in a mixture. This method is very useful for determining the concentration of one amino acid in a large number of different samples, but is not useful for determining each of the amino acids in a hydrolyzate. There are specific chemical reactions or enzymatic assays for some of the amino acids, and these again are useful when information is desired about only a few amino acids in a large number of samples.

Amino acids, or some derivative of the amino acid, can be separated by partition chromatography on paper. Since the amino acids are very similar, it has not been possible to obtain a chromatographic system that will satisfactorily separate all the amino acids in one developing solvent. The chromatogram can be developed in one direction, dried, and redeveloped in a second solvent system run at right angles to the first system. In such a two-dimensional system (Figure 4-17), it is possible to get relatively good separation, and after spraying the chromatogram with a reagent

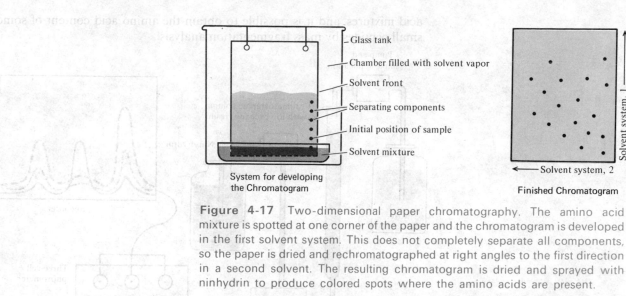

System for developing
the Chromatogram

Glass tank

Chamber filled with solvent vapor

Solvent front

Separating components

Initial position of sample

Solvent mixture

Solvent system, 1

Solvent system, 2

Finished Chromatogram

Figure 4-17 Two-dimensional paper chromatography. The amino acid mixture is spotted at one corner of the paper and the chromatogram is developed in the first solvent system. This does not completely separate all components, so the paper is dried and rechromatographed at right angles to the first direction in a second solvent. The resulting chromatogram is dried and sprayed with ninhydrin to produce colored spots where the amino acids are present.

which reacts with the amino acids, the spots on the paper can be eluted off and the amount of each amino acid determined. This method is not as accurate as is the separation of amino acids by cation exchange chromatography on sulfonated polystyrene columns.

In what is now the standard method of amino acid analysis, a low pH solution of the amino acid mixture is placed on the top of a column containing a polystyrene resin. The functional groups of the resin are negatively charged sulfonates, and the positively charged amino acids are tightly held to them. A buffer, which is gradually increased in pH and ionic strength, is pumped through the column, and the most acidic amino acids are eluted from the column first, followed by the neutral, and eventually the basic, amino acids. The concentration of amino acids in the column effluent is usually determined by use of the ninhydrin reaction, the the type of elution diagram obtained is illustrated in Figure 4-18. This method was first developed as a practical procedure by Moore and Stein, and there are commercially available "amino acid analyzers" which will automatically regulate the buffer changes, mix the effluent with ninhydrin reagent to develop the color, determine the absorbance of the effluent, and plot it on a recorder. An amino acid analysis can now be routinely carried out, in a period of ninety minutes, on the hydrolyzate from 100 micrograms (μg) of protein with an accuracy of ± 5 percent for each of the amino acids. If desired, the analysis can be done on less than 10 μg of protein, but a great deal of care must be taken to insure that the protein sample does not become contaminated with amino acids from other sources. The development of this type of apparatus has made possible rapid advances in the determination of the structures of the proteins.

Even though amino acid analysis can now be obtained routinely, efforts to perfect simpler or faster methods have continued. Progress has been made in adapting gas phase chromatography to the analysis of amino acid mixtures, and it is possible to obtain the amino acid content of some small peptides by mass fragmentation analysis.

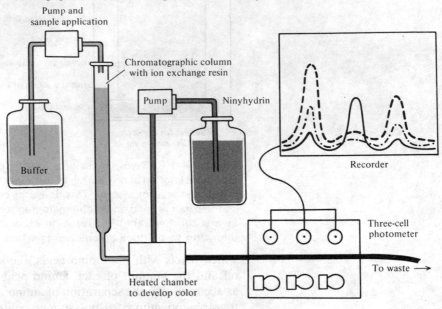

Components of an automated amino acid analyzer

Figure 4-18 Automated amino acid analyzer. The diagram at the top indicates the essential components of such a system. Three photometers are used so that it is possible to measure the optical density of effluent at two different wavelengths and two different levels of sensitivity. The curve at the bottom illustrates the type of separation of amino acids that can be achieved by this method.

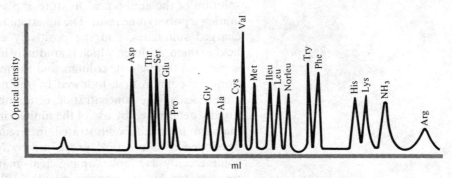

Chromatogram of an amino acid mixture

DETERMINATION OF AMINO ACID SEQUENCE

The determination of the order of each amino acid residue in a protein, the amino acid sequence, is the first step toward an understanding of protein structure. Although this is certainly not a routine laboratory exercise, the problem has now been approached often enough so that the general steps involved are clear.

1. *Preparation of a pure protein.* There will be no chance of success unless it can be shown by rigorous physical and chemical methods that the

protein chosen can be obtained in a completely homogeneous state.

2. *Determination of amino acid composition.*

3. *Determination of molecular weight.* A minimal molecular weight can be calculated from the amino acid composition and the determination of the molecular weight by various physical methods will establish what multiple of this minimal value the true molecular weight is.

4. *Determination of a number of peptide chains and their separations.* On the basis of the reactions to be discussed, the number of amino terminal (N-terminal) amino acid residues, and therefore the number of peptide chains, can be determined. If there is more than one peptide chain, the disulfide bond holding the chains together will be split and the two or more peptides which are obtained can be separated by chromatographic procedures.

5. *Determination of the sequence of each peptide.*

6. *Location of disulfide cross-links.*

Methods for Determination of the Amino Acid Terminal Residue

The method which has been used most extensively is the procedure utilizing 2,4-dinitrofluorobenzene, which was successfully applied by Sanger in his original studies on the primary sequence of insulin. The approach is illustrated in Figure 4-19. The bond formed between the free α-amino group and the dinitrophenyl group is not an amide and is stable to subsequent acid hydrolysis of the resulting derivatized protein. Following hydrolysis, the only amino acid that does not carry a positive charge at an acid pH will be the one which was on the amino terminal end of the molecule, and it can therefore be isolated by extracting it into an organic solvent. The bright yellow dinitrophenyl derivative can then be identified by paper chromatography. During the treatment with Sanger's reagent, the ε-amino groups of lysine and to some extent the phenol, thiol, and imidazole groups in the molecule will also become derivatized. However, only the derivative of the N-terminal residue will be uncharged in the acid solution following hydrolysis and will go into the organic phase during extraction.

1-Dimethylaminonaphthalene-5-sulfonyl chloride, commonly called the "Dansyl" group or DNS-Cl, will react with free amino groups to form a derivative with properties similar to those formed by Sanger's reagent. It has an advantage in that the derivative formed has a strong visible fluorescence when activated by ultraviolet light, and can be detected at about 1 percent of the concentration of the dinitrophenyl derivatives.

If a peptide is treated with phenylisothiocyanate, a phenylthiocarbamyl peptide is formed which, under acid conditions, can be cleaved with the eventual formation of the phenylthiohydantoin of the amino terminal residue (Figure 4-20). This derivative can be identified, or alternatively, the amino acids in the peptide that is left can be determined. The procedure, which was developed by Edman, has an advantage over other methods in

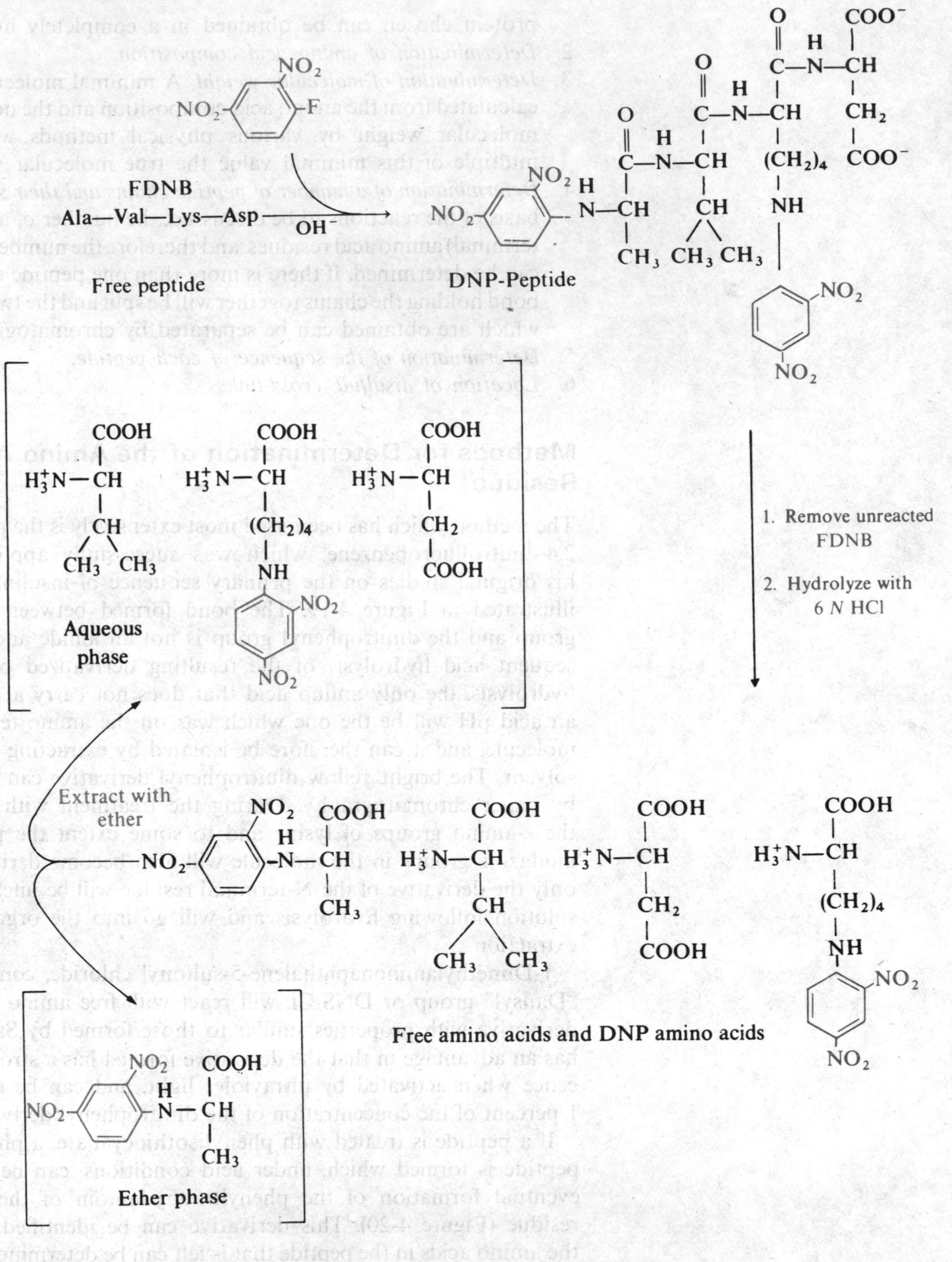

FDNB
Ala—Val—Lys—Asp
Free peptide

DNP-Peptide

1. Remove unreacted FDNB

2. Hydrolyze with 6 *N* HCl

Aqueous phase

Extract with ether

Free amino acids and DNP amino acids

Ether phase

Figure 4-19 The chemical reactions involved in determining an amino terminal residue by the use of Sanger's reagent.

that the entire sequence of reactions can be repeated on the remaining peptide with the identification of the new amino terminal residue during the next cycle. The Edman degradation can be repeated six to eight times before poor yields in the various steps make the results ambiguous. This procedure has now been automated, and it is possible to sequence as many as fifty to sixty residues of a protein by repeated and automated application of this method.

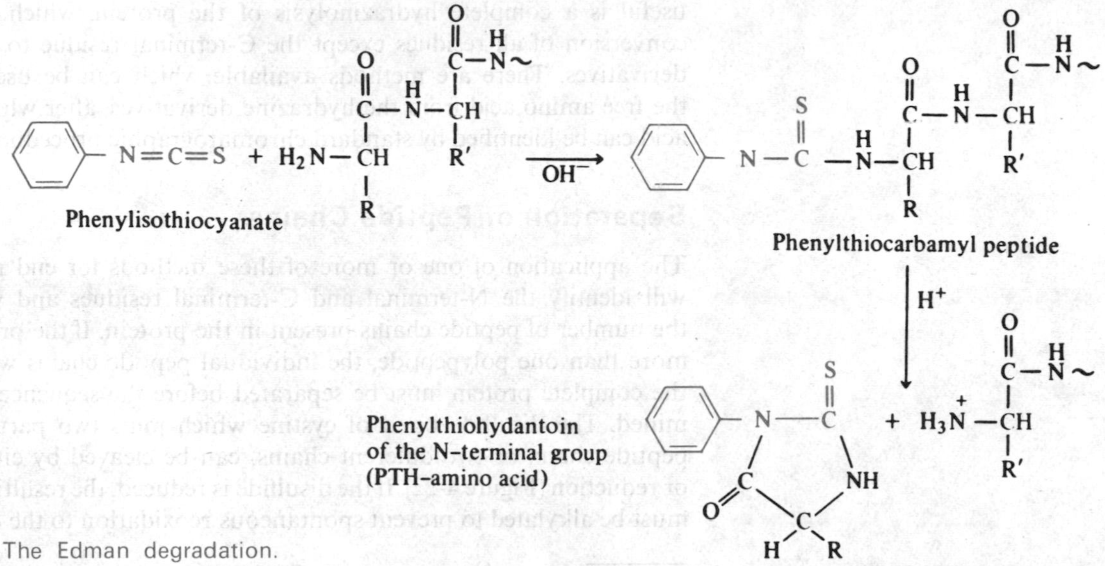

Phenylisothiocyanate

Phenylthiocarbamyl peptide

Phenylthiohydantoin of the N-terminal group (PTH-amino acid)

Figure 4-20 The Edman degradation.

The enzyme leucine amino peptidase will catalyze the hydrolytic cleavage of leucine, and most other amino acids, from the amino terminal end of a protein or a peptide. If a peptide is incubated with the enzyme, and the amino acid which is liberated identified, it is possible to determine the amino terminal group. The method is complicated by the fact that as soon as the N-terminal amino acid is cleaved off, a new amino terminal group is formed, and the enzyme can cleave either that or the original amino terminal group on a second molecule. If the rate of cleavage of the second residue is much greater than that of the first, the two amino acids will be released almost simultaneously, and it will not be possible to tell which one was the original N-terminal group.

Methods for Determining Carboxyl Terminal Residues

Carboxypeptidases are available which will sequentially degrade peptides from the C-terminal end. The interpretation of the data obtained from this procedure is subject to the same difficulties as were indicated for the use of leucine amino peptidase. The data in Figure 4-21 illustrate the problems which may be encountered. In this particular case, if glutamic acid rather than phenylalanine had been the C-terminal amino acid, the rate of cleavage of the second residue would have been so great that glutamic acid and phenylalanine would have appeared at roughly the same rate.

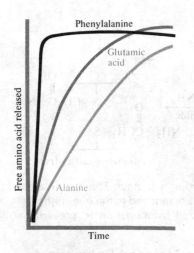

Figure 4-21 Action of the enzyme carboxypeptidase on a protein to determine the carboxy terminal residue.

Chemical methods for determining the carboxyl terminal residue of a peptide have not been as successful as those used for amino terminal analysis. The method most commonly used (the Stark method) utilizes an addition of ammonium thiocyanate to the carboxyl terminal residue to form a thiohydantoin derivative. Using the same type of approach as that in the Edman degradation, this modified residue can be cleaved from the rest of the peptide and identified. A second method that has proved useful is a complete hydrazinolysis of the protein, which results in the conversion of all residues except the C-terminal residue to the hydrazine derivatives. There are methods available which can be used to separate the free amino acid from the hydrazine derivatives, after which this amino acid can be identified by standard chromatographic procedures.

Separation of Peptide Chains

The application of one or more of these methods for end group analysis will identify the N-terminal and C-terminal residues and will determine the number of peptide chains present in the protein. If the protein contains more than one polypeptide, the individual peptide chains which make up the complete protein must be separated before the sequence can be determined. The disulfide bond of cystine which joins two parts of the same peptide chain, or two different chains, can be cleaved by either oxidation or reduction (Figure 4-22). If the disulfide is reduced, the resulting sulfhydryls must be alkylated to prevent spontaneous reoxidation to the disulfide form.

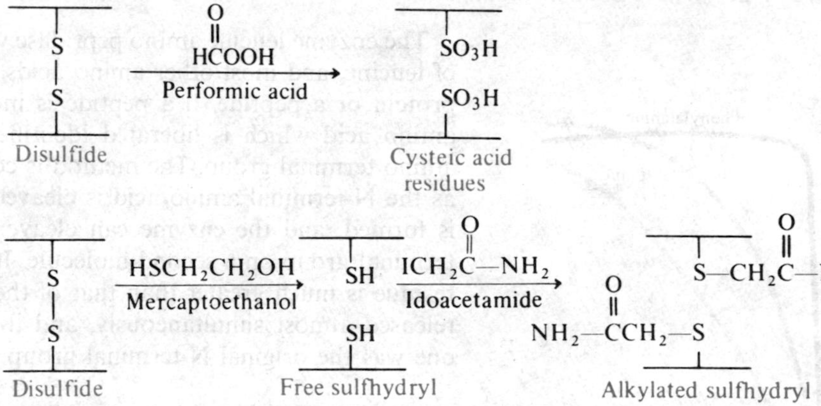

Figure 4-22 Cleavage of the disulfide bond in a protein. The disulfide can be oxidized to the stable cysteic acid derivatives or reduced to the free sulfhydryl. In the latter case the group must be stabilized by alkylation to prevent reoxidation.

Determination of Sequence of the Isolated Polypeptide

Ideally, a polypeptide could be sequentially degraded by the Edman method and the entire sequence determined. In practice it has almost always been necessary to cleave the parent peptide into a number of smaller fragments, separate these fragments, and characterize each of them. If relatively small peptides are obtained, the problem of determining their sequence will be greatly aided. An example of the type of data actually obtained in such a study is illustrated in Figure 4-23.

Correct sequence of two peptides

Peptide A

Val–His–Glu–Ser–Leu

Peptide B

Lys–Glu–Thr–Ala–Ala–Ala–Lys

Amino acid composition of the peptides

Amino acid	Moles of a.a./mole peptide	
	A	B
Ala		2.95
Glu	1.09	1.03
His	0.98	
Leu	1.02	
Lys		2.01
Ser	0.93	
Thr		1.00
Val	0.98	

Determination of the sequence of peptide A

Leucine aminopeptidase for 24 hr released: Val 0.95, Glu 1.04, Ser 1.01
Leu and His determination was lost.

Carboxypeptidase for 20 min released: Leu 1.00,
for 24 hr released: Leu 1.00, Ser 0.04.

Edman degradation

Step	Yield	Val	His	Glu	Ser	Leu
1	90%	0.09	0.83	1.11	1.03	1.03
2	89%	0.02	0.23	1.06	1.00	0.95
3	36%	<0.01	0.25	0.41	0.99	1.01
4	49%	<0.01	0.025	0.31	0.40	0.58

Determination of the sequence of peptide B

DNP-end group: Lys 0.70 residue

Hydrazinalysis: Lys 0.41 residue

Carboxypeptidase for 6 hr released: Lys 1.00, Ala 3.15, Thr 0.65, Glu 0.00.

Figure 4-23 Determination of the sequence of two peptides from ribonuclease. An example of the type of data actually obtained. In the Edman degradation of peptide A, the amount of each amino acid remaining in the peptide was determined rather than identifying the phenylthiohydantoin which was cleaved off. The amino acid underlined after each of the Edman degradation steps was the one assumed to have been removed. Note that in the last step, the results are rather ambiguous, but that a decision between serine and leucine can be made on the basis of the carboxypeptidase data.

Cleavage to Small Peptides Procedures are available which will selectively cleave specific peptide bonds to give the necessary small fragments from large peptides. The earliest attempt to solve this problem made use of the observation that certain peptide bonds are more susceptible to acid hydrolysis than others, and that a partial hydrolysis will yield some rather large peptides constituting a specific length of the chain. In general, this method lacks specificity and has not proved very useful. The most common method used is to subject the protein to the proteolytic action of the endopeptidase trypsin. The specificity of this enzyme is such that it hydrolyzes peptide bonds where the carbonyl carbon is furnished by the basic amino acid arginine or lysine. The enzymes pepsin and chymotrypsin have been used to obtain fragments which overlap those produced with trypsin. They show some preference for cleavage at residues involving aromatic rings, but they lack the specificity of trypsin. New enzymes with a high degree of specificity are continually being sought; a bacterial enzyme, thermolysin, and the plasma protease, thrombin, have been used successfully to generate specific peptides.

The protein may be chemically modified to either increase or decrease its reactivity to trypsin. Treatment with a reagent which will form the trifluoroacetyl derivatives of the free amino group of the lysine residues would leave only the arginine residues susceptible to tryptic attack and would decrease the number of peptides formed. Treatment of a protein with ethyleneimine introduces an amino group at a cysteine residue which trypsin cannot distinguish from lysine, and therefore causes a new point of cleavage and increases the number of peptides.

A number of chemical methods for the specific cleavage of polypeptides have been tried. One which seems to have some general value (Figure 4-24)

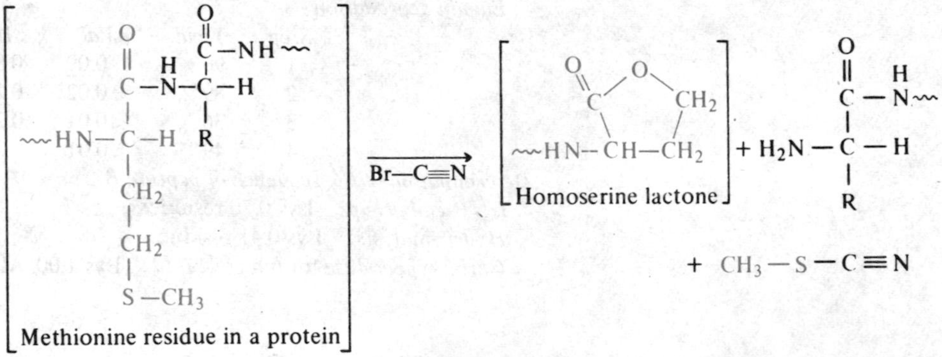

Figure 4-24 Specific cleavage of proteins at a methionine residue by the use of cyanogen bromide.

is a reaction with cyanogen bromide to cause a cleavage at methionine residues. A large number of other specific chemical or enzymatic modifications have been successfully used to obtain fragments from various proteins, and some of these will undoubtedly be developed into more general procedures in the future.

Ordering the Peptide Fragments   Assuming that a peptide can be broken down into a number of small fragments that can be easily sequenced, this information is still not sufficient to obtain the full structure of the original peptide, because the order of these fragments in the original peptide is not known. This problem can only be solved by obtaining a new set of fragments that overlaps the sequences determined in the first set. This approach is illustrated in Figure 4-25. If the complete amino acid sequence can be determined,

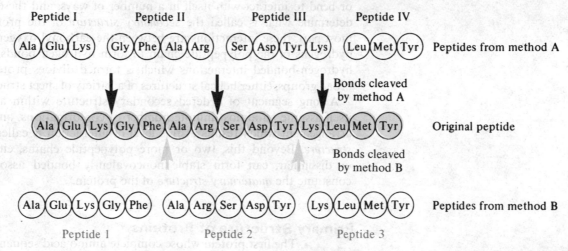

Figure 4-25   Hypothetical cleavage of a peptide by two methods. Assuming that all the peptides obtained by cleavage of the original peptide by method A could be sequenced, it would not be possible to determine the original sequence. The data would not indicate what the order of peptides II and III should be, although peptides I and IV could be placed from a knowledge of the amino and carboxyl end groups of the original peptide. In the case of method B, peptide I could be placed at the amino terminal end, but peptide 2 and 3 have the same carboxy terminal residue and they could not be placed unambiguously. However, inspection of the data from both methods will indicate that they can be ordered in only one way to satisfy the sequence of all the peptides. The points of cleavage by method A are those which would be expected from trypsin treatment; the points of cleavage by method B are those expected from chymotrypsin.

the detailed structure of a protein is not known until the correct position of the various intra- and interpeptide disulfide bonds is determined. This can be achieved by partial chemical or enzymatic degradation of the protein prior to cleavage of the disulfide bonds. After the short peptides which were held together by a disulfide bond are sequenced, their position in the intact protein can be determined.

PROTEIN STRUCTURE

The structure of proteins is usually discussed in terms of the various levels of complexity of the interactions involved in the determination of a unique structure. From this viewpoint, the organization of proteins can be considered at the level of the primary, secondary, tertiary, or quaternary structure. The *primary structure* of a protein refers to the unique sequence of amino acid residues in the polypeptide chain which are covalently linked by peptide bonds, and to the position of the disulfide bonds which form cross-links within or between peptide chains. Although this primary sequence determines a unique chemical structure, the structure can twist or bend to interact with itself in a number of ways, and these interactions determine what is called the *secondary structure* of the protein. For the most part, these interactions are between the carbonyl oxygens and amide hydrogens of the polypeptide chain to form hydrogen bonds. The type of hydrogen-bonded interactions which is formed divides proteins into two broad groups, either helical structures of a variety of sheet structures.

A long segment of ordered secondary structure within a protein can itself assume an almost infinite number of conformations, and the unique arrangement which results at this level of interaction is called the *tertiary structure*. Beyond this, two or more polypeptide chains, either identical or dissimilar, can form stable noncovalently bonded associations that constitute the *quaternary structure* of the protein.

Primary Structure of Proteins

Insulin The first protein whose complete amino acid sequence was determined was bovine insulin in the early 1950s. The structure was determined at Cambridge University in England by Frederick Sanger. Insulin is a small protein hormone produced by the β cells of the pancreas which has a physiological influence on glucose utilization by other tissues. This small protein is readily purified, and most determinations yield a molecular weight of about 12,000. The minimal molecular weight calculated from amino acid analyses is 6000; it can be shown by other methods that the protein has a true molecular weight of 6000 and that the 12,000 value represents an easily formed but rather stable dimer. Treatment of the protein with fluorodinitrobenzene indicates that there are two amino terminal residues and therefore two polypeptide chains. These chains may be separated by performic acid oxidation of the disulfide bonds holding them together. The two polypeptides were separated by Sanger, and an A chain containing 21 amino acids with a glycine as the amino terminal residue, and a B chain ending in phenylalanine and containing 30 amino acids were obtained. The individual chains were then subjected to mild acid hydrolysis to obtain a large number of small peptides suitable for sequence determination. Acid hydrolyses of the B chain produced a complex mixture which was subjected to separation by electrophoresis and various chromatographic techniques until 22 dipeptides, 14 tripeptides, and 12 longer fragments were eventually obtained. The sequence of each of the small peptides was determined by use of fluorodinitrobenzene as a reagent for detecting the amino terminal end of each. When all the small peptides were sequenced, it was possible to deduce the sequence of five large peptides. An example of the information available to obtain the sequence of one of these peptides is shown in Figure 4-26. The order of the large peptides

---

Phe-Val	Glu-His	CySO₃H-Gly
Val-Asp	His-Leu	
	Asp-Glu	Leu-CySO₃H
Phe-Val-Asp		Leu-CySO₃H-Gly
	Glu-His-Leu	
Val-Asp-Glu		
		His-Leu-CySO₃H
Phe-Val-Asp-Glu		
		His-Leu-CySO₃H-Gly
Phe-Val-Asp-Glu-His		
		Glu-His-Leu-CySO₃H

Phe-Val-Asp-Glu-His-Leu-CySO₃H-Gly

Sequences deduced
from the above peptides

Figure 4-26 Peptide sequences from insulin. The sixteen small peptides whose sequences were used to establish the sequence of the first eight residues at the amino terminal end of the β-chain of insulin. The glutamic acid and aspartic acid residues at these positions were later shown to be present as the corresponding amides (Gln, Asn).

within the B chain could not be determined until some overlapping peptides were obtained by other methods. The enzymes pepsin, trypsin, and chymotrypsin were used to obtain these fragments. Insulin has two disulfide (cystine) bonds holding the A and B chains together, as well as an intrapeptide disulfide loop in the A chain. The positions of these linkages were established by hydrolyzing the protein prior to performic oxidation, isolating and characterizing the fragments containing the cystine, and

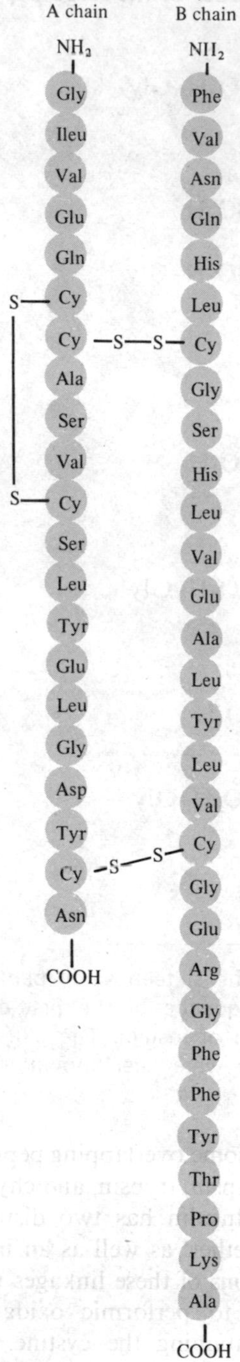

Figure 4-27 Primary structure of bovine insulin.

determining where they fit in the previously established sequence of the oxidized A and B chains. The complete structure of beef insulin is shown in Figure 4-27.

Other Proteins The determination of the structure of insulin was the beginning of the widespread use of these general principles for establishing the amino acid sequences of proteins. The second major advance in this field was the determination of the sequence of the enzyme ribonuclease by Hirs, Moore, and Stein of the Rockefeller Institute in New York, and by Anfinsen at the National Institutes of Health. This protein contains 124 amino acids, and its enzymatic function had been extensively studied before its sequence was determined. Following the determination of the structure of ribonuclease, a large number of proteins were sequenced. Some of the earlier ones completed were cytochrome *c*, myoglobin, hemoglobin, lysozyme, trypsinogen, and the coat protein from tobacco mosaic virus. The available information in this field is now accumulating very rapidly, and a yearly publication lists all the published protein sequences. A recent edition contains the sequences of forty-four species of cytochrome *c* and a large number of hemoglobin variants. A large number of peptide hormones and many enzymes have been sequenced. It is now possible to obtain the primary sequences of rather large proteins, and those with molecular weights in excess of 50,000 have been reported.

The partial acid hydrolysis method of obtaining short fragments used by Sanger has been almost completely replaced by specific enzymatic and chemical cleavages, and Sanger's method for sequencing the small peptides has been largely replaced by the use of the Edman degradation. The availability of commercial apparatus for the automated sequencing of peptides by the Edman degradation method has greatly reduced the effort involved in these procedures and will undoubtedly lead to increased availability of amino acid sequence data.

Structural Generalities A sufficient number of protein sequences have now been determined that it is possible to draw some rather firm generalizations about the primary structure of proteins. First, the covalent bonds found between amino acid residues in soluble proteins are peptide bonds and the disulfide bond of cystine which holds two chains together. The only exceptions to this general rule are some covalent bonds which link peptide chains of fibrous proteins. Second, a protein has a unique sequence and is not a mixture of molecules with small variations in amino acid sequences. There are, however, differences in amino acid sequences of the same protein from different species, even if the protein is very similar. These genetic variations in structure are usually confined to certain regions of the protein, whereas other regions appear to have the same sequence in all species (Table 4-4). Presumably these latter regions are more critical to the function of the protein, and less variability in sequence can be tolerated. Third, the sequences of amino acids within proteins appear nearly random. That is, there are no sequences which are found to be repeated over and over in a large number of proteins. It has been shown, however, that large segments of proteins which are of the same evolutionary origin are very similar in structure.

Table 4-4 Species Variation in Amino Acid Sequence of Cytochrome *c*

	Residue Number[a] in the Cytochrome *c*										
Species	55	56	57	58	59	...	70	71	72	73	74
Man	Lys	Gly	Ile	Ile	· Trp		Asn	Pro	Lys	Lys	Tyr
Duck	Lys	Gly	Ile	Thr	Trp		Asn	Pro	Lys	Lys	Tyr
Yeast	Lys	Asn	Val	Leu	Trp		Asn	Pro	Lys	Lys	Tyr
Pumpkin	Arg	Ala	Val	Ile	Trp		Asn	Pro	Lys	Lys	Tyr

[a] Residues 70–74 are near the heme group and apparently any mutation in this region would result in a nonfunctional protein. Residues 55–58 are apparently in a less sensitive part of the molecule and some alteration can be tolerated. Note, however, that the changes are from one polar or nonpolar residue to a similar residue.

Secondary Structure of Proteins

Although there are a number of ways in which a polypeptide chain could associate with itself to form some type of secondary structure, only a few forms have been found. Most of the initial evidence for the nature of these associations was obtained from the interpretation of x-ray diffraction studies of simple crystalline peptides by Pauling and Corey. On the basis of these studies, they were able to set definite criteria for what would be the most stable secondary structures of proteins. These were: (1) that the maximum number of amide hydrogens and carbonyl oxygens would be involved in hydrogen bonding; (2) that the hydrogen involved in such a bond and shared between the two electronegative groups would be on a direct line between the two groups; (3) that each residue would contribute an equal amount to any change in rotation around a central axis; and (4) that the peptide bond would be planar and would have the same bond angles that it would have in a simple peptide. This last point was a crucial one and is the result of a sharing of electrons between the carbonyl double bond and the carbon-nitrogen bond. This effectively reduces the free rotation around the carbon-nitrogen bond (Figure 4-28). Two types of structures which have been identified as occurring in proteins will satisfy these criteria, the α helix and the pleated sheet.

The α-Helix A number of possible helical structures were considered by Pauling and Corey as satisfying the criteria they proposed for stable protein structures. Only one, the right-handed α helix, fully satisfied the conditions imposed. In this structure the hydrogen on the peptide nitrogen is shared with the carbonyl group of the third amino acid residue from it along the polypeptide. The groups on the α carbon of the amino acid are perpendicular to the main axis of the helix, and there are 3.6 amino acid residues per turn of the helix (Figure 4-29).

This helical arrangement of amino acid residues within a polypeptide chain has now been shown to occur in a large number of soluble proteins,

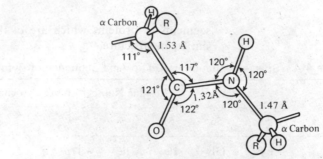

Basic dimensions of the peptide bond

Figure 4-28 Bond angles and resonance forms of the peptide bond. The top structure shows the bond angles and lengths, in angstroms, involved in the bond. The resonance form on the left would put a pure double bond between the C and O and would permit free rotation around the C—N bond. However, there is a sufficient contribution from the other resonance form to restrict C—N bond rotation and to keep the portions of the bond in the shaded area in a single plane.

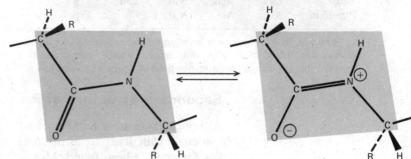

Resonance forms of the bond

although there are probably no proteins which are composed completely of α-helix regions. Because they contain optically active L-amino acids, all proteins will demonstrate optical activity. The presence of a helical region within a protein also contributes to and increases the optical rotation, and this increase can be used as a measure of helical content. Measurements of this type are improved by determining the rotation at many wavelengths. This technique is called optical rotatory dispersion.

Another method used as an indication of helical content makes use of the degree to which the potentially exchangeable hydrogens of the protein are able to exchange rapidly with D_2O. The hydrogens that exchange slowly are considered to be those which are protected from the solvent by the hydrogen bonding that stabilizes the α helix. Knowledge of the complete tertiary structures of some proteins has recently been made available, and this allows a direct calculation of the extent of helix formation in the protein. Charged residues, bulky residues, and glycine residues tend to destabilize an α-helix, and most neutral, less bulky residues favor a helical structure. Sufficient information is now available to allow rather accurate predictions of helical content of proteins and peptides to be made from a knowledge of the amino acid content and sequence.

Pleated Sheet Structures   The second readily identified, hydrogen-bond stabilized structure found in proteins is the pleated sheet. This type of secondary structure, which can exist in either a parallel or an antiparallel form, differs from the α helix in that the hydrogen bonds are not parallel to the long

Figure 4-29 Secondary structure of proteins. (a) The α helix. Model of a right-handed α helix. There are three amino acids in one hydrogen-bonded loop, 3.6 amino acids per turn, a pitch of 5.4 Å, and a diameter of 10.5 Å. All peptide carbonyl groups and amide nitrogen hydrogens form hydrogen bonds. All R groups point away from the helix. (b) Antiparallel pleated sheet arrangement of polypeptide chains. The two chains can also run in the same direction to form a parallel pleated sheet.

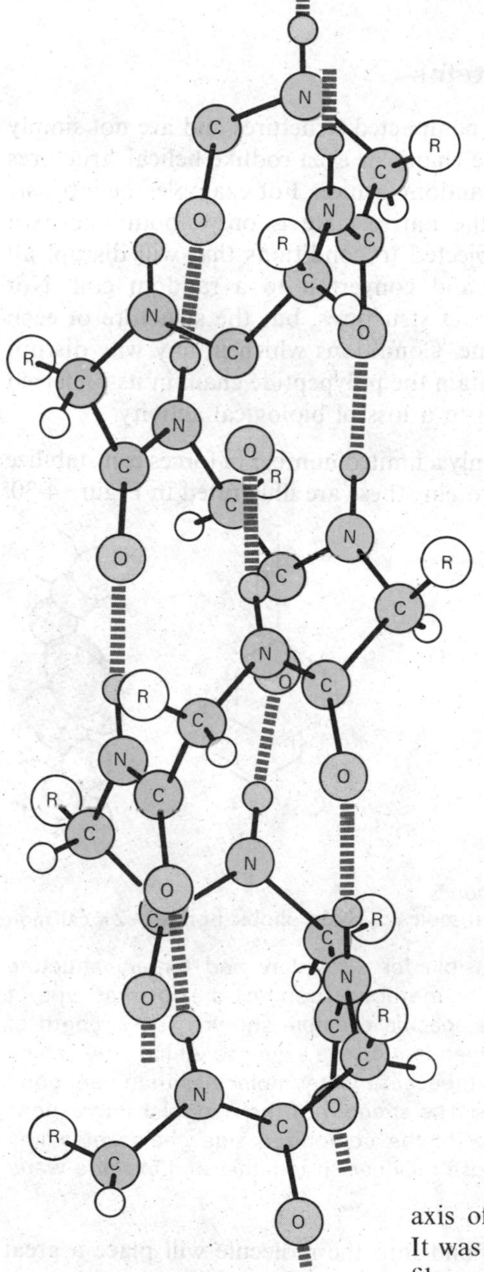

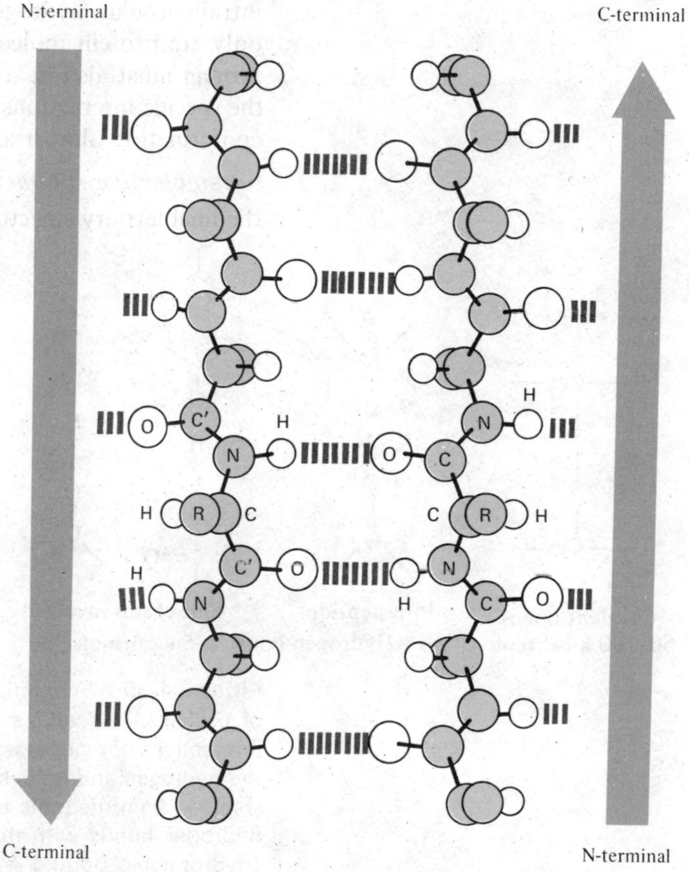

axis of the polypeptide chain, but are perpendicular to it (Figure 4-29). It was long thought that pleated sheet forms were found only in insoluble fibrous proteins such as silk, and that the hydrogen bonds were always intermolecular. More recently, however, short segments of pleated sheet regions have been seen in soluble proteins. In these cases the hydrogen bonding which stabilizes the pleated sheet is intramolecular rather than being between two different polypeptide chains. The residues involved in this type of interaction may be separated by long distances along the peptide chain.

Tertiary Structure of Proteins

Most soluble proteins are tightly compacted structures and are not simply long, randomly coiled polypeptide chains or even rodlike helical structures which are themselves coiled in a random fashion. For example, the intrinsic viscosity of serum albumin in the native state is only about one-sixth that which is seen when it is subjected to conditions that will disrupt all intramolecular hydrogen bonds and convert it to a random coil. Not only are protein molecules compact structures, but the structure of each protein must also be a unique one. Conditions which in any way disrupt the specific interactions that maintain the polypeptide chain in its preferred conformation almost always lead to a loss of biological activity.

Intramolecular Interactions   Only a limited number of forces can stabilize the final tertiary structure of a protein; these are illustrated in Figure 4-30.

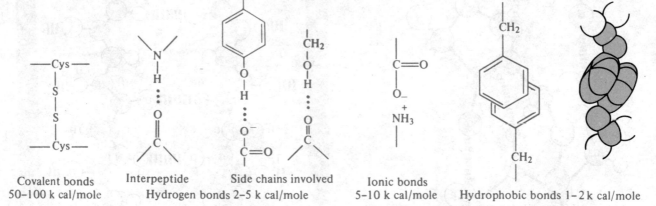

| Covalent bonds 50–100 k cal/mole | Interpeptide Hydrogen bonds 2–5 k cal/mole | Side chains involved | Ionic bonds 5–10 k cal/mole | Hydrophobic bonds 1–2 k cal/mole |

Figure 4-30 Interactions responsible for secondary and tertiary structure of proteins. The values indicated for the bond strengths are for that type of interaction, not necessarily for the specific example shown. The strength of the hydrogen and ionic bonds is dependent on the extent to which they can be shielded from the ionic environment, because water molecules form very good hydrogen bonds with these groups. The strength of the nonpolar interactions (hydrophobic bonds) between any of the nonionized side chains of amino acids is also dependent on how close they can fit together and exclude water molecules from the region.

The introduction of a disulfide bond into the molecule will place a great deal of constraint on the movement of the rest of the molecule and is undoubtedly of critical importance in determining the final structure. Although covalent bonds are very rare, they have definitely been shown to link a number of peptide chains in fibrous proteins such as collagen, fibrin, and elastin. Hydrogen bonds that utilize the hydrogen on the peptide bond nitrogen can occur not only in the formation of an α-helix or pleated-sheet region, but can also be formed when any two suitable groups come in close proximity. The hydroxyl hydrogens on a tyrosine or serine residue can also participate in hydrogen bond formation, and the various charged groups on the protein can participate in ionic attractions. How effective these rather polar interactions will be in stabilizing the structure of a protein will depend to a large extent on how well this portion of the protein chain is protected by the rest of the molecule from the aqueous environment. Water molecules will compete very effectively for both the hydrogen bonds and ionic bonds which could contribute to stabilization of the protein.

It has become increasingly apparent that the hydrophobic interactions between the nonpolar amino acid side chains of the protein are of extreme importance in stabilizing the conformation. In many proteins these groups tend to form a nonpolar region in the center of the molecule when water is excluded, while the amino acid residues with polar side chains are directed toward the aqueous surface of the protein.

X-ray Diffraction of Proteins   Detailed knowledge of the three-dimensional structure of proteins has been made available by the application of x-ray diffraction methods to proteins. If an x-ray beam is directed at a crystalline molecule, the regular array of atoms within the crystal scatters the x-rays. An x-ray diagram which shows this regular scattering can be obtained. From the differences in intensities of the scattered x-rays it is possible to produce an electron-density map or Fourier projection (Figure 4-31). Although it was realized some time ago that protein crystals would give x-ray diffraction patterns with sufficient detail to interpret the structure down to a resolution of 2 Å, the complexity of the patterns obtained made interpretation impossible. The problem was solved by Kendrew for the respiratory pigment, myoglobin. He and Perutz found that it was possible to prepare heavy metal derivatives of myoglobin which did not distort the structure of the molecule, and to use these so-called isomorphous replacements for the x-ray diffraction studies. When this was done, the heavy metal provided a point of orientation of the diffraction pattern which was obtained, and allowed the calculation of the Fourier synthesis of extremely complex molecules. Kendrew originally determined the structure of whale myoglobin (153 amino acids, with a molecular weight of 17,500) at a resolution of 6 Å, where it was possible to see the general direction and orientation of the main side chains, and later at a resolution of 1.5 Å, where individual amino acids could be recognized. A number of features about the fine structure of this protein were immediately apparent when the three-dimensional structure was available (Figure 4-32). First, the molecule is very compact and there is room for no more than four water molecules to gain access to the interior of the folded peptide. Second, almost all the amino acids with polar side chains are on the surface and many have water molecules bound to them; the interior of the molecule consists of a large number of tightly packed nonpolar amino acid residues.

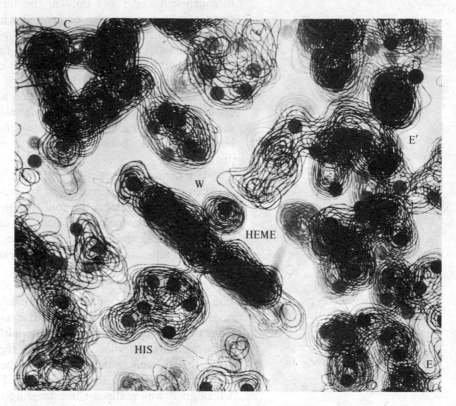

Figure 4-31 Data obtained from the examination of protein crystals are a series of spots on a photographic plate, or a direct measurement of x-rays scattered at different angles from the crystal. From these data, a Fourier synthesis or three-dimensional representation of electron distribution can be obtained. The photograph above is a contour map of electron densities drawn on stacked sheets of clear plastic, showing a portion of the myoglobin molecule. The heme group is seen edge on. His is an amino acid subunit of histidine attached to the iron atom of the heme group. W is a water molecule linked to the iron atom. The region between E and E′ represents amino acid subunits arranged in an α helix. C is an α helix seen end on. The black dots mark atomic positions. (Dr. John C. Kendrew)

Third, eight different helical regions in the molecule can clearly be seen, and they contain 77 percent of all of the amino acid residues. The amino acid proline cannot fit in a helix; therefore, the presence of a proline residue causes an interruption in a helical region and a bend in the structure. Fourth, the chemical nature of the heme-protein interaction can be seen to be due to the binding of the iron of the heme group to a histidine residue of the protein.

Although for some time a large number of generalities about protein structure were made on the basis of the information gained from this one protein, it is now clear that myoglobin is not a very "typical" globular protein. Few proteins have been found that have as high a helical content as myoglobin, and the absence of disulfide bonds in the molecule also makes it a rather atypical protein.

The structure of hemoglobin, the respiratory pigment found in the blood of higher animals, was elucidated by Perutz a couple years after the structure of myoglobin was determined. Hemoglobin consists of four heme-containing polypeptide chains, each of which is similar in size to

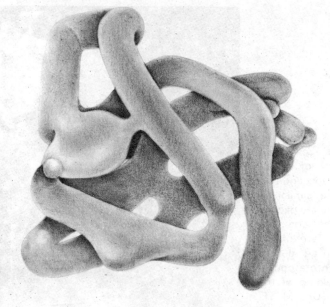

Figure 4-32 Three-dimensional structure of myoglobin. The structure at the top shows what could be learned from the first 6 Å resolution study of the molecule. The main pathway of the helical segments can be seen to fold around the heme group. A more detailed drawing of the structure is shown at the bottom, and the various helical and nonhelical regions are identified. The table on the right indicates the number of amino acid residues in each of these segments. (From *The Structure and Action of Proteins* by Richard E. Dickerson and Irving Geis. Reprinted by permission of the authors and Harper & Row, Publishers, Inc.)

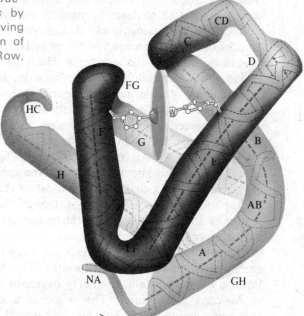

Helix	Length	Nonhelix	Length
A	16	NA	2
B	16	AB	1
C	7		
D	7	CD	8
E_1	10		
E_2	10	EF	8
F	10	FG	4
G	19	GH	5
H	26	HC	4
Total length	121	....	32

the myoglobin molecule and each of which can transport a molecule of O_2. The complete protein is formed from two α chains and two slightly different β chains. Each of the chains has a tertiary structure which very

Figure 4-33 Low resolution model of the structure of hemoglobin. Four chains, two α and two β each of which are similar in structure to myoglobin, are packed together to form the complete quaternary structure. Two of the four heme groups can be seen as gray discs which fit into a pocket resulting from the folding of the polypeptide chain. (Dr. M. F. Perutz)

closely resembles the structure of myoglobin, and these four chains are closely associated to form a compact multiunit structure (Figure 4-33).

Since the pioneering work of Kendrew and Perutz at Cambridge, a number of research groups have been actively involved in structural determinations. Some of the structures which have been solved are indicated in Table 4-5. Certainly one of the most interesting of these has been the determination of the structure of the enzyme lysozyme by Phillips. Lysozyme is an enzyme which has a specificity for certain β-1,4-glycosidic linkages found in bacterial cell walls. The cleavage of these cell-wall components by the enzyme results in the lysis and destruction of the bacterial cell. The structure of lysozyme appears somewhat more complex than that of myoglobin. Only about 45 percent of the amino acid residues are involved in helical segments, and although most of them are in the interior, some hydrophobic groups are on the surface of the molecule. There is a short region of the molecule where the polypeptide chain is folded back on itself to form an antiparallel pleated sheet, although this region is not so extensive as that found in carboxypeptidase, where 20 percent of the residues are involved in this type of bonding. The important advance made by Phillips was the crystallization of lysozyme in the presence of a trisaccharide of *N*-acetyl glucosamine. This sugar is a competitive inhibitor in the enzymatic action of the enzyme, and would therefore be expected to be bound to the protein at the enzymatically active site. Examination of the protein-inhibitor complex did in fact reveal the location of the inhibitor specifically bound to a cleft in the structure of the molecule. The importance of this discovery cannot be overemphasized. It provided the first real demonstration of how substrate molecules could be specifically bound

Table 4-5 Some Proteins Studied by X-Ray Crystallography

Protein	Molecular Weight	Biological Activity	Metal or Prosthetic Group	Resolution, Å
Myoglobin	18,000	Oxygen carrier	Heme	1.4
Hemoglobin	64,500	Oxygen carrier	Heme	5.5
Lysozyme	14,600	β-1, 4-Glycosidase	None	2.0
Ribonuclease A and S	14,000	Phosphodiesterase	None	2.0
α-Chymotrypsin	25,000	Endopeptidase	None	2.0
Carboxypeptidase	34,600	Exopeptidase	Zinc	2.0
Carbonic anhydrase	30,000	CO_2 Hydratase	Zinc	2.0
Papain	22,000	Endopeptidase	None	2.8
Cytochrome c	12,400	Electron carrier	Heme	2.8
Insulin	34,500[a]	Hormone	Zinc	2.8
Lactic dehydrogenase	140,000	Pyruvate reduction	None[b]	2.5
Ferredoxin	6,000	Electron carrier	Fe and S	2.5
Staphlococcal nuclease	16,800	Phosphodiesterase	Calcium	2.0

[a] Crystalline structure contains 3 insulin dimers.
[b] Forms associations with NAD^+ and NADH.

to proteins by interactions with carefully positioned amino acid residues and a rational explanation of how amino residues at the active site participate in the catalytic function. Although much of what was observed had been postulated by more indirect methods, final proof had been lacking. The usefulness of this model in an explanation of the detailed mechanism of the catalytic function of this enzyme will be discussed in Chapter 7.

Recent interest in x-ray crystallography has tended to be in the direction of large multisubunit enzymatically active proteins. Of particular interest has been the determination of the structure of lactic dehydrogenase, and the direct visualization of the coenzyme, nicotinamide adenine dinucleotide, bound to the protein. It now appears that the nucleotide-binding site of the protein is similar even in enzymes which do not appear to have similar primary structures. This similarity of structure of the subunit of *lactic dehydrogenase* and *glyceraldehyde-3-phosphate dehydrogenase* is illustrated in Figure 4-34. The similarity suggests that these proteins have evolved from the same protein, but there is no conclusive evidence of this at the present time.

From all the information now available, some generalizations about the tertiary structure of proteins can be drawn.

1. The molecules are extremely compact.
2. The interior of the molecules are distinctly nonpolar. If polar groups are found in this hydrophobic core, they have often been found to have a catalytic function.
3. There are α-helix regions in proteins, but they may constitute a rather minor portion of the total chain, and they may be distorted from the dimensions predicted by the Pauling-Corey model.

4. Pleated sheet regions do exist in globular proteins.
5. Binding of small molecules to the protein can change its structure. These changes are usually on the order of 0.5—2 Å, but in the case of carboxypeptidase, the binding of the substrate has been shown to cause a 14 Å shift in position of a phenolic side chain.
6. The active site of enzymatic proteins can be made up of amino acid residues which are far removed from one another along the polypeptide chain. They can, however, be brought into close proximity by folding of the chain.

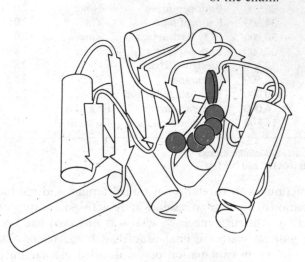 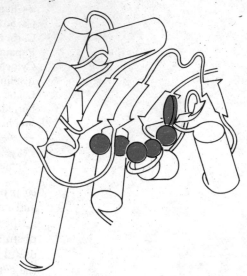

Figure 4-34 Diagrammatic comparison of the structure of two enzymes which bind NAD⁺, glyceraldehyde-3-phosphate on the left, and lactate dehydrogenase on the right. The structures are positioned to show the coenzyme-binding portion of each protein. The cylinders represent helical regions of the proteins, the broad arrow represents pleated sheet regions, and the colored balls represent the bound NAD⁺. (Buehner et al. *Proc. Nat. Acad. Sci.* USA, Vol. *70*, p. 3052, 1973)

The structures that have been determined by x-ray crystallography depict the three-dimensional structure of the protein in the crystalline state. There is, however, reason to believe that this is the same form which is present in solution. The gross measurements of size and shape of protein molecules which can be made on protein solutions are consistent with the established crystalline structure. The ability to crystallize enzymatic proteins with a substrate or an inhibitor bound to it suggests that the fine structure of the enzyme is retained in the crystal. In fact, some of these crystals can be shown to possess enzymatic activity if the substrate is given sufficient time to diffuse into the crystal.

Formation of Tertiary Structure   If each protein represents a unique three-dimensional configuration, a logical question is: How is this structure biologically determined? It has always been considered that two general situations could explain this uniqueness of structure. First, the primary sequence of the protein could determine what the secondary and tertiary structure would be, and second, the tertiary structure could be determined and directed by some step in the biosynthetic process. The available evidence would indicate that the first explanation holds, that is, that the protein assumes a thermodynamically preferred conformation, and that this may then be further stabilized by the formation of disulfide bonds. This does not mean, however, that there cannot be an enzymatic process which will increase the rate of this reaction during synthesis of the protein. An enzyme that will catalyze this disulfide bond formation has been discovered in liver microsomes.

The best evidence for this hypothesis has come from studies where proteins which contain a large number of disulfide bonds are carefully reduced to the sulfhydryl form with a loss of biological activity. They are then allowed to reoxidize with a re-formation of the disulfide bonds and a restoration of biological activity. If the disulfide bonds were rejoined by a random process, it would be expected that only a few percent of the molecules would be in the correct form and would be capable of showing enzymatic activity. However, if the experiments are carefully done, full activity can be restored (Table 4-6). Obviously, in this case, the information available in the molecule itself is sufficient for the correct structure to be formed.

Table 4-6 Renaturation of Disulfide-Bond-Containing Proteins[a].

Protein	Disulfide Bonds	Recovery of Activity (%)	Random Recovery (%)
Ribonuclease	4	95–100	1
Lysozyme	4	50–80	1
Taka-amylase A	4	48	0.3
Insulin (govine)	3	5 10	6.7
Alkaline phosphatase	2	80	33

[a] In these experiments, the disulfide bonds of the proteins were reduced to the free sulfhydryls in 8 M urea and then allowed to reoxidize. The percentage of the molecules that would be expected to re-form the correct disulfide bonds on a purely statistical basis is shown in the last column.
Source: C. J. Epstein et al., *Cold Spring Harbor Symp. Quant. Biol.*, 28 439 (1963).

Denaturation of Proteins

The term denaturation is rather loosely used to describe a number of conformational changes in proteins. It is most commonly used in a more restricted sense to designate a change in structure which causes a loss of biological activity without a cleavage of the peptide chain. The term can be defined as a change in the tertiary, and generally the secondary, structure of the protein from its native state to a more disordered arrangement. This change may be either a reversible or an irreversible arrangement, and a random coil structure for the protein would represent the completely denatured state.

Extreme heat or extremes of pH will usually result in irreversible denaturation of most proteins, while an 8 *M* urea solution is commonly used as an agent which will reversibly denature many proteins by competing for the intramolecular hydrogen bonds which stabilize the structure. Other agents which have been used with some success as reversible denaturing agents are guanidinium chloride, lithium bromide, and various organic solvents and detergents.

Quaternary Structure of Proteins

Many proteins are composed of more than one polypeptide chain and are held together by some force other than a covalent bond. The association of more than one polypeptide chain to form a stable, larger unit is called the quaternary structure of the protein. One example of this, hemoglobin, has already been mentioned as containing a total of four subunits, two α and two β chains. The individual subunits of a protein may or may not be active when dissociated from the large unit. An example of this is the behavior of the enzyme glutamic dehydrogenase. In this case a large molecule can be dissociated into smaller enzymatically active subunits, but the dissociation of these subunits into a smaller polypeptide by conditions that cause unfolding of the polypeptide chains leads to a loss of activity. The presence of a subunit structure makes available methods of metabolic control, which will be considered in later chapters.

Isozymes The multipolypeptide nature of some proteins also provides an explanation for the existence of isozymes (isoenzymes). Isozymes are different forms of the same enzymatic activity found in the same cell or tissue. They are usually detected by electrophoresis of what otherwise might appear to be a homogeneous protein preparation. One of the most extensively studied examples of this is the lactic dehydrogenase system from various species. This enzyme catalyzes a conversion of pyruvate to lactic acid. In some tissues five distinct forms of the enzyme can be electrophoretically identified. It has been shown that the enzyme consists of four subunits, and that two very different polypeptides are involved. One type, the H form, contains all of one polypeptide and is the form found in heart muscle; the M- or muscle-form contains the other polypeptide. The other three types consist of the other three possible combinations of the H- and M-polypeptides to form a tetrameric structure (Figure 4-35). The biological significance of these forms appears to be related to their apparent function in different tissue. The M form is found in anaerobic tissues and presumably functions by producing lactate at a rapid rate in the presence of high concentrations of pyruvate. The H form predominates in aerobic tissue and is inhibited by pyruvate. Its function is probably to oxidize

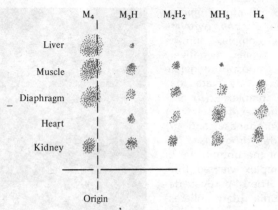

Figure 4-35   Distribution of lactate dehydrogenase isozymes in the tissues of an adult white rat, as determined by zone electrophoresis of tissue extracts in a starch block. The extracts are applied at the line marked *origin*. Four of the isozymes move to the anode and one to the cathode at the pH chosen. The size and density of the spots reflect the amount of each isozyme. (Fine, Kaplan, and Kuftinic, *Biochemistry*, Vol. 2, p. 116, 1963)

lactate to pyruvate which can be utilized aerobically, rather than to convert pyruvate to lactate. Examples of a number of other proteins with subunit structures are shown in Table 4-7.

Table 4-7   Examples of Quaternary Protein Structure

Protein	Molecular Weight	Number of Subunits
Ribonuclease	13,700	1
Chymotrypsin	24,500	1
Bovine serum albumin	66,500	1
Yeast enolase	67,000	2
Liver alcohol dehydrogenase	78,000	2
Hemoglobin	68,000	4
Lactic dehydrogenase	140,000	4
Aspartic transcarbamylase	310,000	8

MULTIENZYME COMPLEXES

Some proteins are involved in interactions even more complex than the quaternary structures seen in multisubunit proteins. In many cases where a series of enzymes acts sequentially on a substrate to carry out a complex alteration, it has been shown that there is a definite association of these enzymes into a large multienzyme complex. The system from *E. coli* which takes pyruvic acid and converts it to acetyl coenzyme A with a loss of CO_2, the pyruvate dehydrogenase complex, is one of the most extensively studied examples. This system, whose enzymatic action is considered in

Figure 4-36 Electron micrographs of the *E. coli* pyruvate dehydrogenase complex and its component enzymes, negatively stained with phosphotungstate (×200,ppp). (*a*) Pyruvate dehydrogenase complex; (*b*) dihydrolipoyl transacetylase; (*c*) pyruvate dehydrogenase; and (*d*) dihydrolipoyl dehydrogenase. The photographs at the bottom are of the entire complex viewed from different sides. The 24 pyruvate dehydrogenase units (dark spheres) and 24 flavoprotein units (light spheres) are distributed in a regular manner along the edges of the transacetylase cube which is in the center. (Reed and Cox, *The Enzymes,* 3d ed., Vol. 1, Paul Boyer, ed., Academic Press)

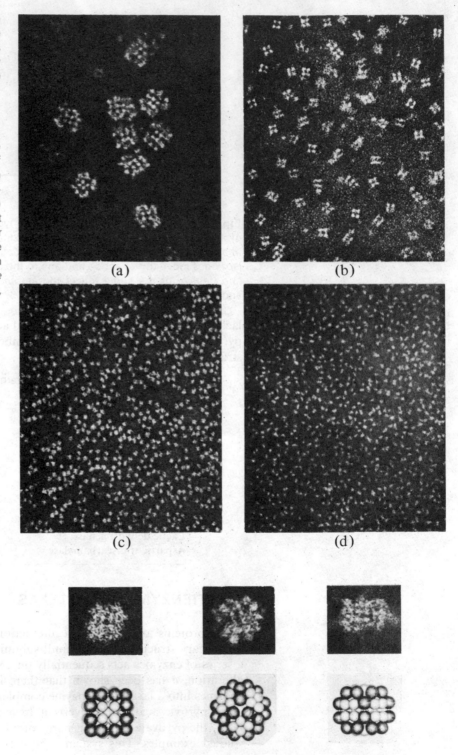

(a)

(b)

(c)

(d)

detail in Chapter 13, consists of three enzymes, a pyruvate decarboxylase, a transacetylase, and a dihydrolipoyl dehydrogenase, which can be separated and purified independently. The entire complex has a molecular weight of about 4 million and consists of 24 transacetylase subunits of molecular weight (MW) 40,000 which are organized into eight trimers that fit together to form a cube. This cube is at the center of a large complex formed by distributing 24 of the flavoprotein subunits, 55,000 MW and 24 of the pyruvic decarboxylase subunits of 90,000 MW around its surface. Electron micrographs of some of these large protein particles and diagrammatic illustrations of how these complex particles might be held together are shown in Figure 4-36.

PROBLEMS

1. Sketch a titration curve for the dipeptides glycylglycine and histidyl-glutamic acid. On each curve, indicate the pI of this compound.

2. Consider this peptide:

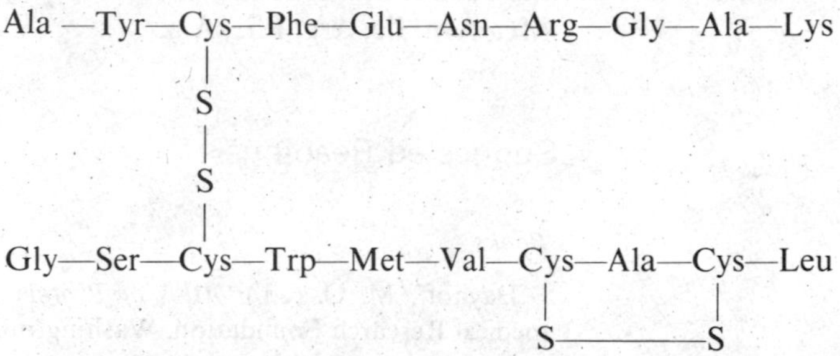

```
Ala —Tyr—Cys—Phe—Glu—Asn—Arg—Gly—Ala—Lys
             |
             S
             |
             S
             |
Gly—Ser—Cys—Trp—Met—Val—Cys—Ala—Cys—Leu
                             S———S
```

If this peptide were subjected to performic acid oxidation, how many peptides would be formed? If it were subjected to the Edman degradation procedure, which amino acid(s) would be present as the thiohydantoin derivative in the first cycle? If it were mixed with insulin and subjected to gel filtration, would it come off the column before or after insulin? In the peptide, draw an arrow (→) at any bonds that would be cleaved by trypsin; a bar (|) through any bonds that would be cleaved by CnBr; a box (□) around the amino acid residue that would contribute most to its absorption at 280 nm; and a circle (○) around any residues that could have a positive charge at pH 6.

3. A pentapeptide obtained from treatment of a protein with trypsin was shown to contain arginine, aspartic acid, leucine, serine, and tyrosine. To determine the amino acid sequence, the peptide was cycled through the Edman degradation procedure three times. The composition of the peptide remaining after each cycling was as follows:

After cycle 1:   arginine, aspartic acid, leucine, serine
After cycle 2:   arginine, aspartic acid, serine
After cycle 3:   arginine, serine

What is the sequence of the pentapeptide?

4. It is possible to obtain much of the information needed to sequence proteins by obtaining large numbers of small peptides by partial acid hydrolysis. The following peptides were obtained from a partial acid hydrolyzate of an oligo peptide which had the composition (Ala,Arg$_2$,Asp$_2$,-Glu$_2$,Leu,Lys,Ser,Thr). Treatment with DNFB indicated that the N-terminal residue was Glu, and treatment with carboxypeptidase A for one hour gave 0.8 Glu and 0.5 Asp. The peptides obtained were:

(Asp,Glu)	(Ala,Asp)	(Arg$_2$,Thr)	(Ser,Thr)
(Glu,Leu)	(Ala,Asp,Ser)	(Arg,Lys)	(Asp,Lys)
(Asp,Glu,Leu)	(Asp,Glu,Lys)	(Ala,Asp,Leu)	(Arg,Thr)

What is the amino acid sequence of the protein? (*Note*: When parentheses are placed around a peptide, it indicates that the order of residues is not known.)

5. Which four of the following residues would you most likely find on the outside of a globular protein at physiological pH's? Leu, Val, Glu, Thr, Met, Asn, Ile, Arg, Phe, Ala.

Suggested Readings

Books

Dayhoff, M. O. (ed.). *Atlas of Protein Sequence and Structure*. National Biomedical Research Foundation, Washington, D. C. (1972).

Dickerson, R. E., and I. Geis. *The Structure and Action of Proteins*. Benjamin, Menlo Park, Calif. (1969).

Haschemeyer, R. H., and A. E. V. Haschemeyer. *Proteins: A Guide to Study by Physical and Chemical Methods*. Wiley, New York (1973).

Meister, A. *Biochemistry of the Amino Acids*, 2nd ed. Academic Press, New York (1965).

Van Holde, K. E. *Physical Biochemistry*. Prentice-Hall, Englewood Cliffs, N.J. (1971).

Articles and Reviews

Anfinsen, C. B. Principles that Govern the Folding of Protein Chains. *Science* **181**:223–230 (1973).

Kendrew, J. C. The Three-Dimensional Structure of a Protein Molecule. *Sci. Amer.* **205**:96–111 (1961).

Merrifield, R. B. The Automatic Synthesis of Proteins. *Sci. Amer.* **218**:56–74 (1968).

Perutz, M. F., 1964. The Hemoglobin Molecule. *Sci. Amer.* 211(5):64–76.

Sanger, F., and E. O. P. Thompson. The Amino Acid Sequence in the Glycyl Chain of Insulin. *Biochem. J.* **53**:353–374 (1963).

Spackman, D. H., W. H. Stein, and S. Moore. Automatic Recording Apparatus for Use in the Chromatography of Amino Acids. *Anal. Chem.* **30**:1190–1206 (1958).

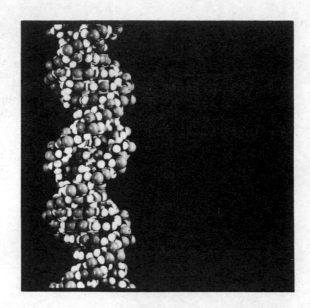

lipids and biological membranes

chapter five

Lipids are cellular components soluble in nonpolar organic solvents. This crude mixture contains neutral fats, glyceral triesters of long-chain fatty acids, and a variety of complex lipids that contain both hydrophobic and hydrophylic groups. Biological membranes are organized structures containing both proteins and complex lipids. These membranes regulate the entry of compounds into cells, separate different portions of the cell, and organize enzymatic activities within the cell. The proteins in membranes are either loosely bound extrinsic proteins or intrinsic proteins which are closely associated with lipid and difficult to remove. Most membranes appear to be composed of a bilayer of lipid with the hydrophobic end of the lipids oriented to the center of this structure and the charged hydrophylic end interacting with the aqueous medium. The membrane proteins float in or span this bilayer, and are held in position by hydrophobic interactions with the nonpolar regions of the lipid.

Lipids are biological materials containing a mixture of compounds of widely differing chemical composition. In general, lipids may be defined as the greasy material which can be extracted from plant or animal tissues by various nonpolar solvents. The amount of this material and its chemical composition varies greatly from one tissue to another. In the animal body, the lipid content may range from a few percent in muscle or liver tissues to near 80 percent in adipose tissues.

NEUTRAL FATS

The simplest class of lipids consists of glycerol triesters, the neutral fats. The acids forming the ester bonds are almost always the long-chain mono-carboxylic acids of even chain length. These are commonly called fatty acids, and in nearly all the naturally occurring lipids, the saturated fatty acids containing 16 or 18 carbons and unsaturated 18-carbon fatty acids predominate. The nomenclature and structure of the more common, naturally occurring fatty acids are shown in Table 5-1.

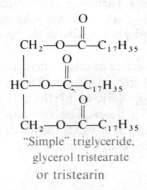

"Simple" triglyceride, glycerol tristearate or tristearin

"Mixed" triglyceride β-oleo-α,α'-stearopalmitin or 1-stearo-2-oleo-3-palmitin

Table 5-1 Structure and Nomenclature of Fatty Acids

No. of Carbons	Common Name	Geneva Name	Structure
Common Saturated Fatty Acids			
4	Butyric	Tetranoic	$CH_3(CH_2)_2COOH$
6	Caproic	Hexanoic	$CH_3(CH_2)_4COOH$
8	Caprylic	Octanoic	$CH_3(CH_2)_6COOH$
10	Capric	Decanoic	$CH_3(CH_2)_8COOH$
12	Lauric	Dodecanoic	$CH_3(CH_2)_{10}COOH$
14	Myristic	Tetradecanoic	$CH_3(CH_2)_{12}COOH$
16	Palmitic	Hexadecanoic	$CH_3(CH_2)_{14}COOH$
18	Stearic	Octadecanoic	$CH_3(CH_2)_{16}COOH$
20	Arachidic	Eicosanoic	$CH_3(CH_2)_{18}COOH$
Common Unsaturated Fatty Acids			
One double bond			
18 C	Oleic	9-Octadecenoic	
		$CH_3(CH_2)_7CH{=}CH(CH_2)_7COOH$	
Two double bonds			
18 C	Linoleic	9,12-Octadeca*dienoic*	
		$CH_3(CH_2)_4CH{=}CHCH_2CH{=}CH(CH_2)_7COOH$	
Three double bonds			
18 C	Linolenic	9,12,15-Octadeca*trienoic*	
		$CH_3CH_2CH{=}CHCH_2CH{=}CHCH_2CH{=}CH(CH_2)_7COOH$	
Four double bonds			
20 C	Arachidonic	5,8,11,14-Eicosa*tetraenoic*	
		$CH_3(CH_2)_4CH{=}CHCH_2CH{=}CHCH_2CH{=}CHCH_2CH{=}CH(CH_2)_3COOH$	

It should be noted that oleic acid and the common polyunsaturated fatty acids have the cis configuration about the double bond. The polyunsaturated fatty acids do not contain a conjugated double-bond system, but rather they are 1,4-polyenes. In addition to those shown in Table 5-1, various other types of fatty acids, such as saturated acids with odd-number carbon chains, branched-chain acids, α-hydroxy acids, acids containing 5- and 3-member carbon rings, trans unsaturated acids, conjugated unsaturated acids, and fatty acids containing carbon-carbon triple bonds, have been found in various sources. However, except in specialized cases, these are of little quantitative importance.

Lipids isolated from natural sources are a complex mixture of various triglycerides; information on the amount of a particular triglyceride in these mixtures is very difficult to obtain. In most cases, all that is known is the fatty acid composition of the mixture, and nothing about the specific triglycerides present.

Because they have the same chemical properties as other esters, neutral fats can easily be hydrolyzed to glycerol and free fatty acids. This can be accomplished by acid- or base-catalyzed hydrolysis or by the action of enzymes which specifically hydrolyze lipids called *lipases*. The base-catalyzed hydrolysis of a lipid is called *saponification* and as it yields the alkali salts or soaps of the fatty acids, the reaction readily goes to completion.

$$
\begin{array}{ccc}
\text{CH}_2\text{—O—}\overset{\displaystyle O}{\overset{\|}{\text{C}}}\text{—R} & \text{CH}_2\text{OH} & \text{R—COO}^-\ \text{K}^+ \\[2ex]
\text{HC—O—}\overset{\displaystyle O}{\overset{\|}{\text{C}}}\text{—R}' \quad\xrightarrow{\text{KOH}}\quad & \text{HC—OH} \quad + & \text{R}'\text{—COO}^-\ \text{K}^+ \\[2ex]
\text{CH}_2\text{—O—}\overset{\displaystyle O}{\overset{\|}{\text{C}}}\text{—R}'' & \text{CH}_2\text{OH} & \text{R}''\text{—COO}^-\ \text{K}^+
\end{array}
$$

PHYSICAL PROPERTIES OF FATTY ACIDS AND TRIGLYCERIDES

Solubility

Because of the essentially hydrocarbon nature of the common fatty acids, they exhibit a very low solubility in water. The low molecular weight fatty acids such as acetic and butyric are miscible with water, whereas fatty acids having more than six carbons, as in caproic acid, are essentially insoluble in water, but are soluble in many nonpolar solvents.

The fatty acids have a pK_a of from 4.5 to 5.0, but with the exception of acetic and butyric acids, their low solubility prevents them from expressing this acidic nature in solution. They will, however, form salts in alkaline solution, and although the Na^+ or K^+ salts are very soluble, many divalent cation salts are rather insoluble. Because of this, most commercial cleaning compounds which at one time were composed of the soluble soaps of fatty acids are now based on "synthetic" detergents, which are various types of alkyl sulfonates. The formation of the Ca^{2+} and Mg^{2+} salts of the fatty acids which occurs in "hard" water areas necessitates the exchange of these cations for Na^+ in a normal "water softening" system.

Melting Point

The extended hydrocarbon chain of the saturated fatty acids exists in a zigzag conformation with a normal tetrahedral bond angle between adjacent carbon atoms of 109.5°. The introduction of a cis double bond into this system causes a considerable deviation from the essentially linear shape of the saturated molecule, which becomes more pronounced as the degree of unsaturation increases and which effects its physical properties.

Both a decrease in chain length, and more importantly, an increase in degree of unsaturation of the hydrocarbon chain will lower the melting point of a fatty acid. Those saturated fatty acids with chain lengths of 10 or longer are solids at room temperature, while oleic acid with 18 carbons and only one double bond has a melting point of 14 °C. The melting point of a triglyceride is a reflection of its complete fatty-acid composition, and the general term *oil* is usually reserved for lipids which are liquids at room temperature, in contrast to fats which are solids.

Micelles

The true solubility of a polar lipid such as the sodium salt of a fatty acid is very limited because of the energy required to order water molecules around the long hydrophobic hydrocarbon chain. As the concentration increases, a point called the *critical micellar concentration* (CMC) is reached at which the polar molecules form into colloidal-sized particles with the polar groups on the surface and the hydrocarbon chains on the inside (Figure 5-1). These particles are in rapid equilibrium with the fatty acid

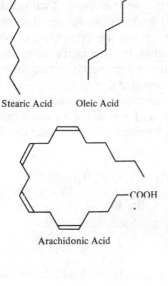

Stearic Acid   Oleic Acid

Arachidonic Acid

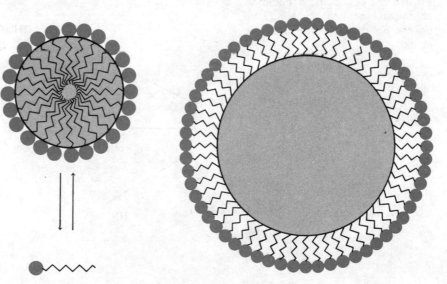

Figure 5-1 Cross-sectional view of a micellar particle (*left*) and emulsified droplet of an oil (*right*). Note that the micellar particle is in equilibrium with free fatty acids that it contains.

salts in solution, and are kept apart by the repulsion caused by negative charges on the surface of each micellar particle. Nonpolar lipids cannot form micelles, but they can be incorporated into the essentially nonpolar interior of the micelles in the form of mixed micelles.

Nonpolar lipids can also be brought into aqueous solutions if they are emulsified by a detergent such as a soap or an alkyl sulfonate. In this case, the detergent molecules can associate at the surface of a lipid droplet and give it a sufficiently high surface charge to repulse other such emulsified droplets and prevent them from coalescing to a larger drop. In contrast to true micelles, these particles are of larger than colloidal size, and are not in equilibrium with lipids in true solution.

CHEMICAL PROPERTIES OF FATTY ACIDS

The carboxyl group of the fatty acids will undergo most of the chemical reactions common to other carboxylic acids and can be esterified and reduced by standard methods. The long alkyl hydrocarbon chain is relatively inert to most reagents. The most important reactions of the naturally occurring fatty acids are those involving the double bond. Two double-bond addition reactions, halogenation and hydrogenation, are illustrated in Figure 5-2. The common vegetable shortenings are composed of plant oils which have been partially hydrogenated by the latter reaction to increase their melting point. During this industrial process a certain amount of isomerization occurs and some of the unnatural trans isomers of the unsaturated fatty acids are formed. This isomerization reaction can be promoted by a number of other catalysts, and results in an equilibrium mixture in which the higher melting trans isomers will predominate. The

Figure 5-2 Some common chemical reactions involving the double bonds of unsaturated fatty acids.

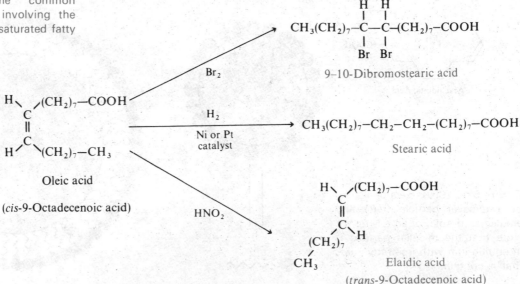

9–10-Dibromostearic acid

Stearic acid

Elaidic acid
(*trans*-9-Octadecenoic acid)

Oleic acid

(*cis*-9-Octadecenoic acid)

conversion of oleic acid to its trans isomer, elaidic acid, by HNO_2 is an example of this transformation.

OXIDATION AND RANCIDITY

The presence of the double bond also renders the fatty acids susceptible to chemical oxidation. Oxidation can be performed under controlled conditions by alkaline permanganate or by ozonolysis to give cleavage products that are useful for characterization, but more importantly, the fatty acids are subject to auto-oxidation. Light, other irradiations, or metal ion contamination will lead to the production of free radicals in lipids. The hydrogen on the methylene carbon atom between the double-bond carbons in the polyunsaturated fatty acids is very easily abstracted by a free radical to yield a new free radical. This new free radical may combine with O_2 to yield a reactive peroxyl radical, or it may first undergo a re-arrangement (Figure 5-3). The peroxyl radicals can form hydroperoxides by abstracting a hydrogen from another fatty acid molecule, and in the process can form a new reactive center. The hydroperoxides can readily be cleaved to form two new radicals, or they may eventually decompose to keto and hydroxyketo acids. It is the accumulation of these compounds which causes an off-flavor to develop in foods. This change, if it occurs in a food product, is called *oxidative* rancidity. Food products can also develop a *hydrolytic* rancidity if partial hydrolysis of the triglycerides releases some of the short-chain volatile noxious smelling fatty acids. To protect against oxidative rancidity, most lipid containing food products are protected by the addition of antioxidants, which are essentially free radical traps to break up the autocatalytic nature of the process. The same type of free radical

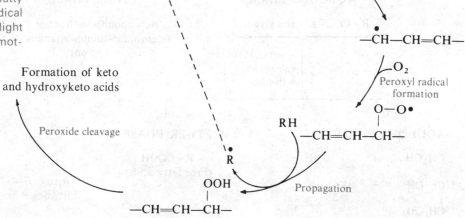

Figure 5-3 Reactions which have been postulated for the auto-oxidation of unsaturated fatty acids. The original free radical (R) arises from the action of light or some other free radical promoting agent on the system.

reaction is important in the polymerization reaction obtained by the use of highly unsaturated oils as "drying oils" in the paint industry.

EXTRACTION AND SEPARATION OF LIPIDS

In natural products much of the total lipid is present in the form of lipoproteins. Although these lipids may be soluble in rather nonpolar solvents such as diethylether or chloroform, the lipid protein interactions are difficult to dissociate in an aqueous environment, and dehydrating solvents such as ethanol, methanol, or acetone are often used for lipid extraction. A particularly effective solvent is a 2:1 mixture of chloroform and methanol, which will dissolve lipids of widely varying polarity from cells and tissues. After extraction, water can be added to this mixture to break it into a two-phase system, and the lipids remain in the chloroform phase.

Crude lipid preparations are readily fractionated by chromatography on silicic acid. Nonpolar lipids do not adsorb to silicic acid, and the more polar lipids can be fractionated by gradually increasing the polarity of the developing solvent.

DISTRIBUTION OF FATTY ACIDS IN LIPIDS

Because little information can be gained about the triglyceride content of a specific fat or oil, these compounds are characterized by their fatty acid composition. The diagram in Figure 5-4 illustrates how free fatty acids can be obtained by the hydrolysis of naturally occurring triglycerides and

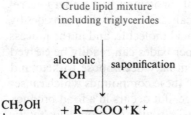

Crude lipid mixture including triglycerides

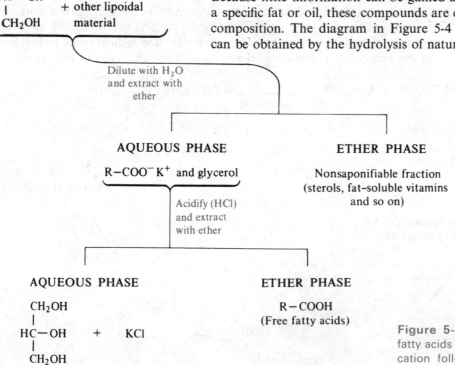

AQUEOUS PHASE

$R-COO^- K^+$ and glycerol

Acidify (HCl) and extract with ether

ETHER PHASE

Nonsaponifiable fraction (sterols, fat–soluble vitamins and so on)

AQUEOUS PHASE

$$CH_2OH$$
$$|$$
$$HC-OH \quad + \quad KCl$$
$$|$$
$$CH_2OH$$

ETHER PHASE

$R-COOH$
(Free fatty acids)

Figure 5-4 Method of isolation of free fatty acids from a crude mixture by saponification followed by selective extraction.

106

isolated from any contaminating lipids on the basis of their different solubilities. Once isolated, the various fatty acids can be separated and the amount of each fatty acid determined. Although separation was first achieved by countercurrent distribution or by partition chromatography, the simplest method, and the one most widely used today, is gas-liquid chromatography.

Before the widespread use of chromatographic methods, it was difficult to quantitatively determine the amount of any particular fatty acid. A number of empirical measurements were developed which would give some idea of the chemical properties of fats and oils and therefore some indication of their fatty acid composition. Two of these are still used to some extent: the saponification number, which is defined as the number of milligrams of KOH needed to saponify 1 g of fat, and the iodine number, which is equal to the number of grams of I_2 absorbed by 100 g of fat. A high I_2 number would therefore be associated with lipids having a high percentage of unsaturated fatty acids. A high saponification number is obtained with triglycerides containing many relatively low molecular weight fatty acids. The predominant fatty acids found in some of the more common fats and oils are given in Table 5-2. In general, lipids obtained from animal

Table 5-2   Fatty Acid Composition, Saponification, and Iodine Number of Some Common Fats and Oils

Source	Percent of Total Fatty Acid as						SAP No.	I_2 No.
	Saturated Acids			Unsaturated Acids				
	C_{14}	C_{16}	C_{18}	C_{16}	C_{18}	C_{20}		
Coconut oil[a]	18	9	1	—	9	—	250	9
Butter fat (cow)[b]	8	22	15	—	38	4	230	30
Lard	1	26	12	2	66	3	198	55
Mutton tallow	5	25	30	—	40	—	193	40
Cottonseed oil	2	20	3	—	75	—	193	110
Sardine oil[c]	6	10	2	13	24	26	191	185
Linseed oil	—	8	8	—	84	—	190	190

[a] Also contains 46 percent C_{12} saturated and 17 percent saturated acids under C_{12}.
[b] Also contains 13 percent saturated acids under C_{14}.
[c] Also contains 19 percent C_{22} unsaturated.

adipose tissue contain less unsaturated fatty acids and therefore have a higher melting point than vegetable oils. Fish oils are very high in polyunsaturated fatty acids; butter fat is characterized by a high concentration of very low molecular weight fatty acids.

COMPLEX LIPIDS

Although the neutral fats are quantitatively the most prevalent class of lipids in most living tissue, they are not necessarily the most biologically

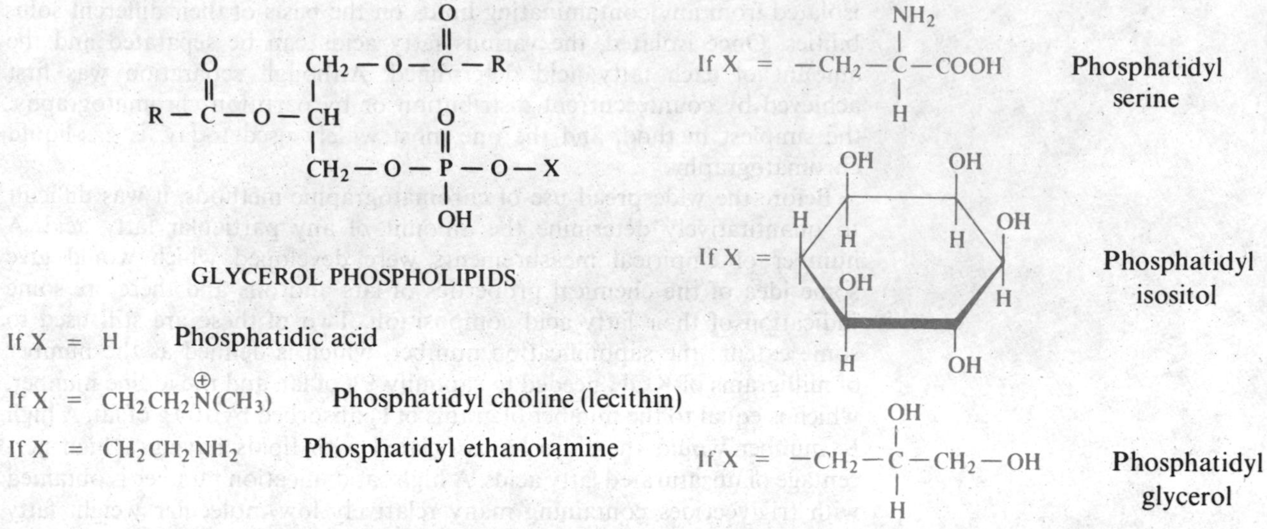

$$
\begin{array}{c}
R-\overset{\overset{\displaystyle O}{\|}}{C}-O-\overset{\overset{\displaystyle CH_2-O-\overset{\overset{\displaystyle O}{\|}}{C}-R}{|}}{\underset{\underset{\displaystyle CH_2-O-\overset{\overset{\displaystyle O}{\|}}{\underset{\underset{\displaystyle OH}{|}}{P}}-O-X}{|}}{CH}}
\end{array}
$$

GLYCEROL PHOSPHOLIPIDS

If X = H Phosphatidic acid

If X = $CH_2CH_2\overset{\oplus}{N}(CH_3)$ **Phosphatidyl choline (lecithin)**

If X = $CH_2CH_2NH_2$ **Phosphatidyl ethanolamine**

If X = $-CH_2-\overset{\overset{\displaystyle NH_2}{|}}{\underset{\underset{\displaystyle H}{|}}{C}}-COOH$ Phosphatidyl serine

If X = Phosphatidyl isositol

If X = $-CH_2-\overset{\overset{\displaystyle OH}{|}}{\underset{\underset{\displaystyle H}{|}}{C}}-CH_2-OH$ Phosphatidyl glycerol

Figure 5-5 Structures of some of the common glycerol phospholipids.

important. There are a group of compounds which, although they differ considerably in chemical composition, are all lipid-soluble surface-active compounds. These compounds, collectively called complex lipids, are found in high concentrations in most biological membranes. In general, their content in various tissues does not vary to the extent that the neutral fat content does. They all contain a hydrophobic group, esterified to either glycerol or sphingosine, and a hydrophylic group, either a phosphate ester or a carbohydrate.

Glycerol Phospholipids

The most common class of complex lipids is the glycerol phospholipids (Figure 5-5) which are substituted diacyl-glycerophosphoric acids (phosphatidic acids). A number of compounds of similar chemical nature are known; the most important are the *phosphatidyl cholines*, *phosphatidyl ethanolamines*, and *phosphatidyl serines*. Because of the hydrophilic phosphate group and charged nitrogen on one end, and the large hydrophobic

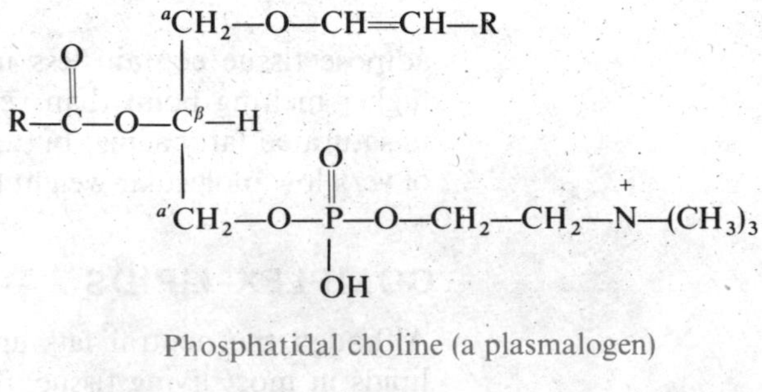

$$
\begin{array}{c}
R-\overset{\overset{\displaystyle O}{\|}}{C}-O-\overset{\overset{\displaystyle ^aCH_2-O-CH=CH-R}{|}}{\underset{\underset{\displaystyle ^{a'}CH_2-O-\overset{\overset{\displaystyle O}{\|}}{\underset{\underset{\displaystyle OH}{|}}{P}}-O-CH_2-CH_2-\overset{+}{N}-(CH_3)_3}{|}}{C^\beta-H}}
\end{array}
$$

Phosphatidal choline (a plasmalogen)

group on the other, the phospholipids are good emulsifying agents; lecithin (phosphatidyl choline) is commercially used for this purpose. The predominant fatty acids in these compounds are highly unsaturated, and care must therefore be taken to prevent oxidation when isolating them. A series of closely related compounds called *plasmalogens* differ from these glycerol phospholipids in that the constituent on the α carbon is a long-chain vinyl ether rather than a fatty acid ester. The plasmalogens which have been characterized have been found to contain either ethanolamine or choline as the nitrogenous base.

Sphingolipids

The second major class of complex lipids is the sphingolipids, for which the basic structural component is an *N*-acyl derivative of the unsaturated fatty alcohol sphingosine (Figure 5-6). As the names of these compounds indicate, they were first isolated from brain and nervous tissue. The *sphingomyelins* are the most widespread of these compounds and the only type which contains phosphate. The similarity of these compounds to the glycerol phospholipids is not readily apparent until the structures are redrawn as illustrated in Figure 5-7. Once they are drawn in this manner,

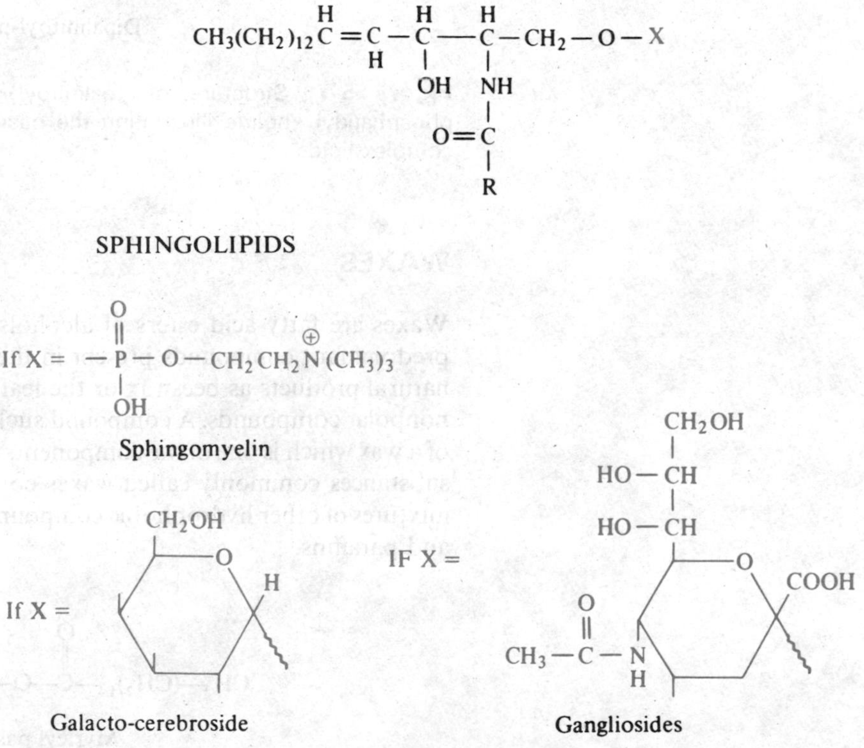

SPHINGOLIPIDS

Figure 5-6   Structures of some of the common sphingolipids.

it is easy to see that both classes of compounds present a long double hydrophobic chain with a strongly hydrophilic group on the other end. In general, the sphingolipids are found to contain saturated fatty acids in contrast to the unsaturated fatty acids found in the various glycerol lipids.

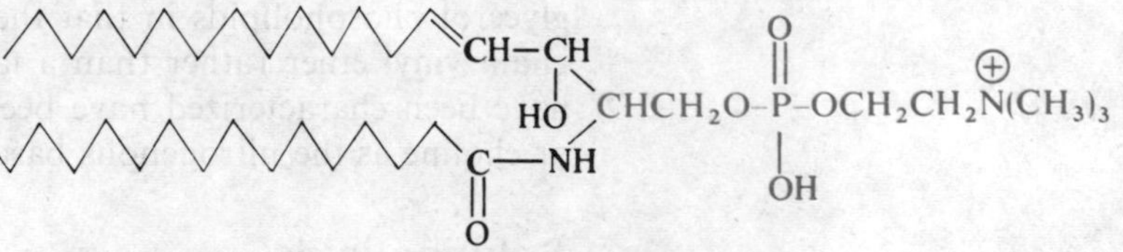

Palmitoyl–sphingomyelin

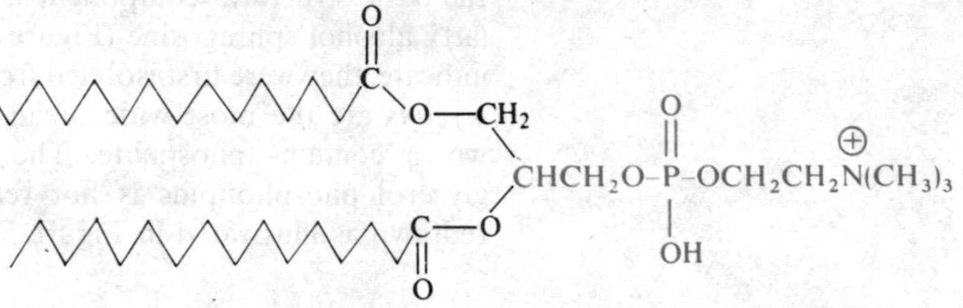

Dipalmitoyl-phosphatidyl choline

Figure 5-7 Structure of (palmitoyl) sphingomyelin and (dipalmitoyl) phosphatidyl choline illustrating the basic similarity of these two classes of complex lipids.

WAXES

Waxes are fatty acid esters of alcohols other than glycerol. They are the predominant compounds present in the mixed lipid which comprises such natural products as beeswax or the leaf cuticle waxes. These are extremely nonpolar compounds. A compound such as myricyl palmitate is an example of a wax which is one of the components of beeswax. The naturally occurring substances commonly called waxes contain not only true waxes, but also mixtures of other hydrophobic compounds such as long-chain fatty alcohols and paraffins.

$$CH_3-(CH_2)_{14}-\overset{O}{\overset{\|}{C}}-O-CH_2-(CH_2)_{28}CH_3$$

Myricyl palmitate

110

ISOPRENOIDS

This class of lipids contains those compounds whose structures can be derived by the condensation of 5-carbon, isoprene units. These compounds (Figure 5-8) range in complexity from a simple 10-carbon compound such

Figure 5-8 Structure of isoprene and examples of two compounds which could be called isoprenoids.

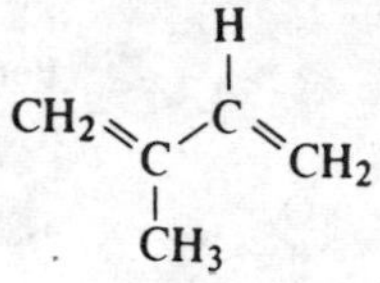

Isoprene (2–Me–1, 3–butadiene)

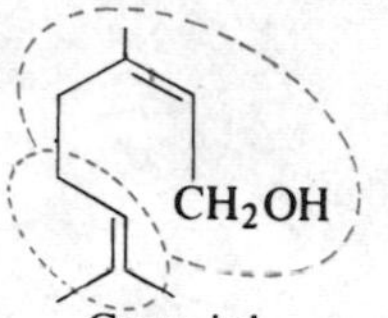

Geraniol

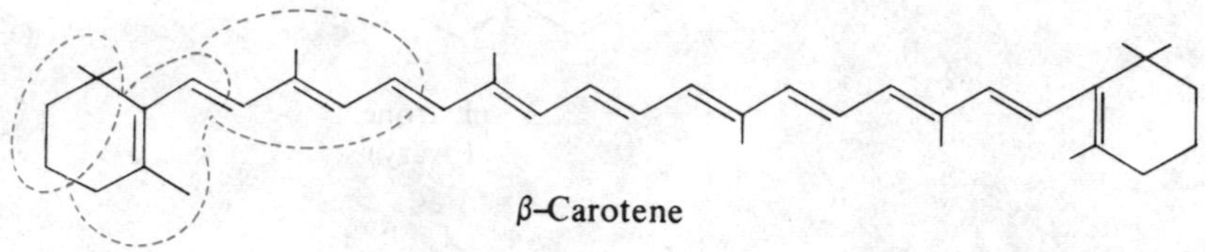

β–Carotene

as geraniol, which is called a terpene, to a tetraterpene such as β-carotene, the common plant pigment which is the precursor to vitamin A.

Steroids

Biochemically, the most important class of isoprenoid lipids is a group called the steroids. Steroids can be considered to be derivatives of the fused, reduced ring system, perhydrocyclopentanophenanthrene (Figure 5-9). Those steroids which have 8–10 carbon atoms in a side chain at position 17 and also a hydroxyl group at position 3 fall in a special class called sterols. The most common animal sterol is cholesterol, which is found in high concentration in brain, nervous tissue, and many membranes. Many of the other common steroids are similar in structure to cholesterol in that they also have an oxygenated substituent on carbon 3 and angular methyl groups at carbons 10 and 13. The steroids can also be classified on the basis of the substitution on carbon 17. Various plant sterols, such as stigmasterol, β-sitosterol, and ergosterol, have an extra methyl or ethyl group and may have an additional point of unsaturation in the hydrocarbon chain at the 17 position. Another important class of steroids is the bile acids, which contain a saturated ring system and a carboxylic acid group on the aliphatic side chain. These compounds, which give the bile its detergent action, are synthesized and degraded in the liver. In the bile, the carboxyl group is usually present as an amide of glycine or taurine to form glycocholic acid or taurocholic acid (Figure 5-9).

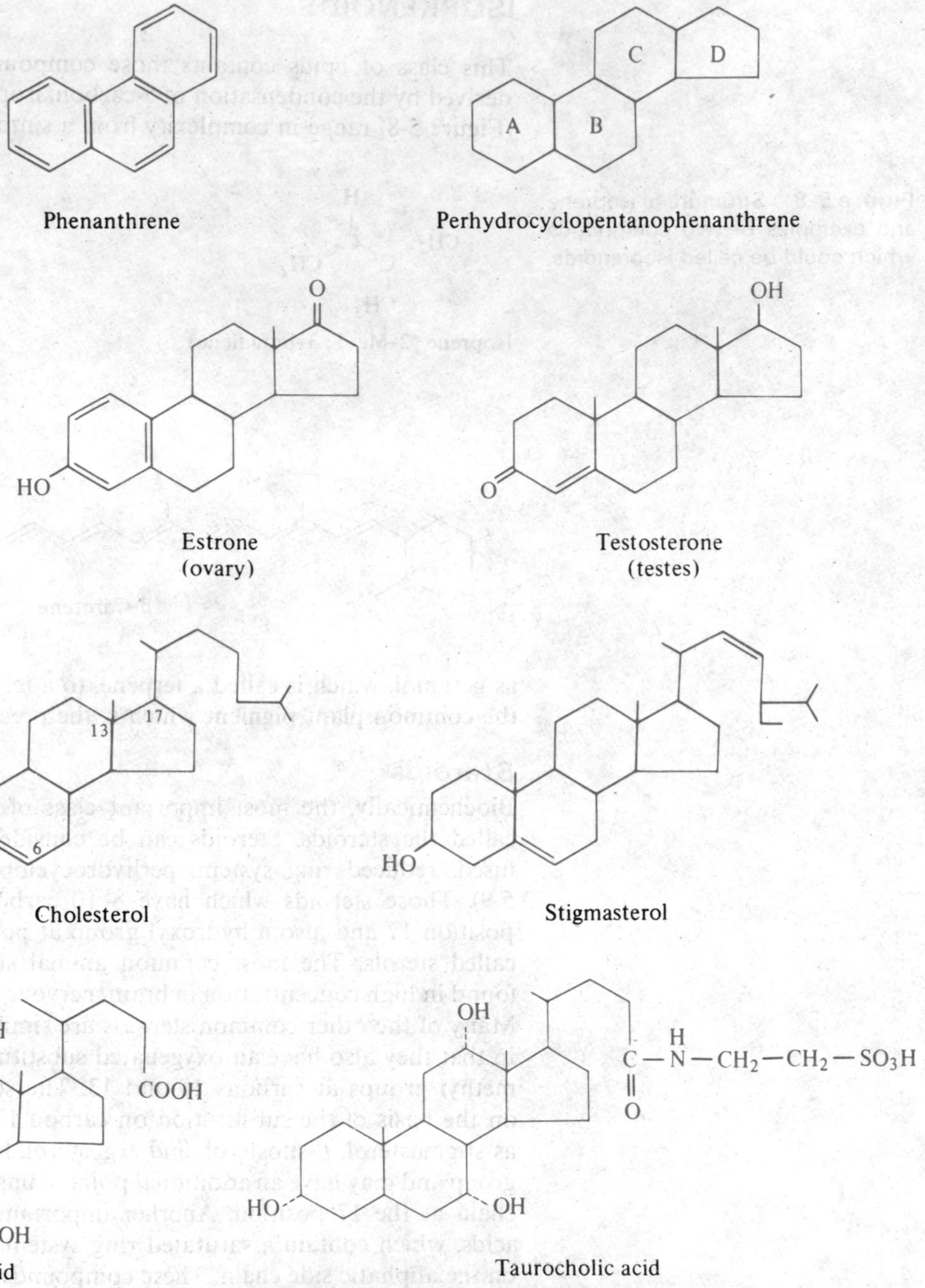

Phenanthrene

Perhydrocyclopentanophenanthrene

Estrone
(ovary)

Testosterone
(testes)

Cholesterol

Stigmasterol

Cholic acid

Taurocholic acid

Figure 5-9 Structures of the fused ring systems that can be considered to be the basis for the sterol structure, and some examples of naturally occurring steroids.

Many of the animal hormones are also steroids; examples of these and the tissue in which they are produced are shown in Figure 5-9. Even though these compounds are present in extremely small amounts, they can exert pronounced physiological responses.

The three, 6-member ring systems which are fused to form the parent sterol structure can exist in the same configurations as any cyclohexane ring system. The junction of the rings can be either cis or trans; and the overall shape of the molecule depends on the configuration. The naturally occurring sterols exist in these different forms, and the configuration of the molecule has a profound influence on the physical properties and reactivity of the molecule.

FAT-SOLUBLE VITAMINS

It will be clear from the discussion of metabolism that animals can derive all the energy they need from carbohydrates or from the carbon skeleton of amino acids. Therefore, they have no specific need for fatty acids as energy sources. Some lipids are, however, essential in the diet of higher animals. The compounds needed are the fat-soluble vitamins (A, D, E, and K) and the essential fatty acids.

Vitamin A

Compounds that possess vitamin A activity can be derived from β-carotene and contain an intact β-ionine ring. The compound most commonly called vitamin A is the alcohol form, or retinol (Figure 5-10). The dietary requirement for this compound was postulated independently by McCollum and Davis and by Osborne and Mendel around 1913–1915, when they demonstrated the need for a fat-soluble dietary factor by growing animals. The compound possessing this activity was isolated, characterized, and synthesized by 1922. The aldehyde form of the vitamin, retinal, has a specific function in the visual cycle. This function involves an isomerization at the 11 position of the trans double bond to the cis isomer. In addition to this function, vitamin A is essential for the maintenance of the integrity of various epithelial tissues. In experimental animals the symptoms of a vitamin A deficiency are poor growth, xerophthalmia, defective bone and tooth formation, and atrophy of many of the epithelial tissues. Poor dark adaption and night blindness are early deficiency signs in humans. Retinoic acid will cure all the other signs of a vitamin A deficiency, but because animals cannot reduce the acid to an aldehyde, it will not prevent the visual defect. The mechanism by which the vitamin exerts its effect is not known, but recent biochemical studies have implicated the vitamin in the reactions involved in the glycosylation of proteins. Whether this effect on the formation of glycoproteins is a primary function of the vitamin, or is secondary to some currently unidentified function, is not yet clear.

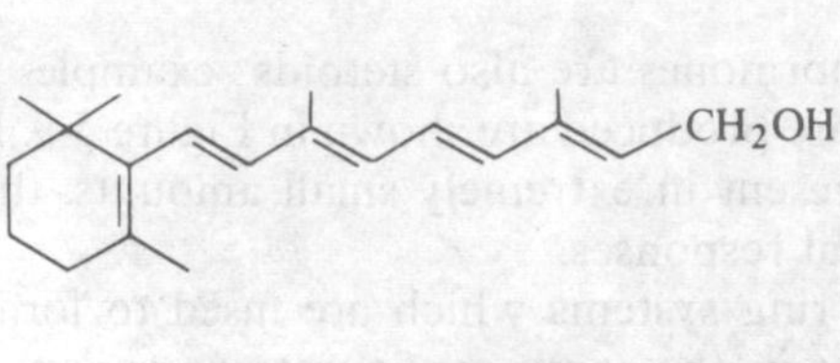

All-trans retinol

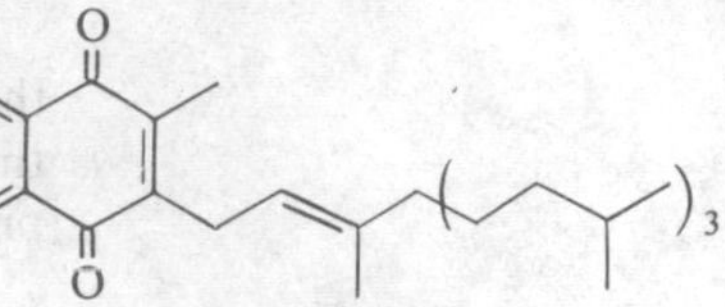

Phylloquinone

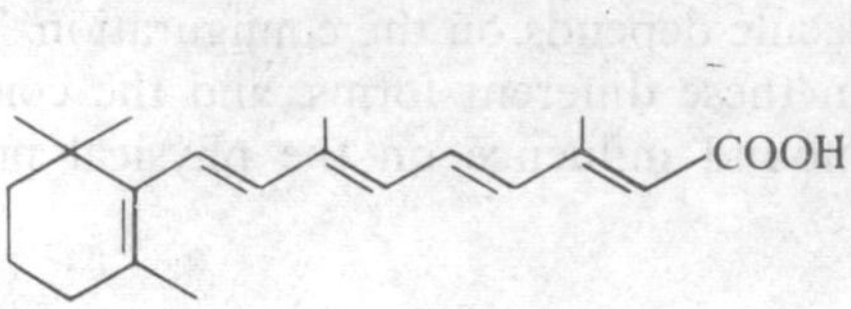

All-trans retinoic acid

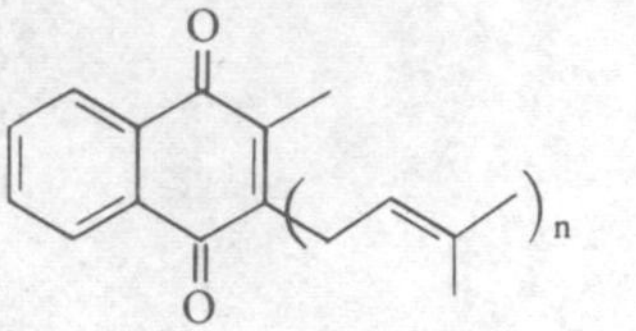

The menaquinones

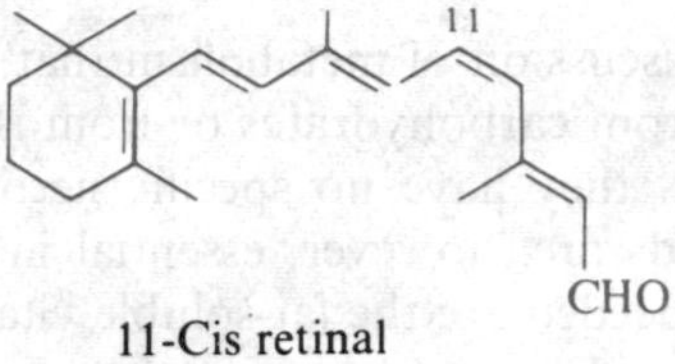

11-Cis retinal

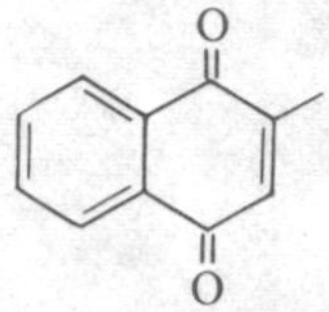

Active forms of vitamin A

Menadione

Active forms of vitamin K

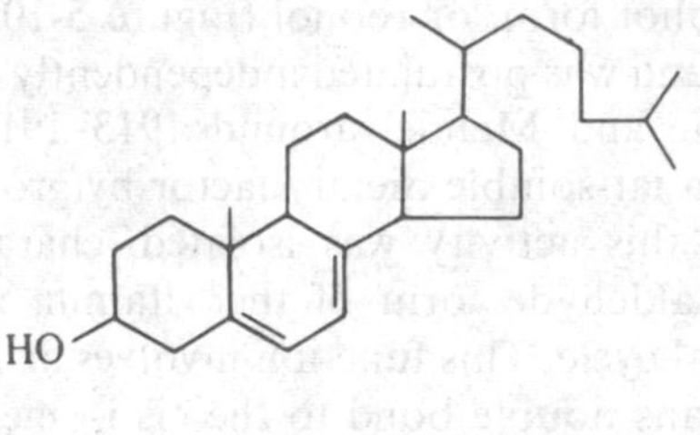

7-Dehydrocholesterol
(Precursor of vitamin D)

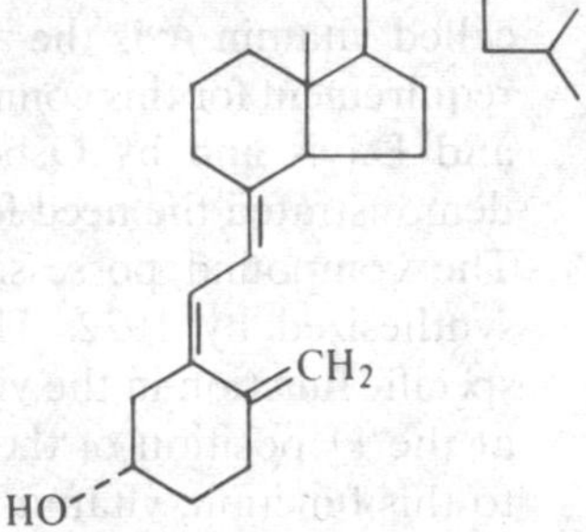

Vitamin D₃

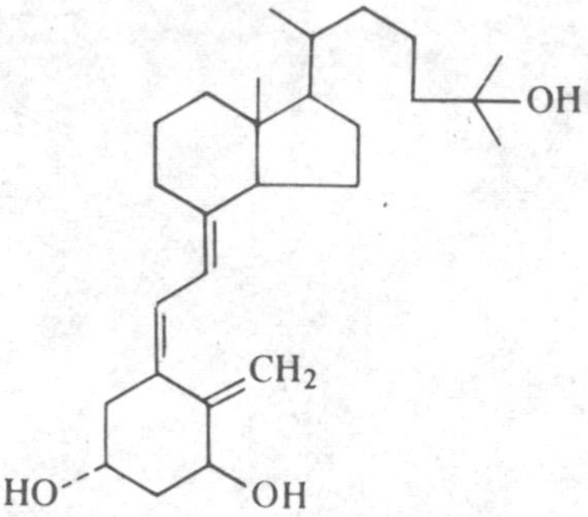

1,25-Dihydroxy vitamin D₃

Precursor and active form of vitamin D

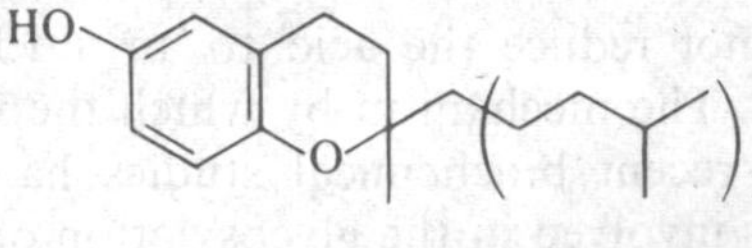

Tocol
(The parent compound for compounds
with vitamin E activity)

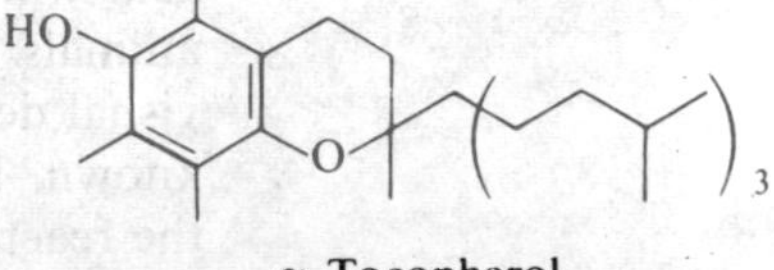

α–Tocopherol
(5, 7, 8, –trimethyltocol)

Vitamin E active compounds

Figure 5-10 Structures of some of the active forms of the four fat-soluble vitamins.

Many green or yellow vegetables contain appreciable quantities of β-carotene, and along with fish liver oils, which contain the vitamin itself, they are good dietary sources for the human.

Vitamin D

Compounds with vitamin D activity are produced from biologically inactive *provitamins* by ultraviolet radiation. Vitamin D_3 is formed by the irradiation of 7-dehydrocholesterol (Figure 5-11), and vitamin D_2, the other common form, is produced by the irradiation of a yeast sterol, ergosterol. A deficiency of vitamin D produces the disease known as rickets, in which the organic matrix of growing bone develops, but calcification does not occur because of low calcium and phosphorus levels in the serum of rachitic animals. Rickets was produced experimentally in 1919 by Mellanby, who later demonstrated that it could be cured by fish liver oils, which are one of the best sources of the vitamin. It was found that sunlight had a curing effect on human rickets, and in 1924 Steenbock demonstrated that the antirachitic potency of many foods could be increased by ultraviolet irradiation. The vitamin was characterized by 1933.

The function of vitamin D is to maintain normal serum calcium and phosphorus concentrations by increasing intestinal absorption and by increasing the mobilization of these minerals from bone. It is now known that vitamin D_3 is metabolized in the liver to 25-hydroxyvitamin D_3 and that this metabolite is further metabolized in the kidney to 1,25-dihydroxyvitamin D_3. The latter metabolite is the active form of the vitamin in the intestine and bone. The mechanism by which the vitamin influences the transport of calcium and phosphorus across cell membranes has not yet been determined.

Vitamin E

In 1922 Evans and Bishop demonstrated the existence of an antisterility factor, which was isolated, characterized, and synthesized by 1936. The compound responsible for the biological action is present in vegetable oils, particularly wheat germ, and is one of a series of naturally occurring trimethyltocols (Figure 5-10). A lack of vitamin E or α-tocopherol causes nutritional muscular dystrophy, sterility, liver necrosis, and erythrocyte hemolysis in various experimental animals. Vitamin E is a potent tissue antioxidant, and many of the symptoms of the deficiency can be explained by an increase in hydroperoxides formed from the oxidation of unsaturated fatty acids in membrane phospholipids. The specificity of the vitamin in preventing these oxidative reactions is, however, not yet clear. It has also been shown that selenium containing enzymes plays a similar role in protecting the cell against oxidative damage.

Vitamin K

Vitamin K, the most recently identified of the compounds commonly classified as fat-soluble vitamins, was observed by Dam in 1934 to prevent a hemorrhagic disease in chicks. The active compounds are substituted

2-methyl-1,4-naphthoquinones (Figure 5-10). In the case of the form found in green plants, phylloquinone, the substitution in the 3 position is a phytyl group. The vitamins found in bacteria, and synthesized in animal tissues from the basic ring structure, menadione, are menaquinones, which have varying numbers of isoprene units in the side chain. The parent compound, menadione, has biological activity, but is probably alkylated in the animal body before eliciting a biological response.

The vitamin is required for the synthesis of plasma prothrombin (factor II) and three other plasma-clotting factors (VII, IX, and X). During the clotting process, prothrombin is converted to thrombin, which then converts circulating fibrinogen to the fibrin clot. The other vitamin-K-dependent clotting factors are involved in the prothrombin activation process. It has now been shown that a protein which is a precursor to prothrombin is produced in the liver, and that this is converted to prothrombin in a metabolic step which requires the vitamin. This alteration, which gives prothrombin the calcium-binding properties needed to be rapidly activated to thrombin, involves the conversion of precursor glutamyl residues to γ-carboxy glutamyl residues in prothrombin.

ESSENTIAL FATTY ACIDS

$$CH_3(CH_2)_4(CH{=}CH{-}CH_2)_3(CH_2)_3COOH$$
6,9,12-Octadecatrienoic acid or
γ-Linolenic acid,

$$CH_3(CH_2)_4(CH{=}CH{-}CH_2)_2(CH_2)_6COOH$$
9,12-Octadecadienoic acid or
Linoleic acid,

$$CH_3(CH_2)_4(CH{=}CH{-}CH_2)_4(CH_2)_2COOH$$
5,8,11,14-Eicosatetraenoic acid or
Arachidonic acid,

Although they are not usually classified as vitamins, a second group of fat-soluble compounds is essential in the diet of many animals. Animals raised with no dietary source of fat develop deficiency symptoms characterized by poor growth and skin lesions. The compounds that were eventually shown to alleviate this syndrome were certain of the polyunsaturated fatty acids: linoleic acid, and two compounds which animals can synthesize from it, γ-linolenic and arachidonic acid. The key feature for biological activity is that the compound must have double bonds at the 6, 7 and 9, 10 positions (counting from the terminal methyl group). The normal form of linolenic acid, which has the final double bond at 3,4 position, from the terminal methyl does not have full biological activity.

PROSTAGLANDINS

Prostaglandins are a class of structurally similar compounds which were originally isolated from seminal plasma. It is now known that they are synthesized not only by the prostate gland, but probably by most cells and tissues in the body from the 20-carbon polyunsaturated fatty acids by a microsomal enzyme system. Although prostaglandins are present in very low concentrations in cells, they have a pronounced and varied effect on metabolism. They can be shown to have marked effects on blood pressure and on smooth-muscle contraction and to mediate a large number of other metabolic effects that are not yet completely understood. Because

Prostaglandin E₁

of their varied pharmacological action, it is thought that prostaglandins will have important clinical applications in the future.

MEMBRANE STRUCTURE AND FUNCTION

Some of the many roles played by biological membranes in living cells were illustrated in Figure 2-1. Membranes represent the primary *permeability barrier* of cells and therefore regulate the flow of substrates into, and products out of, cells. A variety of membrane-bound organelles, each of which represents a separate compartment, are found within the cell. These membranes separate the cytoplasm of the cell from the surrounding medium and divide it into several compartments, each of which plays a specialized role in the economy of the cell. In addition to organizing cellular compartments and separating them from one another, membranes also *organize and orient enzymes* involved in complex reaction sequences such as electron transfer and photosynthesis. If the product of one enzyme is the substrate for the second, the probability of the two-step sequence involving both enzymes is increased when the enzymes are fixed in juxtaposition at the membrane surface. For example, the cytochromes and dehydrogenases responsible for electron transfer to oxygen in mitochondria are membrane-bound (Chapter 8), as are the cytochromes involved in electron transfer to $NADP^+$ in chloroplasts of photosynthetic organisms (Chapter 14). The ATP synthesis associated with electron transfer to oxygen in mitochondria or with light absorption in chloroplasts is an example of membrane-associated *energy transductions*. Biological membranes are also responsible for *signal transductions* in which information is passed from the outside of the cell to the inside, or is passed between the cell compartments. The receptors for many hormones (insulin, for example) and for sensory input such as light (in the rod cell of vertebrate eye) are membrane proteins, which recognize specific stimuli outside a cell and transduce them into appropriate responses inside the cell. A specialized case of signal transduction is the passage of nerve impulses down the axon, which occurs as a result of changes in neuronal membrane permeability to ions. The surface membrane of a cell is the only part of that cell directly exposed to the rest of the world, and, as such, it is the part of the cell recognized in *cell-cell interaction*. It is now clear that antibody production and contact inhibition are mediated at the level of the cell membrane, and thorough understanding of these processes will be achieved only when the structure and function of the membranes involved are better understood.

Microscopic Appearance of Membranes

In the light microscope, membranes appear simply as a boundary of interface between the cytoplasm of a cell and the suspending medium. In the electron microscope, cross sections after appropriate staining show three distinct layers of characteristic dimensions, totaling about 75 Å in cross section (Figure 5-11). This characteristic *trilaminar structure* is seen in

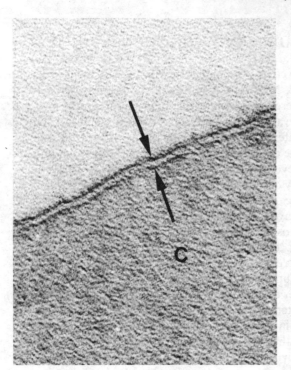

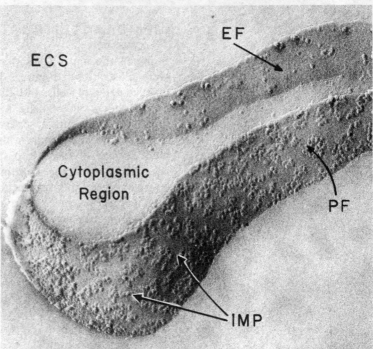

Figure 5-11   Electron micrographs of a red blood cell membrane. (*left*) An electron micrograph of a cross section of a human red blood cell. The cell portion is indicated (C) and the arrows indicate the three layered (dense-light-dense) structure of the plasma membrane. X 250,000 (Courtesy of J. D. Robertson). (*right*) Replica of a freeze-fractured human red blood cell ghost membrane. Freeze-fracturing has hemisected the red blood cell membrane so that Face (PF) originates from the inner leaflet of the membrane and Face (EF) from the outer leaflet. Both fracture faces are studded with 70 Å intramembrane particles (IMP) which are believed to represent integral membrane proteins. The extracellular space (ECS) is devoid of substructure. X 94,000 (Courtesy of R. S. Weinstein).

the membranes of both plants and animals, and is typical not only of the cytoplasmic membrane, but also of the nuclear membrane, the mitochondrial membrane, and the endoplasmic reticulum. Even bacterial membranes have the same general appearance and dimensions when seen in a cross section with the electron microscope.

MEMBRANE COMPOSITION

To study the composition of membranes it is necessary first to isolate them. Membranes are water-insoluble, relatively tough structures of very high molecular weight, and if cells are disrupted by some mechanical means and centrifuged at relatively small gravitational fields, the sediment consists primarily of cell membranes. This sediment can be further fractionated (see Chapter 9) to yield mitochondrial or cytoplasmic membrane, endoplasmic reticulum, and so on. Membranes differ in their lipid content (and thus in buoyant density), in the level of certain enzyme "markers," and in their detailed appearance in the electron microscope, and these differences often allow the biochemist to obtain a preparation greatly enriched for a particular type of membrane. It must be remembered that there is no absolute criterion of membrane purity, such as exists for soluble proteins and nucleic acid. Nonetheless, the composition of a number of fractionated membranes has been studied with results that allow certain generalization regarding the composition of biological membranes. Purified membranes consist almost exclusively of lipid and protein (Table 5-3); most membranes also contain a certain amount of carbohydrate, which is covalently linked either to lipid or to protein, in the form of glycoprotein or glycolipids.

Membrane Lipids

The ratio of lipid to protein is widely variable in membranes from different sources (Table 5-3). In the myelin sheath of nerve tissue, which apparently functions as an electrical insulator, the proportion of lipid is very high, about 80 percent, whereas in the membranes of bacteria or mitochondria, which carry out active metabolism, the proportion of protein is much higher. The type as well as the proportion of lipid varies considerably among membranes from different sources. Animal cells contain large amounts of cholesterol, which is completely absent in bacteria, and plant cells are very rich in sulfolipids, which are not found in bacteria or in animals cells. Bacterial lipids are almost exclusively glycerol phosphatides, and gangliosides, which make up a substantial proportion of the lipids found in nervous tissue of animals, are completely missing in bacteria. Within a given eucaryotic cell, different organelles contain different proportions and different types of lipids, as can be seen by comparing the compositions of mitochondrial and plasma membranes of an animal cell

(Table 5-3). From these examples, it should be apparent that the uniform appearance of biological membranes in the electron microscope cannot be explained simply on the basis of uniform lipid composition.

Table 5-3 Chemical Composition[a] of Cell Membranes

Type of Membrane	Protein (percent)	Lipid (percent)	Carbohydrate (percent)
Myelin	21	79	(5)
Red blood cell ghost	53	47	(8)
Plasma membrane (liver)	58	42	(5–10)
Endoplasmic reticulum	50	50	
Mitochondrial membrane			
Outer	52	48	(2–4)
Inner	76	24	(1–3)
Chloroplast lamellae	44	56	
Gram-positive bacterium (plasma membrane)	75	25	

[a] Compositions of lipid and protein are given as a percentage of total membrane weight. Carbohydrate is present at approximately the levels shown, as glycoprotein or glycolipid.

Membrane Proteins

The proteins associated with biological membranes can be divided roughly into two types called peripheral or extrinsic and integral or intrinsic. *Extrinsic membrane proteins* are less tightly attached to membranes, and can be released by relatively mild treatments, such as a change in the ionic composition of the medium, the removal of divalent cations, or a change in pH. When extrinsic membrane proteins have been released from the membrane, they behave like typical soluble proteins. They usually bind little or no lipid, and they can be purified by standard enzymological techniques. *Intrinsic membrane proteins*, on the other hand, are very difficult to release from membranes; their release generally requires the presence of detergents, organic solvents or denaturants or some combination of these. Once released from the membranes, intrinsic proteins remain tightly associated with lipids and with each other, so that the detergents must be kept present throughout all purification procedures. When lipid and/or detergent is removed, these proteins have a strong tendency to aggregate and precipitate, which makes their purification much more difficult than that of soluble proteins, and accounts for the fact that only a few intrinsic membrane proteins have been thoroughly purified and characterized. The chemical basis for the remarkably strong association of intrinsic membrane protein with lipids and membranes has not been definitely established. At least a few membrane proteins contain a disproportionately large number of hydrophobic amino acid residues, which might account at least in part for their association with membrane lipids. However, many membrane proteins have total amino acid composition not very different from those of soluble proteins. In a few cases, the sequence of intrinsic membrane proteins is known, and there are long sequences of primarily nonpolar amino acid residues. That portion of the protein which contains these nonpolar regions would presumably have a relatively high affinity for lipid and might represent the anchor which holds the membrane protein into the lipid region of the membrane. Furthermore, some membrane proteins contain covalently bound carbohydrates; at least a few membrane proteins are known to have covalently bound lipid, and these nonprotein moieties no doubt influence the interactions of intrinsic membrane proteins with lipids and membranes. Although the number of major lipid components in most membranes is small, the number of different protein species is often very large, a fact that is not surprising in view of the many specific enzyme activities and transport systems known to be present in membranes. Comparison of membrane proteins from a variety of different membranes of animals, plants, and bacterial origin has revealed no single protein species common to all membranes; thus, there is no evidence for a structural protein that can account for the similar appearance in the microscope of membranes from all these sources. One of the major challenges of membrane biology has been the development of a membrane model which accounts for the similarity in the appearance of membranes as viewed in cross section while allowing for the wide variations in both lipid and protein compositions known to exist in these various membranes.

MEMBRANE STRUCTURE: SOME REASONABLE MODELS

In 1925 Gorter and Grendel found that the amount of lipid in red blood cell membranes was just enough to form a monolayer whose surface area was twice that of the surface of the cell, suggesting that the red blood cell membranes might consist of a bilayer of lipid. Although more recent experiments have called into question the precise numbers obtained by Gorter and Grendel, the existence of a lipid bilayer in biological membranes has been confirmed by several other experimental approaches, and the lipid bilayer (Figure 5-12*a*) remains the prominant feature of most current models of membrane structure. To the basic lipid bilayer of Gorter and Grendel, Davson and Danielli added two layers of protein, one at each of the surfaces of the bilayer, held to those surfaces by primarily ionic interaction (Figure 5-12*b*). According to the original formulation, these proteins were spread in an extended conformation (such as the β-pleated sheet), and their interaction with lipids was exclusively with the polar head groups and not with the fatty-acid side chains in the nonpolar regions of the lipid. The kinds of protein absorbed at the two surfaces of the bilayer could be different in this model, accounting for the sidedness of biological membranes. Robertson generalized this model on the basis of his observation with the electron microscope, which showed that membranes from all sources (animal, plant, and bacteria) possessed virtually the same trilaminar structure. Although the Davson-Danielli-Robertson membrane model was appealing for its simplicity and universality, recent developments of the study of membrane composition and structure have made necessary certain basic alterations in the model. It is apparent now that the bonds which hold together the protein and the lipids of membranes are primarily nonionic; raising the salt concentration of the medium in which membranes are suspended does not result in the release or solubilization of membrane proteins, as would be expected if the protein were held to the lipid by strong ionic bonds.

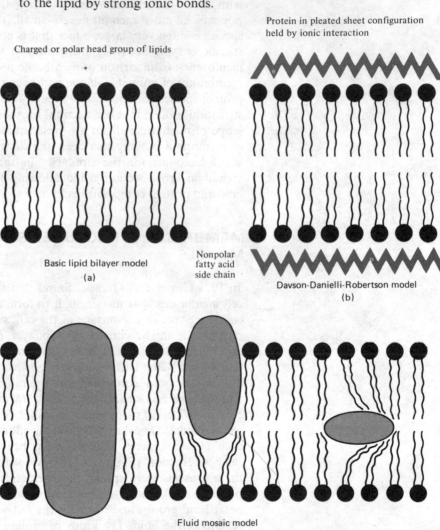

Charged or polar head group of lipids

Basic lipid bilayer model
(a)

Nonpolar fatty acid side chain

Protein in pleated sheet configuration held by ionic interaction

Davson-Danielli-Robertson model
(b)

Fluid mosaic model
(c)

Figure 5-12 The three basic models of membrane structure.

Spectroscopic studies of membrane proteins have indicated that very little if any β-pleated sheet structure exists in membrane proteins; rather, there appears to be a substantial amount of α-helical regions in membrane proteins. A third line of evidence which is at variance with the Davson-Danielli-Robertson model is that obtained by the techniques of electron microscopy and freeze etching, which allows one to see membrane surfaces in "bas relief." Such studies show (Figure 5-11) that membrane surfaces are granular, not smooth. A large number of approximately spherical granules with diameters 50–100 Å appear to float in a featureless matrix. Treatment with proteolytic enzymes removes the granules, suggesting that they are globular proteins. With the discovery of a wide variety of transport systems and the enzymatic activities associated with membranes, it has become apparent that membranes are functional mosaics consisting of many different proteins inserted into the basic structure. Finally, studies of the lipids in biological membranes and in model lipid bilayers using several physical techniques have given strong evidence for mobility within the lipid phase of the membrane. Lipids in one face of the bilayer diffuse rather rapidly in the lateral direction, although movement from one face of the bilayer to the other face proceeds slowly if at all.

As a result of these new observations a substantially altered model for the structure of biological membranes has evolved. In this new *fluid mosaic model* (Figure 5-12c) the basic structural unit is a lipid bilayer. Any of the number of lipids can contribute to this lipid bilayer by arranging themselves with their hydrophobic ends pointing toward the center of the bilayer and their hydrophilic head groups interacting with water on the inside or the outside of the membrane. This basic lipid bilayer is penetrated by globular proteins which may reach part way through the bilayer or may actually span the membrane. In either case, the interaction between protein and lipid is primarily hydrophobic; that is, the nonpolar side chains of the membrane protein have hydrophobic interactions with the nonpolar regions of the lipid. The protein lipid interactions, though rather strong, are not covalent. Within the plane of the lipid bilayer, each individual molecule is free to diffuse laterally. The result is that the protein finds itself floating in a sea of lipid, so that the protein as well as the lipid possesses a certain degree of lateral mobility.

Studies of the model membranes composed of lipid bilayers containing no protein have shown that the lipid bilayer is highly impermeable to charged or polar solutes. Thus, the lipid bilayer of a biological membrane is probably responsible for the impermeability of biological membranes, and specific membrane proteins catalyze the transmembrane flux of specific charged or polar compounds. One of the remaining unsolved problems of membrane biology is the precise mechanism by which membrane proteins catalyze active transport (see Chapter 8).

PROBLEMS

1. State two differences between micellar particles and emulsified droplets.

2. Which of the common complex lipids contain a polar but uncharged group as their hydrophylic function?

3. What is a major physiological disturbance or disease associated with a lack of each of the fat-soluble vitamins in the diet?

4. Biological membranes contain two major classes of compound. What are they?

5. What are some of the properties that distinguish intrinsic membrane proteins from extrinsic proteins?

6. What is the essential difference in the location of protein in the "fluid mosaic" model of membrane structure and the Davson-Danielli-Robertson model?

Suggested Readings

Books

Ansell, G. B., J. N. Hawthorne, and R. M. C. Dawson (eds.). *Form and Function of Phospholipids,* 2nd ed. Elsevier, Amsterdam (1973).

Deluca, H. F., and J. W. Suttie (eds.): *The Fat-Soluble Vitamins,* University of Wisconsin Press, Madison, (1970).

Deuel, H. J. *The Lipids,* vols. I, II, III. Wiley, Interscience, New York (1951, 1955, 1957).

Gurr, M. I., and A. T. James. *Lipid Biochemistry: An Introduction.* Cornell University Press, Ithaca, N.Y. (1971).

Articles and Reviews

Bergström, S., and B. Samuelsson. The Prostaglandins. *Endeavour* **27**:109–113 (1968).

Fox, C. F. The Structure of Cell Membranes. *Sci. Amer.* **226**:30–38 (1972).

Singer, S. J., and G. L. Nicolson. The Fluid Mosaic Model of the Structure of Membranes. *Science* **175**:720–731 (1972).

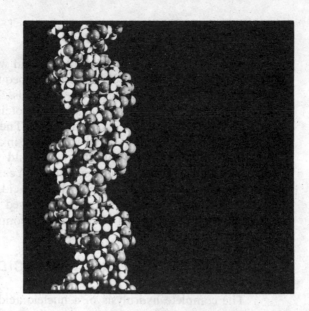

nucleic acid chemistry

chapter six

Nucleic acids are macromolecules composed of pyrimidine and purine bases attached to a phosphodiester linked polymer of ribose (RNA) or deoxyribose (DNA). The basic structure of DNA is a double helix held together by hydrogen bonds between purines and pyrimidines on opposite chains and hydrophobic interactions between these stacked base pairs. The basic structure of RNA is a single-stranded structure which can associate with itself by the same type of hydrogen bonds stabilizing DNA. The DNA of the cell is a very high molecular weight compound that transfers genetic information from one generation to the next. Most cellular RNA is present in the form of ribonucleoprotein particles called ribosomes, while a smaller portion is found as transfer and messenger RNA molecules. All three classes of RNA are involved in the protein synthetic capacity of the cell.

The material now known as nucleic acid was first isolated in a crude form by Miescher in the 1860s. He accomplished this isolation from a preparation of crude nuclei obtained from pus cells. The nuclei were extracted with salt solutions to remove much of the protein, and the nucleic acid was then precipitated under acidic conditions. The material which was obtained had a high molecular weight, was soluble in dilute base, contained a relatively large amount of phosphorus, and could also be obtained from salmon sperm and thymus gland preparations. Yeast cells were found to be another source of a slightly different nucleic acid, and for some time the terms *thymus* and *yeast nucleic acid* were used for what we now know to be deoxyribonucleic acid (DNA) and ribonucleic acid (RNA), respectively.

CHEMISTRY OF NUCLEIC ACIDS

The complete hydrolysis of a nucleic acid preparation yields phosphate, a sugar, a pair of pyrimidine bases, and a pair of purine bases (Table 6-1).

Table 6-1 Hydrolytic Products of RNA and DNA

	Ribonucleic Acid	Deoxyribonucleic Acid
Acid	Phosphoric acid	Phosphoric acid
Sugar	D-Ribose	D-2-Deoxyribose
Bases		
Purines	Adenine	Adenine
	Guanine	Guanine
Pyrimidines	Cytosine	Cytosine
	Uracil	Thymine

Both of the sugars found in nucleic acids are D-pentoses, and both are present in the β-furanose form (Figure 6-1). The structures of the three pyrimidines and the two purines commonly found in nucleic acids are also shown in Figure 6-1. Those bases containing a ring oxygen can undergo a tautomerization between a keto and an enol form. The form which predominates in this equilibrium will depend on the pH, and at physiological pHs the form of the free base indicated in Figure 6-1 will be the predominant one. In the nucleic acid molecule, the chemical binding to the ring nitrogen also pushes the equilibrium to the keto form.

High molecular weight DNA and RNA differ considerably in their susceptibility to hydrolysis under alkaline conditions. Although DNA is rather stable to dilute basic conditions, RNA is readily broken down to nucleotides—building blocks of the nucleic acids. Nucleotides contain one mole each of ribose, a purine or a pyrimidine, and phosphate. In the nucleotide unit, the ribose is joined to the purine or pyrimidine by an N-glycoside bond to the N-9 of a purine or N-1 of a pyrimidine, and the phosphate is present as an ester on the 3′ or 5′ hydroxyl of the ribose. It can

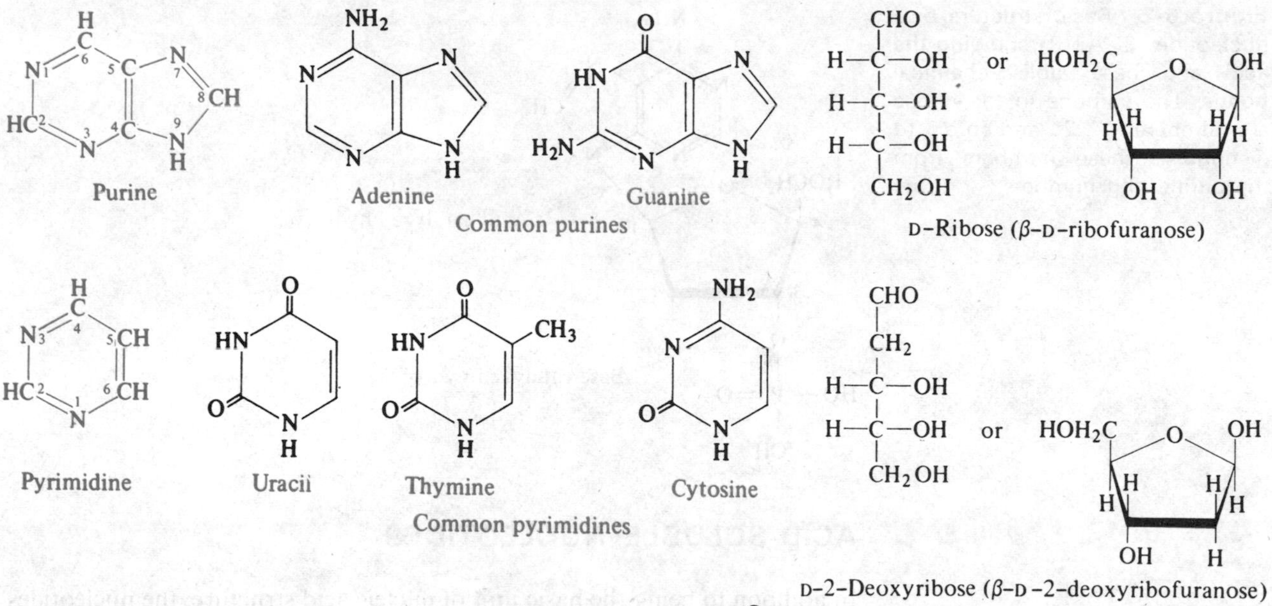

Purine

Adenine Guanine

Common purines

D–Ribose (β–D–ribofuranose)

Pyrimidine Uracil Thymine Cytosine

Common pyrimidines

D-2-Deoxyribose (β–D–2-deoxyribofuranose)

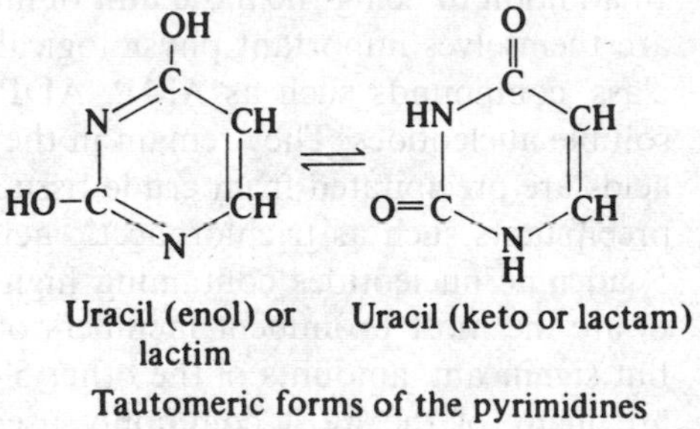

Figure 6-1 Structures of the sugars, pyrimidines, and purines found in nucleic acids. The predominant tautomeric form of a pyrimidine or purine is dependent upon pH and upon the presence of other chemical bonds on the molecule.

Uracil (enol) or lactim Uracil (keto or lactam)

Tautomeric forms of the pyrimidines

be further degraded by alkaline hydrolysis to phosphate plus a nucleoside, which is a pyrimidine or purine plus the sugar. Alternatively, acid hydrolysis will convert the nucleotide to a sugar phosphate plus the free base (Figure (6-2). Table 6.2 gives the nomenclature of nucleosides and nucleotides.

Table 6-2 Nomenclature of Nucleotides and Nucleosides

Base	Ribose Nucleoside	Deoxyribose Nucleoside	Ribonucleotides (5′ as an example)	Deoxyribonucleotides (5′ as an example)
Adenine	Adenosine	Deoxyadenosine	Adenosine-5′P[a] (AMP)[b]	Deoxyadenosine-5′P (dAMP)
Uracil	Uridine	Deoxyuridine	Uridine-5′P (UMP)	Deoxyuridine-5′P (dUMP)
Cytosine	Cytidine	Deoxycytidine	Cytidine-5′P (CMP)	Deoxycytidine-5′P (dCMP)
Guanine	Guanosine	Deoxyguanosine	Guanosine-5′P (GMP)	Deoxyguanosine-5′P (dGMP)
Thymine	Ribosylthymine	Thymidine	Ribosylthymine-5′P (TMP)	Thymidine-5′P (dTMP)

[a] Can also be named as a series of acids, for example, 5′-adenylic acid for AMP, 3′-cytidylic acid for 3′-CMP.
[b] AMP means 5′-AMP; any other ester must be specified, for example, 3′-dAMP for deoxyadenosine-3′-phosphate or 2′-UMP for uridine-2′-phosphate. The di- and triphosphates are abbreviated the same way, for example, GTP, for guanosine-5′-triphosphate.

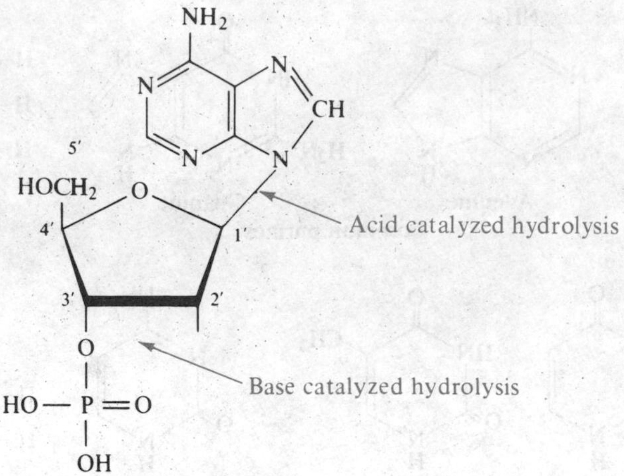

Figure 6-2 Basic structure of a nucleotide (3'-AMP) showing the acid and base labile chemical bonds. The carbons in the ribose are numbered 1', 2', and so on, to distinguish these numbers from the purine ring numbers.

ACID-SOLUBLE NUCLEOTIDES

In addition to being the basic unit of nucleic acid structure, the nucleotides are themselves important physiologically active compounds. As a general class, compounds such as AMP, ADP, and ATP have been called acid-soluble nucleotides. They remain in the supernatant if proteins and nucleic acids are precipitated from crude tissue homogenates by common protein precipitants such as trichloroacetic acid or perchloric acid. Although the 5'-adenine nucleotides containing high-energy phosphate bonds (Chapter 8) are the most prominent members of this class, tissues contain smaller, but significant, amounts of the other 5'-nucleotides. Included among these are many of the most commonly used coenzymes such as the pyridine nucleotides, the flavins, and coenzyme A. A large number of intermediates of carbohydrate and complex lipid biosynthesis are also nucleotides, and they fall in the same general class.

PHYSICAL PROPERTIES OF NUCLEOTIDES

The nucleotides, because they are typical phosphate esters, would be expected to act as strong acids in solution. Upon titration they show a pK_a of about 1.0, owing to the primary dissociation of the phosphate group, and a pK_a around 6.0, owing to the secondary dissociation of the phosphate. Titration of AMP, CMP, and GMP will also show the presence of an ionizable group with a pK_a of between 2.0 and 4.0, and UMP and GMP, a group with a pK_a of from 9 to 10. These pK values are associated with the removal of protons to the ring nitrogens or amino groups of the bases of the ring nitrogens. The most useful physical property of the nucleotides is probably the ultraviolet absorption contributed by the nitrogenous bases. Although there are slight differences in the absorption maximums of the different bases, and the peak is shifted somewhat by pH changes, an

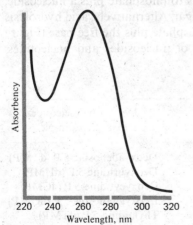

Figure 6-3 Absorption spectrum of nucleic acids showing a maximum in 260 nm region.

128

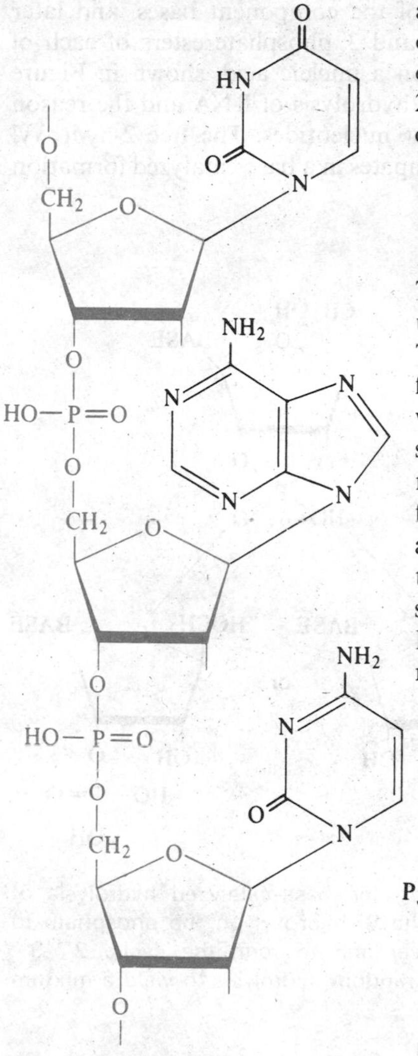

absorption maximum at about 260 nm (Figure 6-3) is a general characteristic of nucleic acids. As the absorption of proteins, which is due to the presence of the aromatic amino acids, has a maximum at 280 nm, the ratio of the 260:280 absorption is often used as a measure of the relative contribution of protein and nucleic acid to a mixture.

PRIMARY STRUCTURE OF NUCLEIC ACIDS

The primary structure of the nucleic acids was determined mainly by the use of specific enzymatic and acid- or base-catalyzed hydrolytic procedures. The high molecular weight nucleic acid molecules were cleaved into smaller fragments, and the knowledge of the structure of these was used to determine what chemical linkages were present in the parent molecule. The basic structure, which is illustrated in Figure 6-4, is composed of a series of nucleosides held together in a long linear chain by a phosphodiester linkage from the 3′ hydroxyl of one sugar to the 5′ hydroxyl of the next. Note that an alternating phosphate-sugar-phosphate-sugar-phosphate chain forms this structure, and that the nitrogenous bases are simply attached to this sugar phosphate backbone. The type of structural formula written in Figure 6-4 is cumbersome to use, and two shorthand designations commonly used are illustrated in Figure 6-5. By convention, the end containing

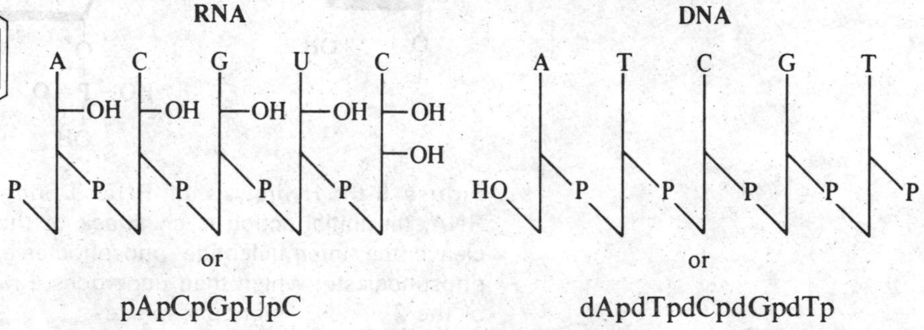

RNA

A   C   G   U   C

—OH —OH —OH —OH —OH
                        —OH

P   P   P   P   P

or

pApCpGpUpC

DNA

A   T   C   G   T

HO   P   P   P   P   P

or

dApdTpdCpdGpdTp

Figure 6-5 Two ways of designating the structure of oligonucleotides.

a 5′-OH or 5′-phosphate is written to the left and the end with a 2′ or 3′ phosphate, or both positions unesterified, is written to the right. These shorthand formulas for oligonucleotides make it possible to discuss efficiently and unambiguously different sequences of nucleic acid segments without involving a cumbersome structural formula.

HYDROLYSIS OF NUCLEIC ACIDS

One of the early observations made about the chemical behavior of the nucleic acids was the extreme alkali lability of RNA compared to DNA. The alkaline degradation of RNA was also shown to result in a mixture

Figure 6-4 Segment of a ribonucleic acid molecule, showing phosphodiester bonds between nucleotide units. A deoxyribonucleic acid molecule would be identical except that the sugars would be deoxyribose, and thymine rather than uracil would be one of the bases.

of two different nucleotides for each of the component bases, and later work revealed that these were the 2'- and 3'-phosphate esters of each of the nucleosides. The action of alkali on a nucleic acid, shown in Figure 6-6, illustrates the basis for the ease of hydrolysis of RNA and the reason for the production of the two different nucleotides. The free 2-hydroxyl adjacent to the 3'-phosphate ester participates in a base-catalyzed formation

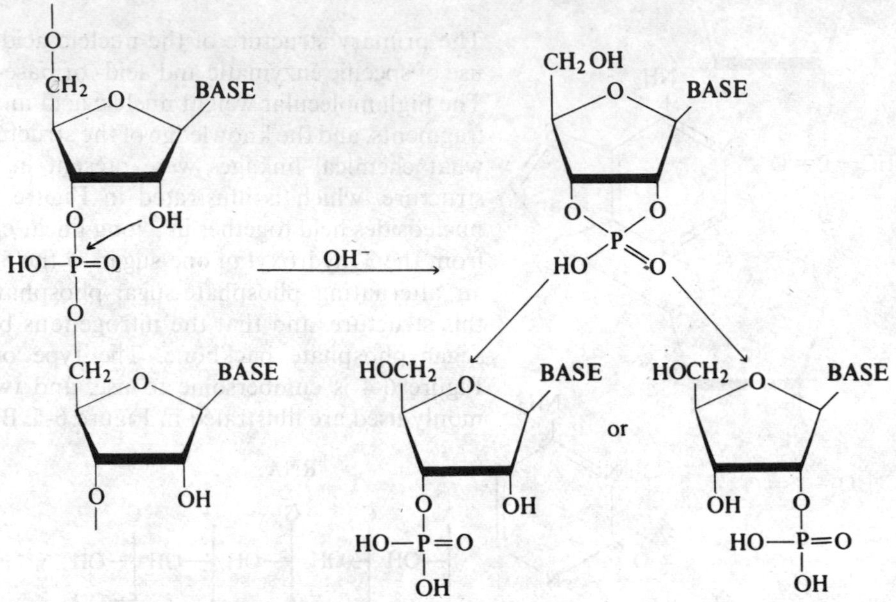

Figure 6-6 Hydrolysis of RNA. During the base-catalyzed hydrolysis of RNA, the initial action is an attack of the 2' hydroxyl on the phosphate to cleave the internucleotide phosphodiester and to form the cyclic 2', 3'-phosphodiester which then undergoes a random hydrolysis to yield a mixture of the 2', and 3'-monophosphates.

of a cyclic diester, which is then further cleaved to yield a mixture of the 2' and 3' phosphates. Although DNA, having no hydroxyl on the adjacent carbon to participate in the hydrolysis, is rather stable to alkaline conditions, it undergoes an interesting reaction under rather mild acidic conditions. The *N*-glycoside bond is acid labile, particularly if it involves a purine, and under appropriate conditions most of the purines can be cleaved from DNA with very little hydrolysis of the sugar phosphate backbone to yield apurinic acid.

A number of enzymes will catalyze cleavages of nucleic acid chains at specific points. These have been very useful in studies on the sequence of nucleic acids, and the specificity of some of those used to hydrolyze RNA is illustrated in Table 6-3.

Table 6-3   Specific Cleavage of Ribonucleic Acid

Method	Type of Cleavage[a]
Dilute alkali	All nucleotides to yield the 2- and 3-nucleotides
Pancreatic RNase	Endonuclease, cleaves at cytosine and uracil to give fragments with a pyrimidine-3′-P
RNase T1	Endonuclease, cleaves at guanine to give fragments with a guanosine-3′-P
Snake venom diesterase	Exonuclease, cleaves from end with a free 3′ OH to give 5′-mononucleotides, will also cleave deoxyribonucleic acid
Spleen phosphodiesterase	Exonuclease, cleaves from the 5′ end to give 3′-mononucleotides, will also cleave deoxyribonucleic acid

[a] *Example*, cleavage of pGpUpCpApGpA by:
Base yields pGp, Up, Cp, Ap, Gp, A
Pancreatic RNase yields pGpUp, Cp, ApGpA
RNase T1 yields pGp, UpCpApGp, A
Snake venom diesterase yields pG, pU, pC, pA, pG, pA

SECONDARY STRUCTURE OF NUCLEIC ACIDS

Although much of the early work on the chemical structure of nucleic acids came from the study of RNA, it was clear to early workers in this field that DNA was a much more ordered structure, and ideas of the three-dimensional structure of nucleic acids developed from its study. The first information providing a clue to the type of structure involved came not from any physical measurements on the molecule, but rather from chemical analysis of the nucleotide composition of DNA from various sources. These data (Table 6-4) revealed that there were considerable differences in the actual percentages of the different nucleotides in the DNA from various sources, but that within experimental error the amount of adenine was always equal to the amount of thymine, and the amount of guanine was equal to the amount of cytosine. A corollary to this is, of course, that the purine content must equal the pyrimidine content.

Table 6-4   Base Composition of DNA from Various Sources

Source	A	G	C	T	$\dfrac{A+T}{G+C}$	$\dfrac{A}{T}$	$\dfrac{G}{C}$
Man	30.9	19.9	19.8	29.4	1.52	1.05	1.00
Calf	29.0	21.2	21.2	28.5	1.35	1.01	1.00
Hen	28.8	20.5	21.5	29.2	1.39	0.99	0.95
Salmon	29.7	20.8	20.4	29.1	1.42	1.02	1.02
Yeast	31.3	18.7	17.1	32.9	1.79	0.95	1.09
E. coli	24.7	26.0	25.7	23.6	0.93	1.04	1.01

At the same time that these chemical measurements were being made, around 1950, a group of x-ray crystallographers in London directed by Wilkins were making various physical measurements of DNA fibers. These measurements led to the conclusion that the molecule was a long thin rod with a diameter of about 20 Å, and that it contained some sort of repeating structural unit with spacings of 3.4 and 34 Å. Crystallographers trying to interpret the x-ray data at the time were concerned with such problems as whether the phosphate groups or the bases were directed toward the interior of this rod and whether two or three chains were somehow coiled together to produce the rodlike structure.

Double Helix

The key observation which allowed the correct postulation of the structure of DNA was made in 1953 by Watson and Crick, who were working in Cambridge, England. They demonstrated that it was possible to arrange the pairs of bases, adenine and thymine, and guanine and cytosine, such that hydrogen bonds were readily formed between them. The hydrogen bonds are formed between a pair of electrons on the keto group or ring nitrogen of one base and a hydrogen atom on a ring nitrogen or amino group on the other (Figure 6-7). The two base pairs which were formed had very similar dimensions and allowed the construction of a double helical structure where the bases extended to the inside of the helix and held it together by their hydrogen-bonded interactions. Although an individual hydrogen bond is very weak, the large number of them formed in the DNA molecule effectively stabilizes the structure. The structure is also stabilized by hydrophobic interactions between the stacked base pairs.

The 3.4 Å spacing which had been seen in x-ray studies was associated with the distance between the base pairs, and the 34 Å spacing corresponded to the distance covered by a single twist of the helix. The structure required that the two strands of DNA run in opposite directions, that is, the direction from 3' to 5' phosphates was different, but posed no restriction on what bases might be present at any point along the chain. An A:T pair had essentially the same dimension as a G:C pair (Figure 6-8). It was clear to Watson and Crick at the time the structure was proposed that the specific base pairing involved in stabilizing the structure of DNA provided the mechanism by which DNA could be accurately duplicated during cell division, and by which a DNA of unique base sequence passed from a mother to a daughter cell. Although other types of base pairing have been suggested, the currently accepted model for the structure of DNA is essentially the same as that originally proposed. It has also been possible to prepare synthetic DNA-like polymers whose properties are those predicted by the model.

RNA STRUCTURE

In contrast to DNA, RNA is essentially a single-stranded molecule whose secondary structure must differ considerably from DNA. Measurements

132

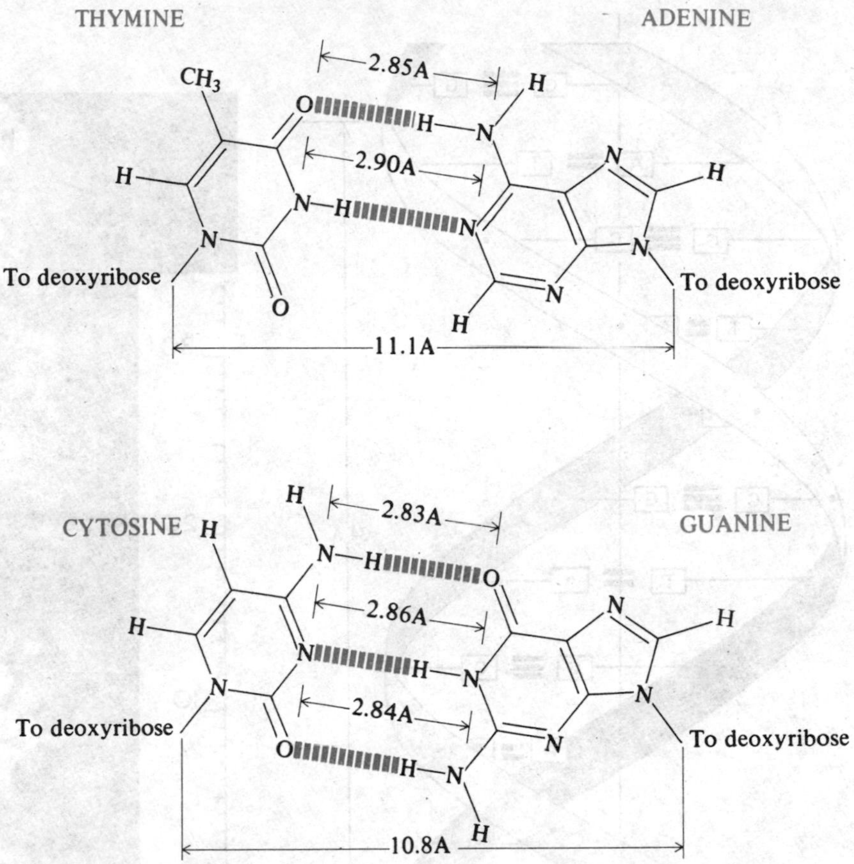

Figure 6-7   Hydrogen bonding between the adenine-thymine and guanine-cytosine bases in DNA. The A:T pair has two hydrogen bonds; the G:C pair has three.

of so-called *melting curves* do indicate, however, that RNA does have regions where base pairing must occur. If the temperature of a DNA solution is raised, hydrogen bonds between bases are broken, and an increase in uv absorption is seen (hyperchromic effect). For native DNA, this occurs over a narrow range of temperature, and the temperature at which 50 percent of the hyperchromic effect is seen is called *the melting point*. The hyperchromic effect is largely a characteristic of the guanine and cytosine content of the molecule. A similar response can be seen with some RNA preparations (Figure 6-9), but the hyperchromic effect is less extensive and the transition not as sharp, indicating that there is less extensive base pairing in the molecule. The present evidence would indicate that there might be regions in the RNA molecule where it is doubled back on itself to form helical areas, and that these areas are separated by extensive non-helical regions where the molecule assumes a random coil type of structure (Figure 6-10).

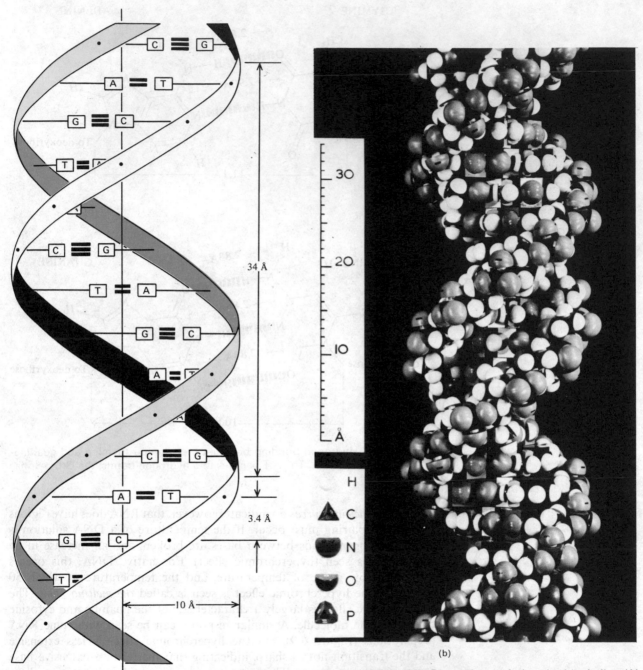

Figure 6-8 Structure of DNA. (*a*) Dimensions of the double helix. (*b*) Space-filled model showing the manner in which the phosphate groups point to the outside of the helix and the bases to the interior. (Professor M. H. F. Wilkins, Medical Research Council, King's College, London)

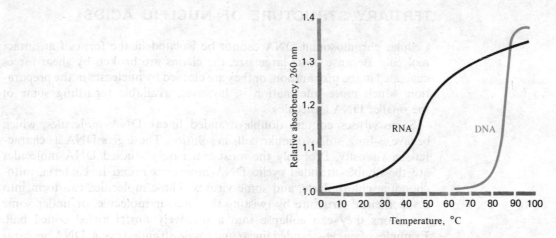

Figure 6-9 Typical melting point curves of a high molecular weight RNA and native DNA. The lower temperature, less extensive absorbancy changes, and less abrupt transition in the curve for RNA indicate that RNA is less extensively base-paired than DNA.

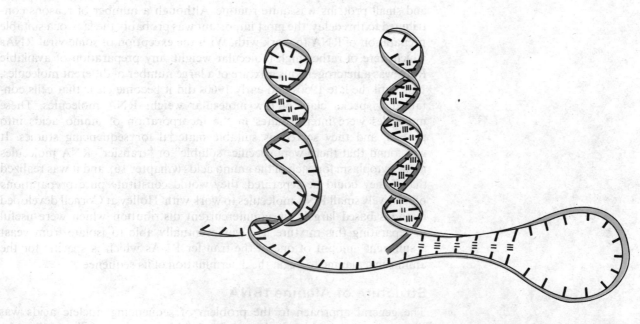

Figure 6-10 Hypothetical model of the type of base-pairing that might occur in single stranded RNA. The molecule forms double stranded and helical regions by doubling back on itself, but it also has large regions where the bases are unpaired.

TERTIARY STRUCTURE OF NUCLEIC ACIDS

Cellular chromosomal DNA cannot be isolated in the form of an intact molecule. Because of its large size, the chains are broken by shear forces generated in the preparation, or they are cleaved by nucleases in the preparation. Much more information is, however, available regarding some of the smaller DNA molecules.

Some viruses contain double-stranded linear DNA molecules, which behave as long, stiff, wormlike coils in solution. These give DNA its characteristic viscosity. Probably the most extensively studied DNA molecules are the double-stranded cyclic DNA molecules found in bacteria, mitochondria, chloroplasts, and some viruses. These molecules can form into a superhelical structure by twisting the circular molecule, or under some conditions they can collapse into a relatively unstructured coiled ball. Examples of single-stranded linear and single-stranded cyclic DNA preparations are also known.

PRIMARY SEQUENCES OF BASES IN RNA

Progress on the determination of the nucleotide sequence of an RNA molecule was delayed beyond the time when the sequencing of peptides and small proteins was quite routine. Although a number of reasons contributed to this delay, the most important was probably the lack of a suitable preparation of RNA to work with. With the exception of some viral RNAs which were of rather high molecular weight, any preparation of available RNA was a heterogeneous mixture of a large number of different molecules. Not until the late 1950s and early 1960s did it become clear that cells contained a special class of low molecular weight RNA molecules. These molecules were intermediates in the incorporation of amino acids into protein, and they served as suitable material for sequencing studies. It was found that there were specific "soluble" or "transfer" RNA molecules in the cytoplasm for each of the amino acids (Chapter 16), and it was realized that if they could be separated, they would constitute pure preparations of relatively small RNA molecules to work with. Holley at Cornell developed methods based largely on countercurrent distribution which were useful in separating this mixture. He was eventually able to isolate from yeast a sufficient amount of one of the transfer RNAs which is specific for the amino acid alanine to begin the determination of its sequence.

Structure of Alanine tRNA

The general approach to the problem of sequencing nucleic acids was the same as that which had been developed and successfully used in the sequencing of peptides and proteins. The molecule was split into large fragments, these were separated, and their nucleotide sequence determined. The major fragments which were useful in the characterization were those which were obtained by ribonuclease cleavage of the molecule at the linkages following pyrimidine nucleotides, to give pyrimidine-3'-phosphates and cleavage with takadiastase-T-1. Under usual conditions the latter enzyme is able to cleave the polynucleotide at all guanosine residues. Holley determined that at low temperature and at appropriate Mg^{2+} concentrations it could be used to cleave only one particularly sensitive linkage and split the molecule into two roughly equal parts. The fragments from the enzymatic digests were separated by various column chromatographic methods, many of which had to be developed for this purpose, and their sequence determined by hydrolytic and enzymatic degradation. The problem of aligning the correctly sequenced fragments was aided tremendously by the fact that the transfer RNA molecules contain not only the four common bases, but also a number of so-called "rare" nucleotides (Figure 6-11). Some of these had not been characterized before Holley

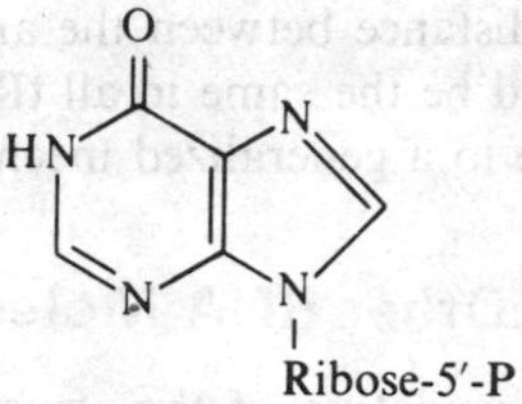

Ribose-5'-P

Inosinic acid (I)

Ribose-5'-P

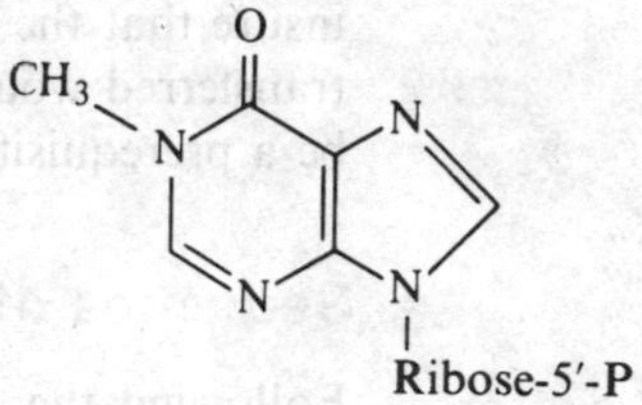

1-Methylinosinic acid (I^m)

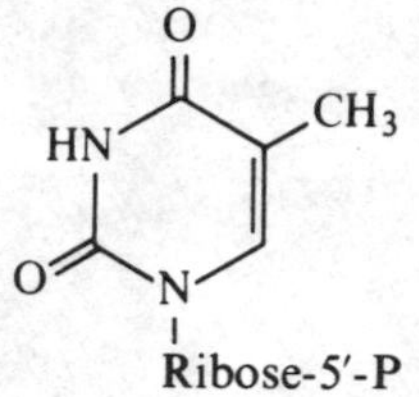

Ribose-5'-P

Ribothymidylic acid (T)

Ribose-5'-P

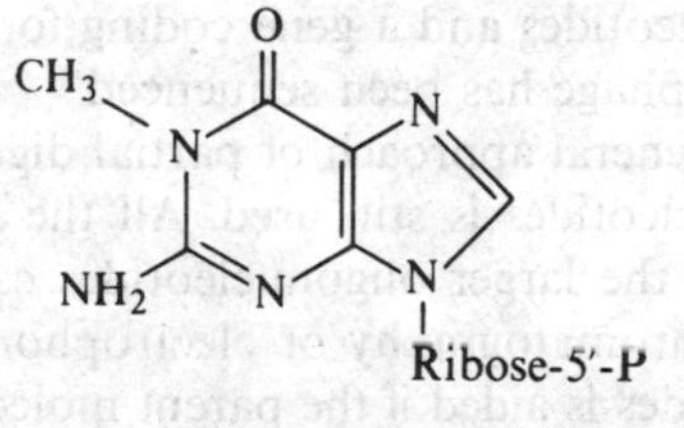

1-Methyl guanylic acid (G^m)

Ribose-5'-P

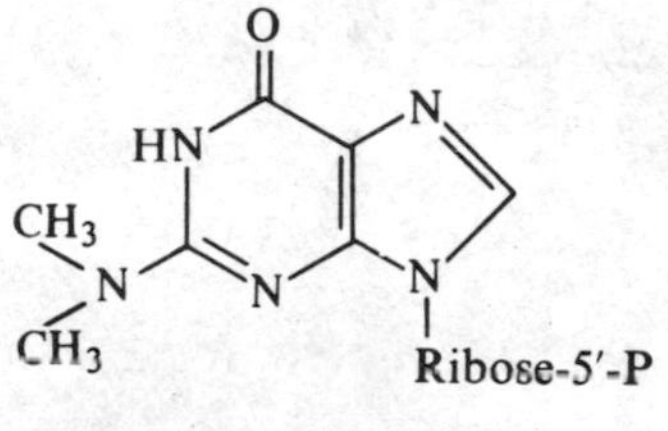

N^2-Dimethyl guanylic acid (G^m)

Ribose-5'-P

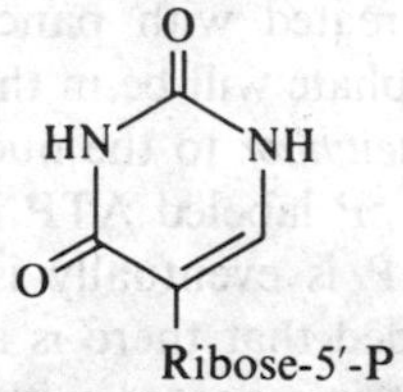

Pseudouridylic acid (ψ)

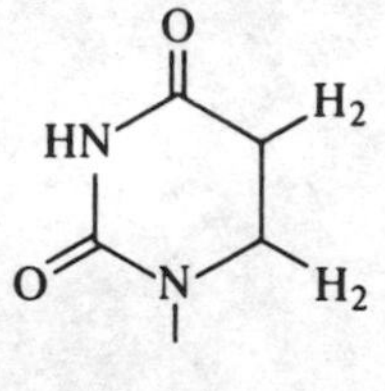

Dihydrouridylic acid (U^h)

Figure 6-11 Structures of some of the nucleotides found in tRNA molecules which are not found in most other ribonucleic acids.

began his study of the structure, but once identified, they provided important markers which could be used to align the small fragments into the appropriate order. In all, there were seven of these "rare nucleotides," and the four common ribonucleotides, in the molecule. The structure of the complete molecule, which contains 77 nucleotides and has a molecular weight of 26,600, is shown in Figure 6-12. Since the determination of the structure of the alanine tRNA by Holley, a number of the other tRNA molecules have been sequenced, and some generalities about the structures are readily apparent (Figure 6-12). It will be seen in Chapter 16 that there must be a three-nucleotide portion of each tRNA (the anticodon) that recognizes a complementary sequence (the codon) on a messenger RNA molecule during protein synthesis. When the structures of a number of tRNAs were available, it was apparent that when they were all written in a cloverleaf fashion which maximized the possibility of internal base pairing, the anticodon occupied the same region of each molecule. This would insure that the distance between the anticodon and the amino acid to be transferred would be the same in all tRNA molecules, and would seem to be a prerequisite to a generalized mechanism of protein synthesis.

Sequence of Other RNA Molecules

Following the elucidation of the structure of some of the tRNAs, larger molecules have been sequenced. The 5S RNA from bacterial ribosomes, which contains 120 nucleotides and no rare bases, was sequenced, and methods are now available to sequence rather large molecules. Particular attention has been given to viral RNAs, and one such molecule containing 474 nucleotides and a gene coding for the synthesis of a coat protein of a bacteriophage has been sequenced.

The general approach of partial digestion and identification of resulting oligonucleotides is still used. All the possible di- and tri-nucleotides and some of the larger oligonucleotides can be identified by two-dimensional paper chromatography or electrophoresis. The sequencing of large oligonucleotides is aided if the parent molecule can be synthesized in an *in vitro* system. If this can be done, the synthesis can be carried out with nucleoside triphosphates labeled in the $5'$-α-phosphate with ^{32}P, and this radioactive phosphate will be incorporated into the phosphodiester bond of the RNA molecule (see Chapter 16). If an oligonucleotide isolated from this radioactive RNA is treated with pancreatic ribonuclease or with alkali, the radioactive phosphate will be in the $3'(2')$ position of the nucleotide, which was the *nearest neighbor* to the nucleotide, originally labeled triphosphate. For example, if ^{32}P labeled ATP is used to synthesize a nucleic acid and ^{32}P labeled GMP is eventually isolated from an oligonucleotide digest, it can be concluded that there is a GpA sequence in the oligonucleotide. If the experiment is repeated with the other three labeled nucleotide triphosphates, a large amount of sequence information can be obtained. An example of this approach is illustrated in Figure 6-13.

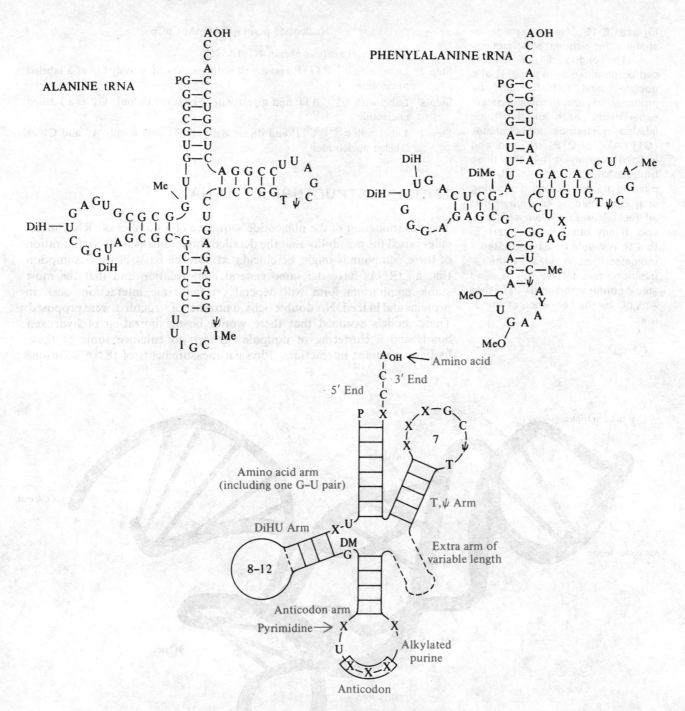

Figure 6-12 Primary structure of alanine tRNA and phenylalanine tRNA, and the generalized "cloverleaf" structure that will fit all the tRNA molecules that have been sequenced.

Figure 6-13 Nearest neighbor analysis for sequence determination. The oligonucleotide shown can be obtained from a digest of a nucleic acid which can be produced *in vitro*. In three separate experiments, each of the three labeled nucleoside triphosphates (GTP, ATP, or CTP) are used, and in each experiment the other three triphosphates in the incubation mixture are not labeled. Following step 2, it could be concluded that all the G residues were at the 5' end. If any other residue were 5' to a G, it would be labeled. Step 3 indicates that A was in the 3' position next to the 4th G, and step 4 confirms this, and establishes ApCpC as the sequence at the 3' end.

Nucleotide pGpGpGpGpApCpCp

Step 1   By chemical analysis obtain 4G : 1A : 2C

Step 2   Label with α ^{32}P GTP and digest with alkali; obtain only Gp as a labeled nucleotide

Step 3   Label with α ^{32}P ATP and digest with alkali; obtain only Gp as a labeled nucleotide

Step 4   Label with α ^{32}P CTP and digest with alkali; obtain only Ap and Cp as labeled nucleotides

TERTIARY STRUCTURE OF tRNA

The determination of the nucleotide sequence of a number of tRNA molecules raised the possibility that the detailed three-dimensional configuration of these compounds might be elucidated. On the basis of the assumption that all tRNAs have the same general conformation and that the most stable equilibrium form will depend on the same interactions seen in proteins and in the DNA double helix, a number of structures were proposed. These models assumed that there would be maximization of hydrogen bonds and a clustering of nonpolar groups to enhance some of these hydrogen-bonded interactions. Physical measurements of tRNA solutions

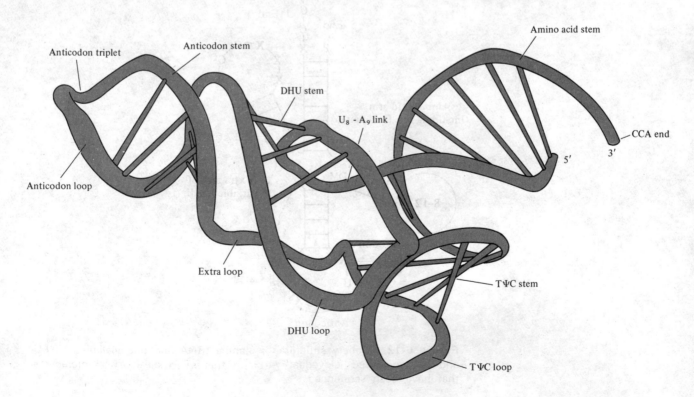

suggested that the molecule had the general shape of a long thin rod, which might easily dimerize.

The possibility that the detailed structure could be determined was enhanced when preparations of specific tRNA molecules were crystallized by Bock and by Rich. These crystals were found to be suitable for x-ray diffraction analyses, and early studies indicated the general structure that must be present. Recently, detailed models for the structure of yeast phenylalanine tRNA have been proposed by groups at MIT, Wisconsin, and Cambridge, England. The molecules are in the form of an L with the amino acid acceptor group at the end of one arm of the L and the anticoden at the other. The cloverleaf structure is twisted so that the four stem regions of the cloverleaf model are arranged into two arms of an L, each of which contains a column of bases stacked roughly perpendicular to the arm. In addition to the base pairs that were postulated from the cloverleaf structure, there are tertiary interactions between some of those bases not involved in the stem regions which hold various parts of the polynucleotide chain in a defined position. Although some disagreement still exists about some of the details of the structure, the proposed model shown in Figure 6-14 is probably essentially correct.

SYNTHESIS OF OLIGONUCLEOTIDES

The chemical synthesis of specific oligonucleotides containing internucleotide 3'-5'-phosphodiester linkages involves essentially the same problems as those encountered in the synthesis of peptides. A phosphomonoester must be activated so that it can phosphorylate the hydroxyl group of another nucleotide, and various functional groups on the molecules must be protected to prevent side reactions. The problems are more complex than those involved in peptide synthesis because of the multiplicity of functional groups present. These include primary and secondary hydroxyl groups on the sugar residues, and amino groups in the purine and pyrimidine rings and, if an attempt is made to synthesize ribopolynucleotides rather than deoxyribopolynucleotides, the 2' hydroxyl on the ribose. An example of the reaction sequence involved in the synthesis of a deoxyribotrinucleotide is illustrated in Figure 6-15. The activation of the phosphomonoester and the condensation between the activated group and the free hydroxyl group on the second nucleotide are carried out as a single step. Dicyclohexylcarbodiimide is used as the reagent, and the activated intermediate is not isolated. It is very difficult and sometimes impossible to purify and crystallize the products from these reaction mixtures. In practice the products are separated by chromatography on modified cellulose columns. These chemical reactions can be combined with the use of an enzyme which will use the oligonucleotides as a template for nucleic acid synthesis to obtain high molecular weight polynucleotides of known composition. It will be shown in Chapter 16 that these products can be used as synthetic messenger

Figure 6-14 (*opposite*) Schematic model of yeast phenylalanine tRNA. The ribose-phosphate backbone is drawn as a continuous cylinder and the various portions of the molecule are related to the typical cloverleaf structure (Figure 6-12). The bars drawn between different portions of the backbone represent Watson-Crick-type hydrogen bonded base pairs that form short helical segments. There are other hydrogen bond interactions (not shown) that are important in holding the molecule in its correct three-dimensional configuration. (Courtesy of M. Sundaralingam).

RNA molecules in *in vitro* protein-synthesizing systems. These techniques, and another enzymatic reaction which joins oligonucleotide fragments, have been used by Khorana to carry out the complete synthesis of the segment of DNA which contains the genetic information for the biosynthesis of alanine tRNA.

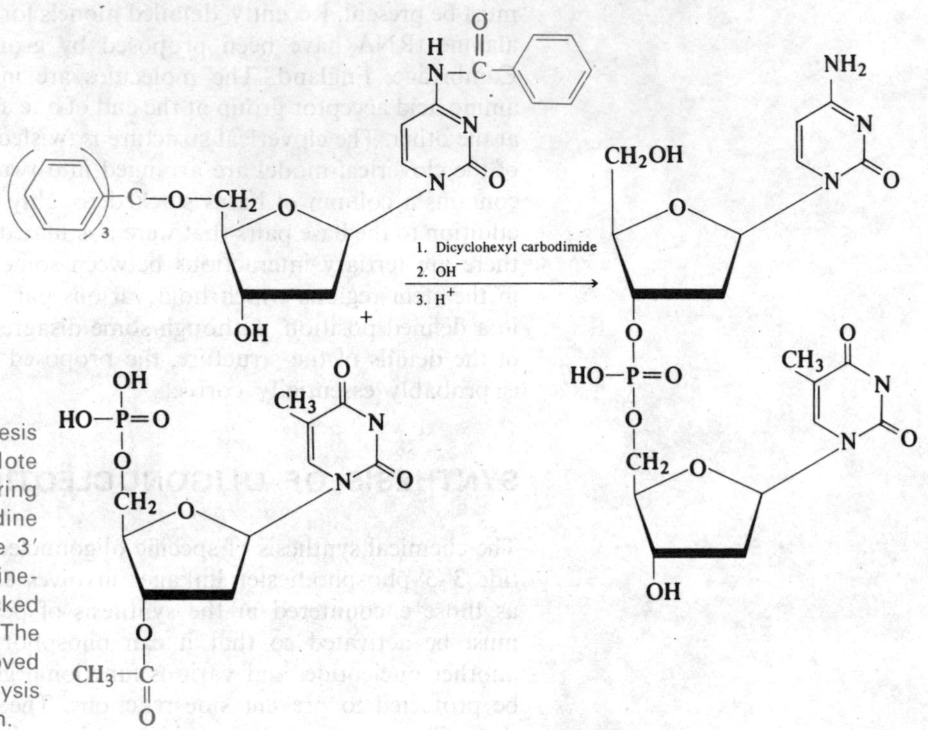

Figure 6-15 Chemical synthesis of a dinucleotide (dCpdT). Note that the free 5′ hydroxyl and ring amino group of the deoxycytidine must be blocked, and that the 3′ hydroxyl of the deoxythymidine-5′-phosphate must also be blocked to prevent side reactions. The blocking groups must be removed by alkaline and acid hydrolysis after the condensation reaction.

It is now known that the tRNA molecules are synthesized as larger precursors which are later cleaved to form the active tRNA molecule. Khorana's group has now synthesized the segment of DNA that corresponds to the gene for the precursor and the initiation sites of a tRNA that might be incorporated into a bacterial genome.

DISTRIBUTION AND SIZE OF NUCLEIC ACIDS

The accepted values for the molecular weight of DNA isolated from various sources have been subject to continual revision. For some time improved techniques of both DNA isolation and molecular weight determination gave estimations of larger and larger molecular weight. It was soon realized that the molecules being studied were of such high molecular weight that they were being degraded, not only by various nucleases present in many crude DNA preparations, but also by the mechanical shear forces involved in handling the solutions. Techniques were eventually developed so that it was possible to show that all the DNA of the bacterium *E. coli* is present as one circular double strand of DNA with a molecular weight of 2×10^9. If fully extended, the length of this molecule would be about 1 mm, even though it is contained in a cell with a diameter of only a few microns. At the present time, the techniques of electron microscopy are sufficiently advanced so that single molecules of DNA can clearly be photographed (Figure 6-16), and in some cases molecular weights can be estimated by simply measuring the length of the molecule.

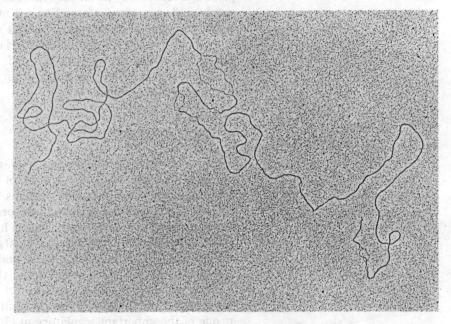

Figure 6-16 Electron micrograph of bacteriophage λ DNA. The preparation has been partially denatured by treatment at high pH, and the denatured sites are clearly seen as loops where the double-stranded DNA has separated into two single strands. The molecular weight of this molecule is about 32 million, and the length of the undenatured DNA is about 17 μ. (Courtesy of R. B. Inman and M. Schnos)

The majority of the DNA in eucaryotic cells is found in the nucleus, where it is associated with nuclear proteins to form chromosomes. The sizes of the individual DNA molecules making up the chromosomal complex are not known, but they are undoubtedly large. More recently, the presence of DNA in the mitochondria has been verified. This DNA is distinctly different from the nuclear DNA and, like bacterial DNA, has been shown to be circular.

The ribonucleic acid in the cell consists of three general types. The majority, 60–80 percent, is present in the form of large ribonucleoprotein particles, the ribosomes, which will be discussed in detail in Chapter 16. Most of the ribosomal RNA is of rather high molecular weight, from 0.6×10^6 to 1.8×10^6 depending on the source of ribosomes, but a small amount is present as the 5S RNA, which contains only 120 nucleotides. What function the 5S RNA plays other than as a structural component of the ribosomes is not clear.

The majority of the remaining RNA of the cell is in the form of the soluble or transfer RNA fraction. This fraction contains a mixture of low molecular weight polyribonucleotides of roughly 25,000 MW. These molecules play a role in protein synthesis by serving as the vehicle by which the activated amino acids are directed to the correct position in the growing peptide chain.

A rather small percentage of the total RNA of the cell is in the form of the messenger RNA fraction. These are the molecules which furnish the link between a sequence of nucleotides in the basic genetic material, chromosomal DNA, and the sequence of amino acids in the gene product, the protein. Because they direct the synthesis of proteins of different molecular weight and may carry information for more than one protein, messenger RNAs constitute a polydisperse fraction varying in molecular weight from 5×10^4 to 5×10^6. Except in very special cases, such as the immature red blood cell, where the vast majority of the protein being synthesized is hemoglobin, it is difficult to obtain from this mixture the messenger RNA for one specific protein.

NUCLEOPROTEINS

Because of the high density of negative charge that they carry at physiological pH, the DNA and much of the RNA of the cell can be tightly associated with various positively charged proteins. Most important of these are the class of low molecular weight basic proteins found in eucaryotic cells called histones. These proteins, which have a molecular weight of from 10,000 to 20,000, are composed of up to 30 percent lysine or arginine residues, and have only a few percent glutamic and aspartic acid residues. The tight association between these proteins and the acidic DNA molecule may be one way of regulating which portions of the DNA molecule are available for transcription. Because of this possibility, histones have been postulated as one of the important regulatory mechanisms in the process of cell differentiation. Another class of basic proteins with a high arginine content, the protamines, are found specifically associated with the DNA of sperm cells. The bulk of the RNA in the cell is also in the form of a specific ribonucleoprotein complex, the ribosome. It has been shown that the protein portion of the ribosome is composed of a large number of different basic proteins which have molecular weights in the range of 20,000.

ISOLATION OF NUCLEIC ACIDS

Because of their tight association with cellular proteins, the isolation of nucleic acids has presented many problems. As are many proteins, the polynucleotides are soluble in dilute salt solutions and are precipitated by the common protein precipitants trichloracetic acid or perchloric acid. The procedure most commonly used to isolate ribonucleic acid is to extract a buffered suspension of the tissue with aqueous phenol. The denatured protein is extracted into the phenol phase or remains at the interface, and the nucleic acids remain in the aqueous phase. The RNA can then be precipitated from this solution by the addition of ethanol. The RNA prepared by this method is a complex mixture and must usually be fractionated before use. These high molecular weight molecules are very susceptible to digestion by various nucleases, and purification is often carried out in the presence of nuclease inhibitors. Crude preparations of RNA are often fractionated on the basis of molecular weight, by centrifugation in a sucrose gradient or, if only small amounts are available, by electrophoresis in polyacrylamide gels. Various column chromatographic methods and countercurrent distribution have also been used to fractionate RNA.

DNA is also obtained by disruption of the protein nucleic acid complex with organic solvents, phenol, or detergents and selectively precipitating the DNA from an aqueous phase. In addition to the problem of nuclease digestion, care must be taken to prevent mechanical shear forces from degrading the extremely long DNA molecules. Different types of DNA molecules can be separated by centrifugation through gradients of sucrose, or CsCl, or by chromatography.

VIRUSES

The most extensively studied class of nucleoproteins are the small infectious particles called viruses. Viruses are parasitic agents and are capable of replication only within a host cell. The basis for the understanding of the chemical nature of the viruses was established in 1935 with the isolation and crystallization of tobacco mosaic virus by Stanley. He was able to demonstrate that this virus consisted of a single ribonucleic acid molecule surrounded by protein subunits which formed a protective coat (Figure 6-17). The infectivity of the virus is a function of the nucleic acid portion only, and the protein serves as a protective device to stabilize the mature viral particle and, in some cases, to aid in the process of infection.

Types of Viruses

The viral particles which have now been studied differ a great deal in the relative amount of nucleic acid and protein they contain, the molecular weight and structure of the nucleic acid present, the number of different viral proteins present, and their morphology. Much of the early work in virology was done with plant viruses which were all single-stranded RNA viruses. The viruses which may infect vertebrates can contain either RNA, usually single-stranded but occasionally double-stranded, or double-stranded DNA. Much of the work on viruses has been done on the viruses that have a bacterial host, the bacteriophages, which are mainly double-stranded DNA-containing viral particles. Viral particles and the size of the nucleic acid in them can vary a great deal. Some examples are shown in Table 6-5. Many of the viruses have much more complicated structures than the simple rod composed of identical subunits that forms the tobacco mosaic virus structure. An example of one of these is shown in Figure 6-17.

145

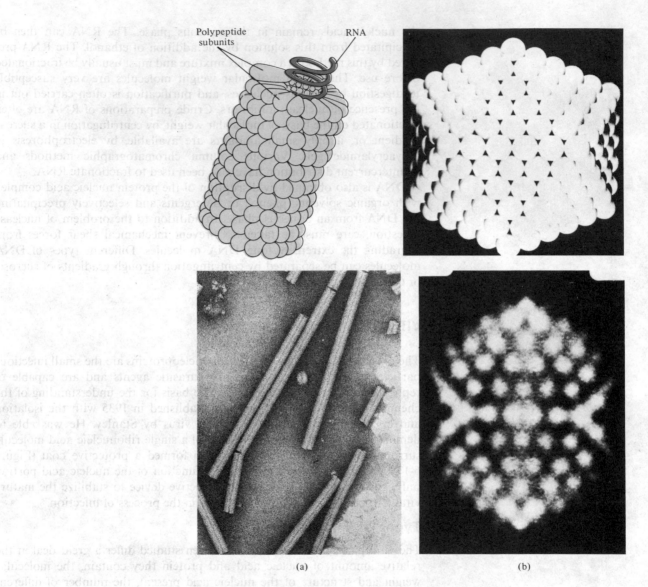

(a)

(b)

Figure 6-17 (*a*) Tobacco mosaic virus particle. Each of the identical subunit proteins has a molecular weight of 18,000. The RNA strand forms a helix which is surrounded by the subunits. The intact particle is a rod which is 160 × 300 Å and has a particle weight of 40 × 10⁶. (Electron micrograph courtesy of R. Williams, the Virus Laboratory, University of California, Berkeley) (*b*) Adeno-virus. This virus consists of 252 protein subunits (capsomers) arranged in the shape of an icosahedron. The DNA is in the middle of this closed shell of identical subunits. The virus is about 700 Å in diameter and has a particle weight of 200 × 10⁶. (R. W. Horne et al., *J. Mol. Biol.*, Vol. 1, 1969, Academic Press)

Table 6-5   Properties of Some Viruses

Virus	Type of Nucleic Acid	Particle Weight Daltons $\times 10^{-6}$	Molecular Weight of Nucleic Acid Daltons $\times 10^{-6}$	Percent of Nucleic Acid
Tomato ring spot	RNA	1.5	0.66	44
Poliomyelitis	RNA	6.7	2.0	22–30
Polyoma	DNA	21.1	3.5	13–14
Tobacco mosaic	RNA	40	2.2	5–6
Adenovirus	DNA	200	10	5
Bacteriophage T2	DNA	220	134	61
Human influenza	RNA	280	2.2	0.8

Viral Replication

The replication of a viral particle can occur only within its cellular host, and involves a rather complex series of events. The specificity of a particular virus is due to its ability to adsorb to specific sites on its host cell, and it will not infect cells not having these specific attachment sites. Following adsorption of the virus to the cell surface, the viral nucleic acid penetrates the cell wall to begin the infective process. In the case of some simple small viruses, the entire virus is taken into the cell by pinocytosis, whereas some of the bacteriophages have elaborate mechanisms for injecting their nucleic acid into the cell.

Once in the cell, the viral nucleic acid directs the synthesis of specific enzymes needed to replicate the viral nucleic acid, and directs the synthesis of the viral coat proteins. The normal synthesis of DNA, RNA, and host cell proteins is usually interfered with while the host is replicating the viral particle. By this mechanism the virus is able to duplicate itself. Many copies of the infective virus are assembled from the newly synthesized compounds in the host cell. In some cases this appears to be a spontaneous process; in other cases a series of virus-directed steps are involved. The progeny virus particles are then released from the host cell by lysis of the cell wall.

PROBLEMS

1. What would be a rapid way of distinguishing a sample of high molecular weight DNA from high molecular weight RNA with a minimum of chemical analysis?

2. Early crystallographic examination of DNA revealed that the molecule contained repeating structural units with spacings of 3.4 and 34 Å. What was the significance of these observations?

3. The oligonucleotide pGpGpApGpApAp is resistant to digestion by pancreatic ribonuclease. Why?

4. If the sequence of a short segment of one strand of a double-stranded DNA molecule is pGpCpTpGpGpA, what is the complementary sequence in the other strand?

5. If a solution of double-stranded DNA is slowly heated, there is an increase in the A_{260} associated with the temperature change. The ability to observe a decrease in A_{260} along the same curve is dependent on the rate of cooling. Why?

6. An oligonucleotide which you know contains two moles of A and one each of U, C, and G has been isolated. Treatment of this compound with pancreatic ribonuclease results in the formation of adenosine and two dinucleotides. Treatment with RNAs T-1 yields a trinucleotide and a dinucleotide. Are these data sufficient to establish a unique structure for the oligonucleotide?

Suggested Readings

Books

Brownlee, G. G. *Determination of Sequences in RNA.* American Elsevier, New York (1972).

Davidson, J. N. *The Biochemistry of Nucleic Acids.* Academic Press, New York (1972).

Watson, J. D. *Molecular Biology of the Gene,* 2nd ed. Benjamin, New York (1970).

Articles and Reviews

Crick, F. H. C. The Structure of the Hereditary Material. *Sci. Amer.* **191**:54–61 (1954).

Holley, R. W., J. Apgar, G. A. Everett, J. T. Madison, M. Marguisee, S. H. Merrill, J. R. Penswick, and A. Zamir. The Base Sequence of Yeast Alanine Transfer RNA. *Science* **147**:1462–1465 (1965).

Horne, R. W. The Structure of Viruses. *Sci. Amer.* **208**:48–56 (1963).

Kim, S. H., F. L. Suddath, G. J. Quigley, A. McPherson, J. L. Sussman, A. H. J. Wang, N. C. Seeman, and A. Rich. Three-Dimensional Tertiary Structure of Yeast Penylalanine Transfer RNA. *Science* **185**:435–439 (1974).

Mills, D. R., F. R. Kramer, and S. Spiegelman. Complete Nucleotide Sequence of a Replicating RNA Molecule. *Science* **180**:916–927 (1973).

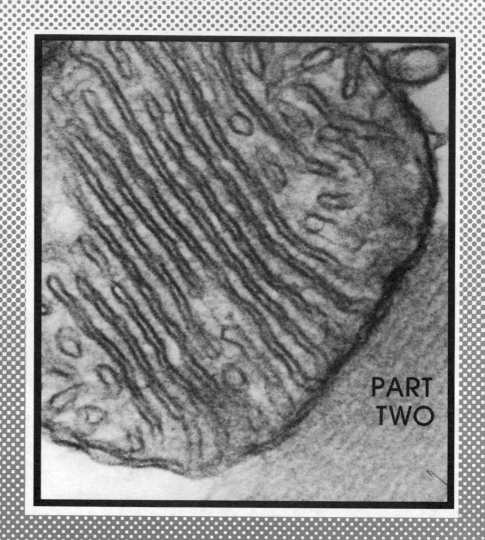

PART
TWO

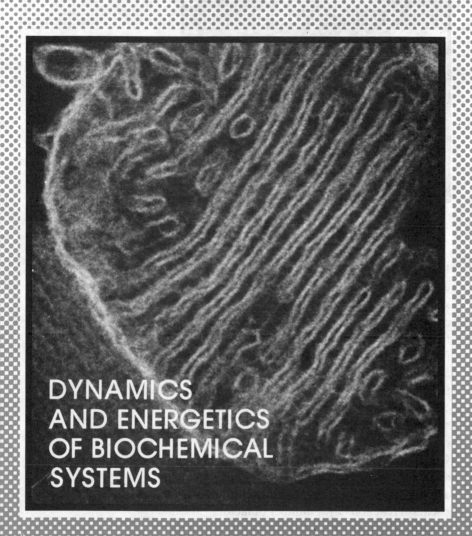

DYNAMICS
AND ENERGETICS
OF BIOCHEMICAL
SYSTEMS

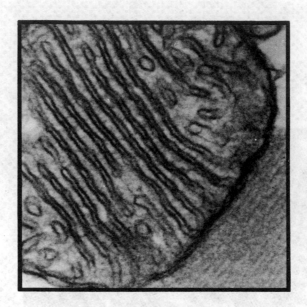

structure and function of enzymes

chapter seven

Enzymes are proteins that function as organic catalysts by combining with their substrate at a specific active site to form an intermediate enzyme-substrate complex and then dissociating to free enzyme and product. The rate of product formation is a function of substrate concentration and can be influenced by temperature, pH, and the presence of activators or inhibitors. Inhibitors can react reversibly with the enzyme at the active site in a competitive manner or at some other site on the protein in a noncompetitive manner. Compounds that form a covalent interaction with some site on the protein can also inhibit enzyme activity. Enzymes are able to increase rates of chemical reactions by a number of factors resulting from the interaction formed between the substrate and the active site amino acid residues of the protein.

For many years the study of enzymes and their function centered around the process of fermentation. The conversion of fruit juices and cereal grains to alcoholic beverages had long interested people, and as early as the seventeenth century, Lavoisier had shown that during fermentation sugar is converted to CO_2 and water. The process was always associated with a living system, and not until the late 1800s was Pasteur's claim that there was no fermentation without life challenged. Büchner was then able to show that a cell-free extract from yeast was capable of bringing about the fermentation of a sugar solution, and the modern study of enzymology began as a study of this yeast extract or zymase. It was 1926 before the first enzyme, urease, was crystallized by Sumner and definitely shown to be a protein. Since that time hundreds of enzymes have been isolated, purified, and crystallized, and as far as is known, only protein molecules are able to function biologically as enzymes.

An enzyme is a catalyst and as such exhibits the general property of being able to accelerate the rate of a chemical reaction without being consumed in it. Enzymes affect only the kinetic, and not the thermodynamic, properties of a reaction; that is, the same chemical equilibrium will be reached with or without the enzyme, although it may not be reached in any reasonably observable time span without the enzyme. An enzyme cannot make a thermodynamically unfavorable reaction go, but it can cause a reaction to proceed rapidly in what is otherwise a very stable system. The enzyme performs its catalytic function by lowering the activation energy barrier for the reaction. As can be seen in Figure 7-1, this means

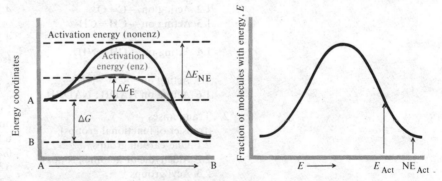

Figure 7-1   Action of a catalytic agent. The curve on the left indicates that for compound A to be converted to B, it must pass through an activation energy barrier, ΔE_{NE}. The activation energy for the same reaction in the presence of the enzyme is much less ΔE_E. The curve at the right indicates that if the molecules in the population show a normal distribution of energy, there will be many more molecules with sufficient energy to overcome the energy of activation in the case of the enzyme-catalyzed reaction.

that at any given temperature, more of the molecules in the population will possess sufficient energy to overcome the activation energy barrier and to undergo the reaction.

CLASSIFICATION OF ENZYMES

Enzyme nomenclature has been rather inconsistent; in the past someone discovering a new enzyme activity was able to name the enzyme pretty much as he pleased. The suffix, *-ase*, has often been used to designate an enzyme activity. A lipase for example would be an enzyme which is active on a lipid, and urease is an enzyme which is active with urea as a substrate. At the present time there are specific and systematic rules for naming new enzymes involving a detailed classification scheme which allows them to be correctly named and given a number which is an indication of their activity (Table 7-1). For example, the enzyme, commonly known as *maltase*, which hydrolyzes a large number of α-D-glucopyranosides is EC 3.2.1.20 and is called α-D-*glucoside glucohydrolase*. *Lactate dehydrogenase* is a more specific enzyme; it is correctly identified as EC 1.1.1.27 and is called L-*Lactate*: *NAD oxidoreductase*. In most cases the systematic name is rather long, and practicing biochemists tend to use the older trivial names for most enzymes.

Table 7-1 International Enzyme Classification

1. Oxido-reductases
   (oxidation-reduction reactions)

   1.1 Acting on —CH—OH

   1.2 Acting on —C=O
   1.3 Acting on —CH=CH—

   1.4 Acting on —CH—NH$_2$

   1.5 Acting on —CH—NH—
   1.6 Acting on NADH; NADPH

2. Transferases
   (transfer of functional groups)
   2.1 One-carbon groups
   2.2 Aldehydic or ketonic groups
   2.3 Acyl groups
   2.4 Glycosyl groups
   2.7 Phosphate groups
   2.8 *S*-containing groups

3. Hydrolases
   (hydrolysis reactions)
   3.1 Esters
   3.2 Glycosidic bonds
   3.4 Peptide bonds
   3.5 Other C—N bonds
   3.6 Acid anhydrides

4. Lyases
   (addition to double bonds)

   4.1 —C=C—

   4.2 —C=O

   4.3 —C=N—

5. Isomerases
   (isomerization reactions)
   5.1 Racemases

6. Ligases
   (formation of bonds with ATP cleavage)
   6.1 C—O
   6.2 C—S
   6.3 C—N
   6.4 C—C

PREPARATION AND PURIFICATION OF ENZYMES

Almost any tissue contains hundreds of proteins with enzymatic activity. The isolation of a particular enzyme from this mixture depends on the discovery of ways of separating this one particular protein from all others, and preferably doing it in such a way that the majority of the desired enzyme is recovered.

Enzyme Assays

Before any attempt can be made to isolate an enzyme, a convenient *assay* for its activity must be developed. An enzyme assay is simply some means of measuring the initial rate of the reaction catalyzed by the enzyme. What is measured can be either the appearance of one of the products of the enzymatic reaction or the disappearance of a substrate with time. Before one of these measurements can be used as an assay, it should be shown that the enzymatic reaction is linear over the time period studied and that the reaction rate is proportional to the amount of enzyme present (Figure 7-2). Although any arbitrary unit of enzymatic activity can be defined

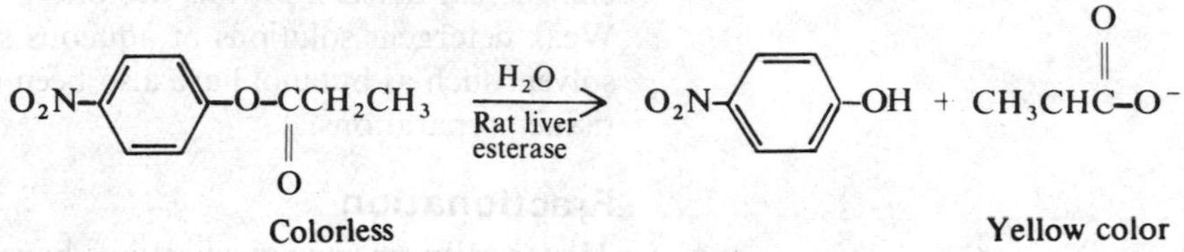

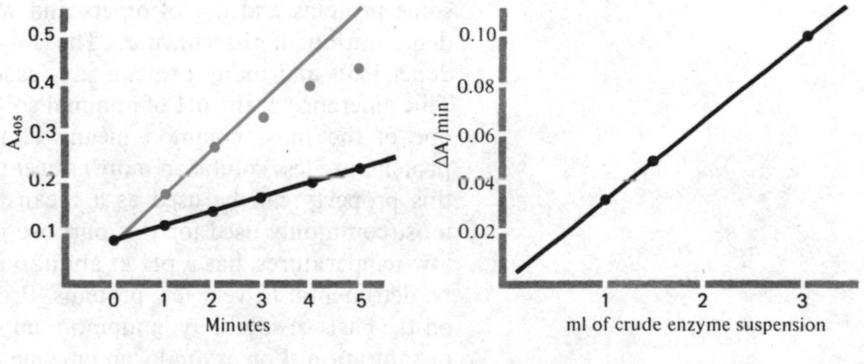

Figure 7-2 Principles of an enzyme assay. In this assay a colorless ester is hydrolyzed to form propionic acid and a colored compound, *p*-nitrophenol. The reaction can therefore be followed by the absorbency increase at 405 nm. The curve on the left shows that the reaction is linear with time when low concentrations of enzyme are present, but that at higher concentrations it is linear for only a short time. The curve on the right shows that for the range of enzyme concentration examined, the rate of conversion of the substrate ($\Delta A/min$) is proportional to the amount of enzyme. This type of curve can then be used as an assay for the activity of the enzyme in other preparations.

and used, the standard and recommended procedure is to define one unit of enzyme activity (U) as the amount of enzyme which will catalyze the transformation of 1 μmole of substrate/minute. The temperature of the reaction must be stated, and although it is usually 30°C, it can be any temperature at which the enzyme is reasonably stable. The other conditions of the assay, such as pH and substrate concentration, should be optimal; if they are not, they should be clearly stated. Specific activities of enzymes are expressed in terms of enzyme units/mg protein.

To purify an enzyme from a mixture of other proteins, it is necessary to determine the specific activity of the enzyme in the mixture and to separate the proteins into two or more fractions by some type of fractionation procedure. The specific activity of the enzyme and total recovery of the protein in each fraction are then determined, and the fraction which contains the majority of the enzyme is further purified. Procedures which leave appreciable amounts of the enzyme in more than one fraction or which give a high specific activity of the enzyme in one fraction but involve a large loss in total activity are of little value.

Extraction

Many enzymes can be solubilized by simply extracting a homogenized or otherwise disrupted tissue with a dilute salt solution. If an enzyme is localized in a particular subcellular particle, a preliminary purification can be made by first obtaining a preparation of this cellular organelle and then beginning the fractionation. Many enzymes are bound to lipid-rich membranes in the cell and because of this association are difficult to extract. In these cases, the tissue may be defatted and dried by acetone extraction, and the enzyme extracted from this acetone powder by an aqueous salt solution. Weak detergent solutions or aqueous solutions saturated with an organic solvent such as butanol have also been used to extract enzymes from crude tissue preparations.

Fractionation

Historically, enzyme purifications have depended on a number of classical treatments of protein solutions which will result in the precipitation of some proteins and not of others and which will not result in irreversible denaturation of the enzymes. The solubility of a protein is usually pH dependent, and many proteins are least soluble at their isoelectric point. This difference in the pH of minimal solubility of various proteins has been one of the most common means of fractionating proteins. In general, proteins are less soluble in more concentrated, than in dilute, salt solutions; this property can be used as a means of protein fractionation. The salt most commonly used for this purpose is $(NH_4)_2SO_4$. It is very soluble at low temperatures, has a pH of about 6 in aqueous solutions, and seems to be detrimental to very few proteins. Protein fractionations are often made on the basis of solubility in ammonium sulfate solutions of varying degrees of saturation. For example, an enzyme may be soluble in a half-saturated ammonium sulfate solution, but would precipitate if the concentration were raised to 0.7 saturation.

Different proteins are heat inactivated at different temperatures, and it is sometimes possible to achieve considerable purification of one enzyme by heating a protein mixture to a temperature that will denature much of the contaminating protein without affecting the activity of the enzyme desired. Although many proteins are irreversibly denatured by treatment with organic solvents at room temperature, aqueous mixtures of solvents such as acetone, ethanol, butanol, and methanol can sometimes be used as a fractionation tool at low temperatures, as can rather high concentrations of some divalent cations such as Zn^{2+}. Specific adsorption of some proteins onto material like Al_2O_3, $BaSO_4$, or charcoal has sometimes been useful, and calcium phosphate is often used as a material for general adsorption of proteins.

During any of these procedures, care must be taken not to denature the enzyme. An attempt is routinely made to keep the temperature as low as possible (2–5 °C), and buffered solutions are used to prevent drastic shifts in pH. Many proteins, particularly if their concentrations are very dilute, are denatured by contact with glass surfaces or at an air-solution interface. Because of this, protein concentrations are usually kept as high as possible in all manipulations, and a protein with no competing enzyme activity, such as serum albumin, is often added to a purified dilute enzyme solution to stabilize it.

Chromatographic Separations

Although enzymes are still isolated by some of the classical techniques based on protein solubility, almost any current procedure for the preparation of an enzyme will utilize a column chromatographic method. The most commonly used methods involve the use of a cellulose support which has been chemically modified to carry either a weakly acidic or a basic group. Of the various types available, DEAE (diethylaminoethyl) cellulose, an anion-exchange material, and CM (carboxymethyl) cellulose, a cation-exchange material, are the most widely used (Figure 7-3). Depending on the pH of the solution, a protein molecule will have an excess of positive or negative charges, and will bind to a column of these materials. By changing ionic strength or pH, various proteins will be selectively dissociated from the charged group and will be separated by being eluted from the column.

Proteins can also be separated on the basis of molecular weight by the use of molecular sieves. These high molecular weight cross-linked polymers can effectively separate proteins of different sizes, and if conditions are carefully controlled, an estimation of molecular weight of the proteins can be obtained. Their use was discussed in more detail in Chapter 5, and an example of the separation of a mixture of proteins by this technique is shown in Figure 5-14. In many cases, ion exchange functional groups such as DEAE are now attached to these cross-linked polymers rather than to cellulose to increase the capacity of the ion exchange material.

A third type of chromatographic procedure that is becoming increasingly useful is the technique of affinity chromatography. In this case a material which will interact strongly with a site on the protein, such as a substrate of the enzyme, is attached to an inert solid support. The mixture of proteins is then poured through a column of this material, and the desired enzyme is held to this material while the rest of the proteins pass through the column.

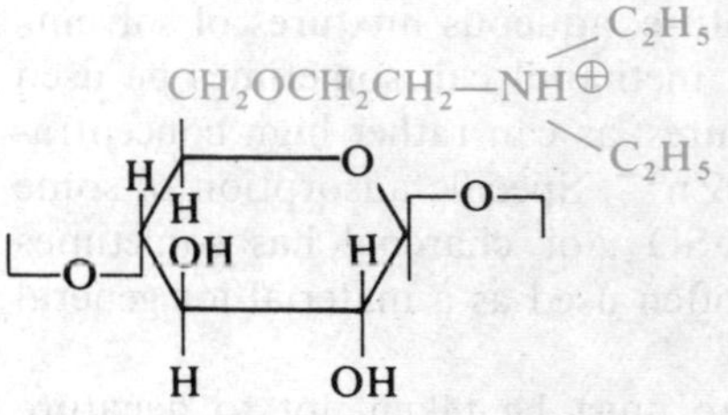

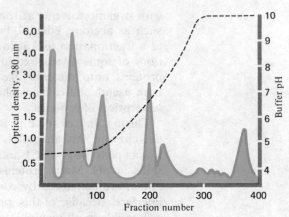

Diethylaminoethyl cellulose (DEAE)

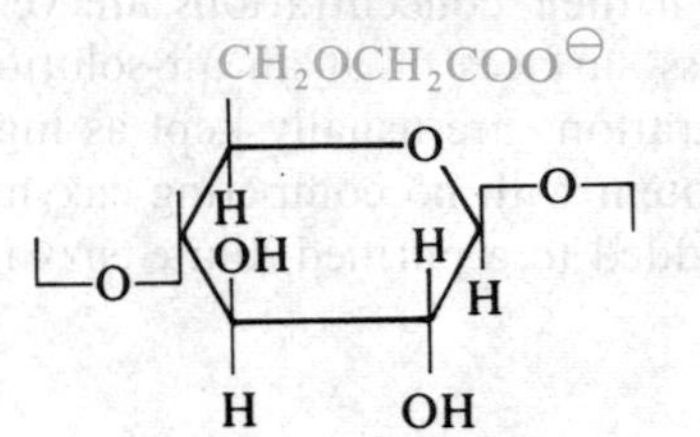

$CH_2OCH_2COO^{\ominus}$

Carboxymethyl cellulose (CMC)

Figure 7-3 Modified cellulose columns. DEAE is an anion exchanger which has its greatest capacity below pH 10. CMC is a weak cation exchanger which has a lower capacity than DEAE and is effective above pH 4.0. The curve on the right is an example of the type of protein separation that might be achieved on CMC. The protein would be put on a column at pH 4.5 and the pH gradually increased to elute different proteins. At high pHs all proteins would have lost their positive charges so they would no longer be attracted to the negative charges on the modified cellulose.

Depending on how tightly the desired enzyme is bound, it may simply be delayed in moving through the column, or it may be necessary to change the ionic composition of the buffer to elute it from the column.

Assay

Purification of a Phosphomonoesterase

$$CH_2OH$$
$$|$$
$$\overset{O}{\underset{\parallel}{}}$$
$$H-\overset{|}{\underset{|}{C}}-O-\overset{O}{\underset{\parallel}{P}}-O^- + H_2O \xrightarrow{\text{Enzyme}} HC-OH + H_2PO_3^-$$
$$\underset{CH_2OH}{\overset{OH}{|}}$$

CH₂OH → CH₂OH, HC—OH, CH₂OH

Figure 7-4 Enzyme purification by classical methods of separation. Two different organic solvent fractionations, $(NH_4)_2SO_4$ fractionation, and charcoal adsorption were all used. An enzyme unit was defined as 1 μg PO_4 released/min at 38°C, pH 9.9.

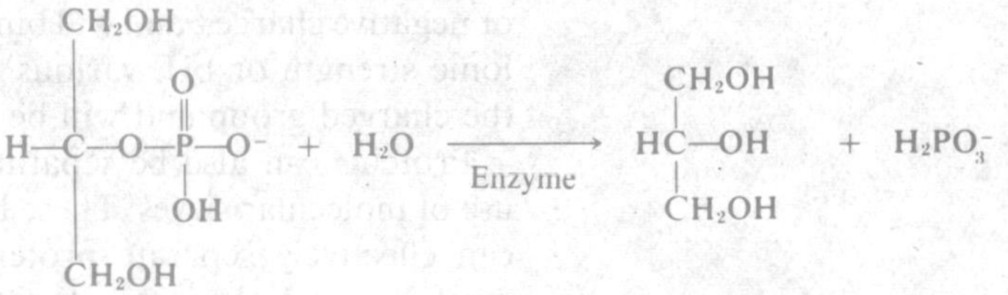

Fraction	Total Units $\times 10^3$	Specific Activity, U/mg N	Recovery, Percent
Buttermilk	2600	28	100
Aq. phase after butanol	1755	325	67
Ether-acetone ppt	1616	910	62
0.63–0.85 sat. $(NH_4)_2SO_4$	1172	4102	45
Charcoal eluate	935	7261	36
Ether-acetone ppt	735	9423	28
Charcoal eluate	510	11900	20
0.69–0.82 sat. $(NH_4)_2SO_4$	348	14200	13
Ether-acetone ppt	268	15314	10

Examples of Enzyme Purification

Figures 7-4 and 7-5 present two actual examples from the literature of protein purifications illustrating an older separation of an enzyme by solubility fractionation and a separation which makes use of modern chromatographic procedures. It can be seen that the specific activity increases during each step of the purification as contaminating proteins are separated from the desired enzyme. The total recovery of enzyme activity usually falls in each step, because none of the separation procedures are completely specific and some compromise between an increase in specific activity and recovery of enzyme must usually be made. If the protein is pure, its specific activity will not be increased by any further separation method, and the attainment of a constant maximum specific activity, along with electrophoretic evidence of a single protein, is often used as a criterion of protein purity.

Figure 7-5 Enzyme purification using chromatographic methods. A NAD-specific glutamic dehydrogenase was isolated from the mold neurospora crassa. The mold mycelium was frozen and homogenized with buffer and glass beads in a blender. This homogenate was centrifuged at 27,000 *g* for 80 minutes and the supernate was used as the crude extract. Nucleic acids in the extract were removed with protamine sulfate and most of the enzyme activity was recovered in a fraction precipitating between 27 and 45 percent ammonium sulfate. This precipitate was further fractionated by DEAE-cellulose and Bio-Gel A-1.5 (gel filtration) chromatography. The enzyme was assayed in the direction of glutamic acid formation by following the disappearance of NADH at 340 nm. (F. M. Veronese, et al., *J. Biol. Chem.*, Vol. 249, p. 7922, 1974)

Purification of Glutamic Dehydrogenase

$$H_2N-\underset{\substack{|\\CH_2\\|\\CH_2\\|\\COOH}}{\overset{\substack{COOH\\|}}{C}}-H + NAD^+ \underset{\textit{dehydrogenase}}{\overset{\textit{Glutamic}}{\rightleftharpoons}} \underset{\substack{|\\CH_2\\|\\CH_2\\|\\COOH}}{\overset{\substack{COOH\\|}}{C}}=O + NADH + H^+ + NH_3$$

Fraction	Protein (mg)	Units $\times 10^6$	Sp. Act (U/mg)	Recovery Percent
Crude extract	34,600	3.4	100	100
protamine sulfate supernate	34,600	3.4	100	100
0.27 sat. $(NH_4)_2SO_4$ supernate	21,600	3.3	157	97
0.45 sat. $(NH_4)_2SO_4$ precipitate	20,000	1.7	85	50
DEAE-cellulose effluent	500	1.5	3000	44
Bio-Gel A-1.5m effluent	180	1.3	8100	38

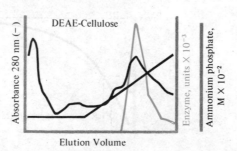

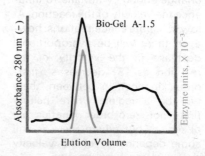

FACTORS AFFECTING ENZYME ACTIVITY

Substrate Concentrations

Rates of enzymatic reactions are expressed, as are rates of any chemical reaction, as changes in the concentration of a substrate or product with time. In enzymology, two types of reactions are of practical importance, those which show zero-order kinetics and those which show first-order kinetics (Figure 7-6). Although an enzyme with more than one substrate can exhibit second-order kinetics, the usual practice is for an enzymologist to make sure that all but one of the substrates is present in a large excess, which makes the system dependent on only one variable. Whether or not an enzyme reaction will exhibit zero- or first-order kinetics will depend on the initial substrate concentration. This important dependence is illustrated in Figure 7-7. It should be carefully noted that this is not a kinetic plot—that is, it does not show the rate of product formation versus time, as does Figure 7-6 but rather a plot of initial velocity (product formed

Figure 7-6 Plots of zero- and first-order reactions. Note that in a first-order reaction the rate is not constant with time, but is proportional to the amount of substrate remaining at any time, while in a zero-order reaction the rate is a constant.

Figure 7-7 Effect of substrate concentration on the velocity of an enzyme reaction. In the region where the enzyme shows essentially zero-order kinetics, the substrate concentration does not change enough with time to influence the velocity of the reaction. At low substrate concentrations, however, there will be a proportional decrease in the velocity of the reaction as [S] decreases (first-order kinetics). In between, there will be a region where neither first- nor zero-order kinetics are observed, but where there is still some dependency of the velocity on substrate concentration.

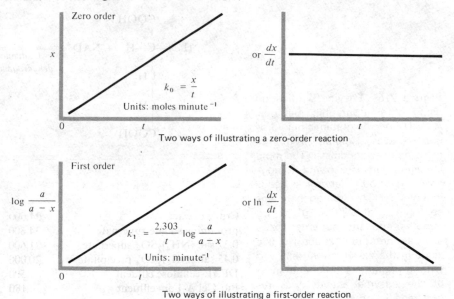

$$k_0 = \frac{x}{t}$$

Units: moles minute^{-1}

Two ways of illustrating a zero-order reaction

$$k_1 = \frac{2.303}{t} \log \frac{a}{a-x}$$

Units: minute^{-1}

Two ways of illustrating a first-order reaction

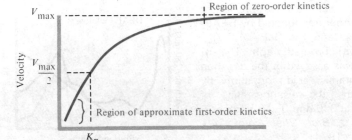

Substrate concentration [S]

160

per unit time) against different substrate concentrations. This plot is based on the following equation for the behavior of a simple enzyme catalyzed reaction in which one substrate is converted to one product.

$$E + S \underset{k_2}{\overset{k_1}{\rightleftharpoons}} ES \xrightarrow{k_3} E + P$$

The expression simply says that the enzyme (E) and substrate (S) react in a reversible manner to form an intermediate enzyme substrate complex (ES) which breaks down to form free enzyme and product (P). This simplified form of the expression assumes that the enzyme is present in a very small concentration compared to the substrate, and that the rate of the reverse reaction, E plus P to regenerate ES, is negligible. For the typical case, this assumption can be made. With this equation in mind, it is easy to understand the nature of the curve in Figure 7-7. At low substrate concentrations, an increase in [S] will shift the E + S equilibrium in favor of ES and the rate will increase. However, as the enzyme becomes saturated, there is no more free E available, so increasing the [S] no longer increases the rate of the reaction. At this point the observed rate is constant; that is, the kinetics are zero-order with respect to [S].

It is possible to derive an equation which relates the velocity of an enzyme reaction to the substrate concentration. This expression is called the Michaelis-Menten equation, which can be written a number of ways, but is most commonly written as

$$\frac{v}{V_{max}} = \frac{[S]}{K_m + [S]} \quad \text{or} \quad v = \frac{V_{max}[S]}{K_m + [S]}$$

where v = velocity of the enzymatic reaction

V_{max} = limiting or maximal velocity at "infinite" substrate concentration

[S] = substrate concentration in moles per liter of solution

K_m = Michaelis constant in moles per liter of solution

The equation is derived by making a number of assumptions about the condition of the enzyme reaction. It is assumed that $[S] \gg [E]$ so that the amount of substrate present as ES can be ignored in the calculations, and that the rate of the reverse reaction, $E + P \rightarrow ES$, is negligible. It is also assumed that [ES] is a constant; that is, it is formed from E and S as fast as it breaks down to E + S or E + P. In the derivation, three forms of E are important: [E], the total enzyme concentration; [ES], the concentration of the enzyme substrate complex; and [E] − [ES], the uncombined enzyme concentration. If [ES] is a constant, then its rate of formation is given by

$$\frac{d[ES]}{dt} = k_1([E] - [ES])[S] \tag{1}$$

and its rate of breakdown by

$$\frac{-d[ES]}{dt} = k_2[ES] + k_3[ES] \tag{2}$$

Under steady-state conditions, where [ES] is constant, the rates of formation and breakdown must be equal, so

$$k_1([E] - [ES])[S] = k_2[ES] + k_3[ES] \tag{3}$$

which can be rearranged to

$$\frac{[S]([E] - [ES])}{[ES]} = \frac{k_2 + k_3}{k_1} = K_m \tag{4}$$

This combination of the rate constants is by definition, the K_m, the Michaelis constant. If k_3 is very small in relation to k_2, then K_m would be approximately equal to k_2/k_1, which would be the dissociation constant (k_s) of the enzyme substrate complex. Therefore, for many enzymes, the K_m is a measure of the affinity of the enzyme for the substrate. It is, however, not correct to assume that $K_m = k_s$ without knowing that $k_3 \ll k_2$. Using Equation (4) the steady-state concentration of ES can be expressed as

$$[ES] = \frac{[E][S]}{K_m + [S]} \tag{5}$$

The initial rate (v) of an enzyme reaction is proportional to the concentration of ES, and the maximum velocity (V_{max}) will be observed when essentially all the enzyme is present as ES; that is, $[ES] = [E]$. Therefore

$$v = k_3[ES] \tag{6}$$

and

$$V_{max} = k_3[E] \tag{7}$$

If the [ES] term in Equation (6) is expressed as in Equation (5),

$$v = k_3 \frac{[E][S]}{K_m + [S]} \cdot \tag{8}$$

which can be divided by Equation (7) to give

$$\frac{v}{V_{max}} = \frac{k_3 \dfrac{[E][S]}{K_m + [S]}}{k_3[E]} \tag{9}$$

When Equation (9) is solved for v, the familiar form of the Michaelis-Menten equation is obtained.

$$v = \frac{V_{max}[S]}{K_m + [S]}$$

An inspection of this equation should make it clear that as [S] becomes large in relation to the K_m, v will approach V_{max}, and when [S] is equal to the K_m, v will be equal to $V_{max}/2$. This, then, is an operational definition of the K_m, or the Michaelis constant: the substrate concentration at which the velocity of the reaction is one-half the maximum velocity. As a general rule, it may be considered that when $[S] \gtrsim 100\ K_m$, then $v = V_{max}$ and zero-order kinetics are observed, and that when $[S] \lesssim 0.01 K_m$, first-order kinetics are observed.

Figure 7-8 A Lineweaver-Burk plot.

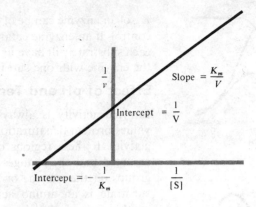

Figure 7-9 Two other straight-line solutions to the Michaelis-Menten equation.

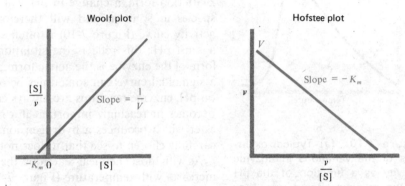

If an attempt is made to determine experimentally the K_m of an enzyme reaction by adding various substrates and measuring the rate of the reaction, it may be difficult to determlne when the maximum velocity has actually been reached. This problem can be solved by rearranging the Michaelis-Menten equation to yield an equation for a straight line. A plot of this equation, called the Lineweaver-Burk

$$\frac{1}{v} = \frac{K_m + [S]}{V_{max}[S]} = \frac{K_m}{V_{max}} \cdot \frac{1}{[S]} + \frac{1}{V_{max}}$$

plot, is shown in Figure 7-8. It has the advantage that measurements of the velocity of the reaction can be made at a number of substrate concentrations and then extrapolated back to infinite substrate concentration at the intercept. This is not the only straight-line solution to the equation, and two other plots which are sometimes used are shown in Figure 7-9.

The measured value for the K_m, which is expressed in molar concentrations, is a characteristic of each individual enzyme, and no broad generalization can be made about the magnitude of this value. Some hydrolytic enzymes, sucrase for example, have very high K_m values (2.8×10^{-2} M), while most enzymes have values more on the order of that for lactate in the lactic dehydrogenase reaction (3.5×10^{-5} M), or even several orders of magnitude lower. As will be discussed in Chapter 17, knowledge of the

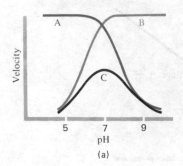

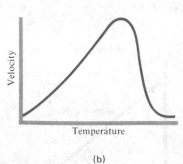

Figure 7-10 (*a*) Typical examples of the variations in enzyme activity as a function of the pH of the medium. Below pH 4 or above pH 11 most proteins are denatured so that their enzymatic activity will be lost; therefore curves *A* and *B* would be expected to drop off rapidly at these points. (*b*) Effect of temperature on the velocity of an enzyme reaction. Once the protein begins to become heat-denatured, the activity of the enzyme drops very rapidly.

K_m of an enzyme can be of considerable value in investigations of metabolic control. If an enzyme catalyzes a reaction between two different substrates, each substrate will have its own K_m, which can be determined by saturating the enzyme with one substrate and varying the other.

Effect of pH and Temperature on Enzyme Activity

Enzyme activity is always found to be pH dependent. At extreme pH values protein denaturation will occur with a concomitant loss of enzyme activity. If these regions of very high or very low pH are avoided, it is assumed that the changes in activity are due to ionizations of functional groups on the enzyme or ionization of the substrate. For example, if the substrate is an amino acid and the enzymatically active species is the zwitterion form, a change in pH will affect the relative abundance of this species in solution and will therefore influence activity. A bell-shaped activity curve (Figure 7-10) is often seen when enzyme activity is plotted against pH; this reflects the situation where an intermediate protonated form of the enzyme is the active form. If only one dissociation is important, a sigmoidal curve can sometimes be observed. Because of this dependence on pH, enzyme reactions are always carried out in buffered solutions. This becomes increasingly important if a reaction such as the hydrolysis of an ester which produces a hydrogen ion is studied. The buffer used must be carefully chosen to see that it does not influence the activity of the enzyme.

As with most chemical reactions, the rate of an enzyme-catalyzed reaction increases with temperature (Figure 7-10); however, in the case of enzymes there is a temperature at which thermal denaturation of the protein causes a loss of activity and the rate begins to slow down. Enzyme assays are commonly run at 30 or 37°C, while the point at which thermal denaturation begins can vary widely for different enzymes but is often around 45°C. From a practical standpoint, it is inconvenient to do assays much under 30°, because it requires a cooling bath rather than a heated-water bath, and reactions are often run at 37° simply because this is near the normal body temperature of man and many laboratory animals.

Effect of Prosthetic Groups, Coenzymes, and Metals

Some enzymes are simple proteins; that is, they contain no nonprotein group bound to them; others contain some type of prosthetic group. If the prosthetic group of a protein is not covalently bound, it can often be dissociated from the protein and removed by dialysis, in which case it would be called a coenzyme. Even if a coenzyme is not covalently bound to the protein, it often has such a tight binding affinity for the enzyme that under normal isolation procedures it will remain with the enzyme. Many of the coenzymes can also be viewed as if they were a second substrate in the enzymatic reaction. Because these few compounds function in a similar manner in a number of different enzymatic reactions, they have been grouped in this special category of coenzymes. Many of the common coenzymes are derivatives of the water-soluble vitamins. A disease resulting

Table 7-2 Common Coenzymes for Enzymatic Reactions

Coenzymes	Action	Corresponding Vitamin
I For oxido-reductases		
Nicotinamide-adenine dinucleotide (NAD) or Diphosphopyridine nucleotide (DPN)	Hydrogen transfer	Nicotinamide
Nicotinamide-adenine dinucleotide phosphate (NADP) or Triphosphopyridine nucleotide (TPN)	Hydrogen transfer	Nicotinamide
Flavin-adenine dinucleotide (FAD)	Hydrogen transfer	Riboflavin
Flavin mononucleotide (FMN)	Hydrogen transfer	Riboflavin
II For transferases		
Adenosine triphosphate (ATP)	Phosphate group transfer	—
Phosphoadenosylphosphosulfate (PAPS)	Sulfate group transfer	—
Cytidine diphosphate (CDP)	Lipid transfer	—
Uridine diphosphate (UDP)	Sugar transfer	—
S-adenosylmethionine (SAM)	Methyl group transfer	—
Tetrahydrofolate (THF)	Formate metabolism and transfer	Folic acid
Biotin Coenzyme	Carboxylations	Biotin
Coenzyme A (CoA-SH)	Acyl group transfer	Pantothenic acid
Thiamine pyrophosphate (TPP)	"Active acetaldehyde" transfer	Thiamine
III For isomerases and lyases		
Uridine diphosphate (UDP)	Sugar isomerization	—
Pyridoxal phosphate	Decarboxylation	Pyridoxine
Thiamine pyrophosphate (TPP)	Decarboxylation	Thiamine
B_{12} coenzyme	Carboxyl displacement	Cobalamine

from a deficiency of the vitamin must ultimately be the result of a lack of a sufficient amount of the corresponding coenzyme for some key reaction. The detailed mechanisms of action of the various coenzymes will be treated when they are first encountered in a metabolic pathway, but Table 7-2 gives a brief summary of types of reactions in which they are involved.

Much the same situation is true for the interaction of proteins with metals. Many enzymes are activated by specific metals which are readily removed from the protein; other proteins contain metals which are so tightly bound that they are always isolated with the protein. These latter proteins are called metalloenzymes. In the case of proteins such as hemoglobin and the cytochromes, the metal, in this case Fe^{2+} or Fe^{3+}, is part of a heme prosthetic group, while in other cases the metal ion is directly bound to the protein. The ions most commonly found in enzymes and an example of each are: Mg^{2+}, most phosphotransferases; Zn^{2+}, alcohol dehydrogenase; Mn^{2+}, arginase; Fe^{2+}, ferredoxin; and Cu^{2+}, cytochrome oxidase.

ENZYME INHIBITION

Any condition that causes denaturation of proteins or reagents that irreversibly react with functional groups on an enzyme will, of course, have an adverse influence on enzyme activity. The term *enzyme inhibition* is, however, usually reserved for less drastic effects. Most of the agents which can reversibly influence the rate of an enzyme-catalyzed reaction can be grouped into two classes: competitive inhibitors and noncompetitive inhibitors. A third general class, uncompetitive inhibitors, is much less common in simple single substrate reactions and will not be considered.

Competitive Inhibition

A competitive inhibitor is one which competes with the substrate for the active site of the enzyme; that is, both the following reactions occur

$$E + S \rightleftharpoons ES \longrightarrow E + P$$

and

$$E + I \rightleftharpoons EI$$

The presence of the inhibitor effectively removes some of the free enzyme and prevents it from forming an enzyme-substrate complex. The rate of the reaction will therefore be decreased. The addition of more substrate will overcome the inhibition, because it will drive the various equilibriums in the direction of the enzyme-substrate complex. The extent of the inhibition will, therefore, depend on

1. Concentration of the inhibitor
2. Concentration of the substrate
3. Relative affinity of the active site for the inhibitor and substrate

The affinity of the active site for the inhibitor is really a measure of the dissociation constant of the enzyme-inhibitor complex, and this term is defined as the K_I, where

$$K_I = \frac{[E]\ [I]}{[EI]}$$

As the maximum velocity at infinite substrate concentration is not changed, the apparent K_m, which is defined as the $[S]$ at half-maximum velocity, will be increased in the presence of the inhibitor. In general, competitive inhibitors are compounds which show a close structural similarity to the substrate. This enables them to bind to the active site, where they block entry of the substrate but fail to undergo the enzymatic reaction. One of the classic examples of this type of inhibition is the inhibition of the enzyme succinic dehydrogenase by malonic acid.

$$
\begin{array}{c}
\text{CH}_2\text{COOH} \\
| \\
\text{CH}_2\text{COOH}
\end{array}
+ \text{Enz} \rightleftharpoons \text{Enz} \cdot
\begin{array}{c}
\text{CH}_2\text{COOH} \\
| \\
\text{CH}_2\text{COOH}
\end{array}
\rightleftharpoons \text{Enz} \cdot \text{H}_2 +
\begin{array}{c}
\text{H} \quad\ \text{COOH} \\
\diagdown\ \diagup \\
\text{C} \\
\| \\
\text{C} \\
\diagup\ \diagdown \\
\text{HOOC} \quad \text{H}
\end{array}
$$

Succinic acid   Enz·sub complex   Fumaric acid

$$
\begin{array}{c}
\text{COOH} \\
\diagup \\
\text{CH}_2 \\
\diagdown \\
\text{COOH}
\end{array}
+ \text{Enz} \rightleftharpoons \text{Enz} \cdot
\begin{array}{c}
\text{COOH} \\
\diagup \\
\text{CH}_2 \\
\diagdown \\
\text{COOH}
\end{array}
\quad \#\!\#
$$

Malonic acid   Enz·inhib complex

Malonic acid, presumably because of the very similar arrangement of its two carboxyl groups, binds very effectively to the succinic acid site on the enzyme. Malonic acid, however, has no carbon-carbon bond to oxidize, so it cannot act as a substrate. Its presence, however, prevents the true enzyme-substrate complex from forming, and it therefore acts as an inhibitor.

Noncompetitive Inhibition

The second generalized case of enzymatic inhibition, noncompetitive, involves the situation where the inhibitor reacts with a site on the enzyme, other than the active site, so it cannot be displaced by the substrate. The presence of the inhibitor may influence either the rate of formation of the IES complex or the rate of dissociation of this complex to IE + P. The inhibition is, therefore, dependent on

1. Concentration of the inhibitor
2. Affinity of the enzyme for the inhibitor

The inhibitor has not influenced the binding of the substrate to the active site of the enzyme, so the apparent K_m is unchanged. Inhibitors of this type are compounds such as heavy metal ions that can reversibly alter the conformation of a protein by interacting with the free sulfhydryl groups of the protein.

Kinetic Plots to Determine Inhibition

Whether or not an inhibitor fits into one of these two basic forms of reversible inhibition can be determined by the use of the Lineweaver-Burk plot. If the rate of an enzymatic reaction at various substrate concentrations is plotted in the presence and absence of a constant amount of inhibitor, the relationship between the two straight lines is an indi-

167

Figure 7-11 Lineweaver-Burk plots of an enzyme reaction in the presence of a noncompetitive and competitive inhibitor. In the case of a noncompetitive inhibitor the V_{max} and slope are changed, but the apparent K_m is the same. For a competitive inhibitor the V_{max} remains the same, but the apparent K_m and slope are altered.

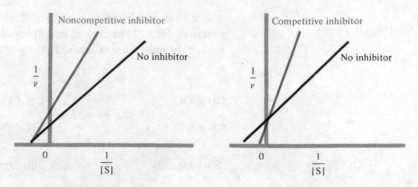

cation of the type of inhibition (Figure 7-11). Not all inhibitors will yield simple kinetic plots such as those shown in Figure 7-11, and other types of effects can be recognized. These two types do, however, illustrate the principles involved. Some of the commonly used enzyme inhibitors, mainly of the noncompetitive type, are indicated in Table 7-3.

Table 7-3 Some Common Reversible Enzyme Inhibitors

Inhibitor	Action
Noncompetitive	
Heavy metals (Ag$^+$, Hg^{2+}, Pb^{2+}, and so forth)	Complex to sulfhydryl groups
Organic mercuricals	Complex to sulfhydryl groups
Metal complexing agents (F$^-$, CN$^-$, N$_3^-$)	Complex to metal ion of metalloproteins
Chelating agents (EDTA, *O*-phenanthroline)	Removal of metal from metalloproteins
Competitive	
Malonate	Competes with succinate
Fluorocitrate	Competes with citrate
Sulfanilamide	Competes with *p*-aminobenzoate
Oxamate	Competes with L-lactate

Irreversible Inhibitors

In addition to inhibition by compounds which readily dissociate from the protein, enzymes can be irreversibly inhibited by agents which covalently

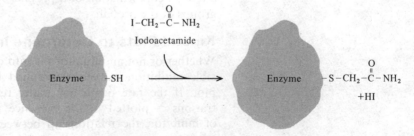

interact with a functional group on the protein. An alkylating agent such as iodoacetamide can form a covalent bond to a free sulfhydryl group of an enzyme and may, therefore, destroy its activity. Active site specific reagents such as diisopropylfluorophosphate (Figure 7-20) are a specialized type of irreversible inhibitor that have been effectively used to study the mechanism of action of enzymes.

ALLOSTERIC ENZYMES

Many enzymes have been found which do not give classical hyperbolic Michaelis-Menten curves when increasing substrate concentration is plotted against the velocity of the reaction. For these enzymes a plot of initial velocity versus [S] gives a sigmoidal curve. These enzymes are often composed of multiple subunits, and the observed curve results from what is called a cooperative effect. That is, the binding of the substrate to one subunit enhances the binding of subsequent substrate molecules to the enzyme.

It is becoming increasingly apparent that the enzyme inhibition which is of the greatest physiological importance in regulating biochemical reactions is of the type called *allosteric control*. In this case, an inhibitor or activator (called a negative or a positive effector) interacts at a site, other than the active site, on a multisubunit protein and thereby changes the conformation of the molecule and its catalytic activity. The type of kinetics seen and a diagrammatic representation of this type of inhibition are illustrated in Figure 7-12. Specific examples of allosteric inhibition and activation will be discussed in the section on metabolic control.

Figure 7-12 Allosteric enzymes. The sigmoidal shape of the control curve is due to the fact that binding of the substrate to one of the catalytic sites increases the affinity of the enzyme for the substrate at the other catalytic site. In the presence of an activator (positive effector) the affinity of the catalytic sites for the substrate is increased, and in the presence of an inhibitor (negative effector) it is decreased. These effects are presumably mediated through changes in the conformation of the proteins involved. There is still some question whether the catalytic site and regulatory site can be on the same subunit of a multisubunit protein, or whether they must be on different subunits. It is likely that there are examples of both possible situations.

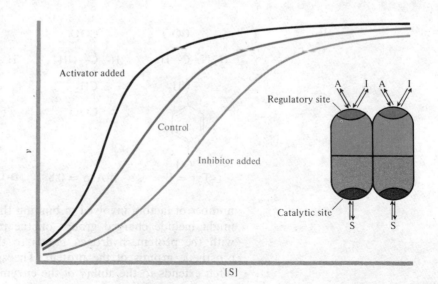

ENZYME SPECIFICITY

Enzymes vary tremendously in the degree to which they exhibit substrate specificity. Some are active with only one, or a small number of, substrates, whereas others will react with a wide range of substrates of the same general type. The specificity of the D-amino acid oxidase system shown in Figure 7-13 is a good illustration of this. The nature of the specificity depends on a

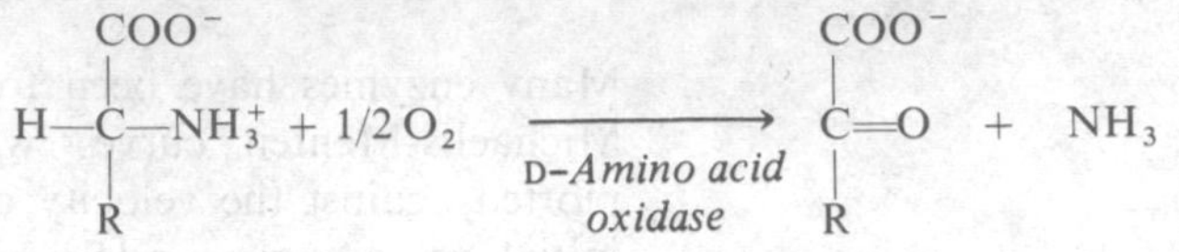

Figure 7-13 Action of D-amino acid oxidase. The enzymatic reaction can be followed by measuring the rate of O_2 consumption. The reaction rates for the various amino acids are expressed relative to the rate of oxidation of D-tyrosine. In general, this enzyme is specific for D-neutral amino acids. It has little or no activity toward L-amino acids, charged amino acids, or peptides. The nature of the nonpolar group does, however, influence the rate of the reaction.

Rate of reaction with different substrates

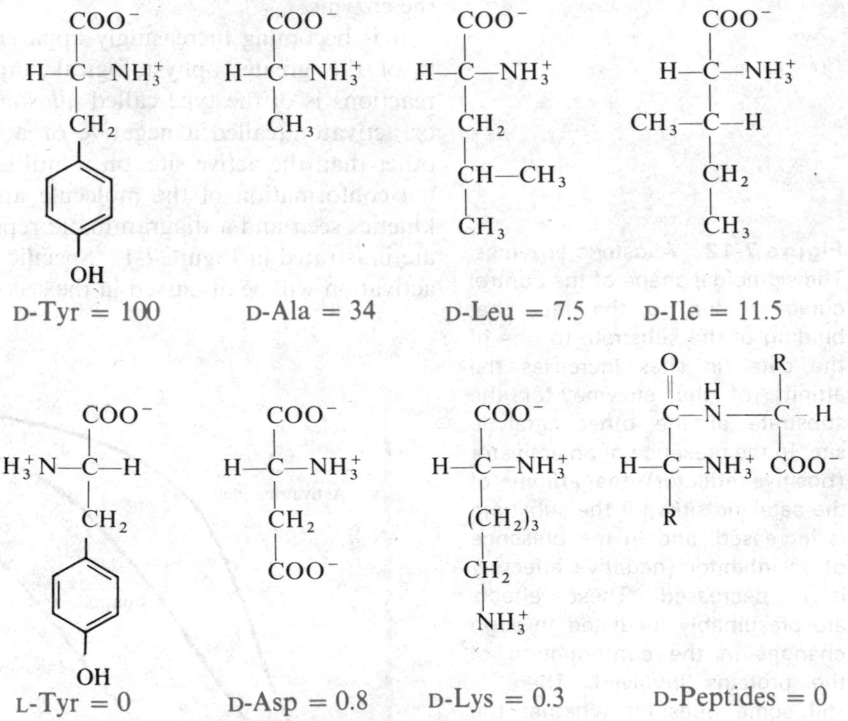

D-Tyr = 100   D-Ala = 34   D-Leu = 7.5   D-Ile = 11.5

L-Tyr = 0   D-Asp = 0.8   D-Lys = 0.3   D-Peptides = 0

number of factors involved in binding the substrate to the enzyme. These might include charged groups on the protein, hydrophobic interactions with the protein, hydrogen bonds to the protein, or interactions with prosthetic groups of the protein. The specificity of an enzyme reaction often extends to the ability of the enzyme to distinguish between the two apparently similar groups on a compound with a "meso" carbon atom.

This was first realized when early studies with radioactive citric acid, a symmetrical compound, indicated that it must have been converted to isocitrate in an asymmetric fashion by an enzymatic process. The problem was solved when Ogston deduced that the two apparently similar groups on the molecule were distinguishable if it was assumed that the substrate had to make contact with the enzymes at three different points (Figure 7-14).

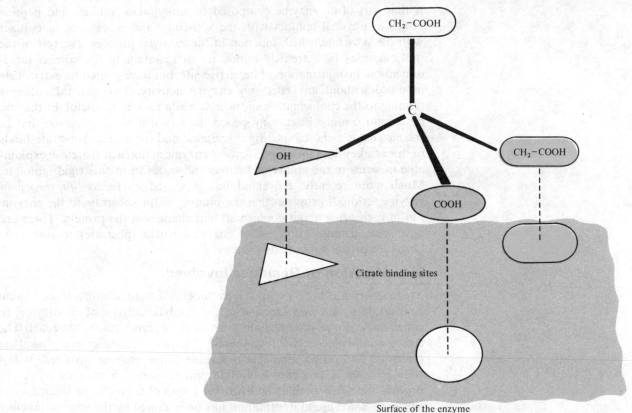

Figure 7-14 Three-point attachment of a symmetrical molecule to an enzyme surface. If there is a specific binding site for the hydroxyl and carboxyl groups, only one of the two hydroxymethyl groups of citric acid can fit the correct binding site for this group.

MECHANISM OF ENZYME ACTION

A complete understanding of the mechanism of action of an enzyme would involve an elucidation of at least the following:

1. The chemical pathway involved in the reaction
2. The amino acid residues of the enzyme which are involved
3. The spatial arrangement of these residues in relation to the substrate
4. A rational explanation of how these residues contribute to a lowering of the activation energy of the reaction

171

Concept of an Active Site

Substrates are small compared to the size of an enzyme; therefore, it is apparent that only a small portion of the amino acid residues of an enzyme is involved in the catalytic action. This observation led early enzymologists to the concept of an *active site*. A fairly general definition of an active site is that part of an enzyme composed of amino acid residues and peptide bonds in physical contact with the substrate, and others not in contact with the substrate, which function in the catalytic process. The rest of the molecule may be extremely important in maintaining the correct three-dimensional conformation of the active site, but it may often be extensively modified without any effect on enzyme activity. Two general concepts relating to the conformation of the active site have been useful. In the late nineteenth century Fischer proposed the *lock and key hypothesis*; that is, the enzyme is disposed as a rigid template and the correct substrate binds to this as a key fits a lock. Specificity of enzyme action was therefore explainable in terms of the ability of various substrates to fit this rigid template. Much more recently, Koshland has proposed an *induced-fit theory* of enzyme action. It proposes that the binding of the substrate to the enzyme can in itself bring about conformational changes in the protein. There are now some definite examples of enzymes where substrates modify conformation which support this theory.

Identification of Residues Involved

Much effort has been expended by biochemists to identify those amino acid residues which do function in the catalytic activity of an enzyme. In some cases, a true intermediate form of the enzyme can be detected. The conversion of glucose-1-P to glucose-6-P requires a transfer of a phosphate group off and on an amino acid residue of the enzyme involved. It has been possible in this case to isolate phosphoserine from a digest of the protein and to show that the hydroxyl group of serine is the intermediate acceptor. Some useful information has been gained by the specific labeling of active site residues with activated substrate analogs. These compounds are similar enough to the substrate to form an enzyme substrate complex, but rather than function as a true substrate, the activated group on the molecule forms a covalent bond with a residue at the active site. One of the best examples of a site-specific reagent has been the use of diisopropyl fluorophosphate to label the active serine of a number of esterases (Figure 7-15). Once this amino acid residue has been covalently labeled, it is possible to identify it subsequently in partial digests of the protein, and to determine rather easily the sequence of the amino acid residues immediately adjacent to the active serine.

Indirect evidence also gives an indication of the group involved at the active site. Changes in enzyme activity with changes in pH suggest that certain groups are being titrated, and treatment with chemicals known to alkylate selectively or otherwise to modify one specific amino acid residue

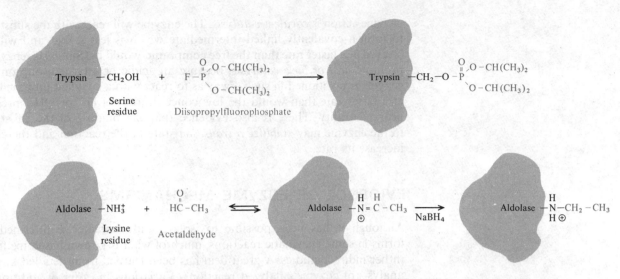

Figure 7-15 Examples of two reagents which have been used to form stable derivatives of an active site residue of an enzyme. In the case of the action of acetaldehyde on aldolase, an equilibrium between free acetaldehyde and its addition product is established. This intermediate can then be reduced to form a stable derivative.

will identify the amino acids involved. The sequence of amino acids in the peptide region around an active residue has now been determined for a large number of enzymes whose complete amino acid sequence is not known. These data have not yielded particularly useful information about catalytic mechanism. It is now apparent that the conformation of these residues and their relationship to other parts of the molecule is of much greater significance than the sequence alone.

Factors Involved in Catalysis

No general mechanism can be stated to explain the action of an enzyme. In general, chemical bond cleavage or formation in an enzymatic reaction will involve the formation of an ionic complex between the substrate and the enzyme to hold it in the correct position, followed by participation of amino acid residues or coenzymes as electron donating or attracting groups. It is possible, however, to determine that a number of distinctly different factors may be involved. In some cases, *approximation* may be important. By holding two reactants on the surface of an enzyme in just the position required for a chemical reaction to occur, the rate of the reaction can be drastically increased. There are regions on the enzyme which are distinctly nonpolar, and if a substrate is held in those regions, its reactivity may be influenced, just as a *solvent effect* is often found if reactions are carried out in solvents of different polarity. This may be particularly important if a transfer of electrons is involved. One of the more common factors in

173

enzyme action is *covalent catalysis.* The enzyme will react with the substrate to form a covalently linked intermediate which is less stable, and will be cleaved at a faster rate than the free compound would be. Similarly, enzymes promote *acid or base catalysis,* because an acidic or a basic group on the enzyme is positioned in such a way as to react with a covalent intermediate at a faster rate than would the low concentration of H^+ or OH^- present near neutrality. There is also evidence that the binding of the substrate to the enzyme may *stabilize a transition state* in the reaction and therefore increase its rate.

EVIDENCE OF ENZYME MECHANISMS

Although it has been possible to isolate and characterize intermediate forms in some enzymatic reactions, much of what is known has come from rather indirect studies. A great deal has been learned from detailed kinetic analysis of enzyme catalyzed reactions. Variations in order of addition of substrates to the enzyme, alterations of temperature, use of a number of similar substrates or various inhibitors, or alterations of the pH or polarity of the solvent can all be used to obtain an indication of the nature of the reaction that is being catalyzed by the enzyme. Much useful information has also been gained by the study of different isotope exchange reactions catalyzed by the enzyme. The use of this type of experiment is illustrated in Figure 7-16.

Examples of Specific Reactions

A large number of enzymatic reactions have now been studied in sufficient detail to permit a reasonable interpretation of the data in terms of a chemical pathway of the reaction. A few examples are sufficient to indicate what types of data are obtained.

Ribonuclease The enzyme ribonuclease has been subjected to extensive study, and the information gained from these studies is typical of what can be learned about the mechanism of enzyme action from various indirect methods. The enzyme catalyzes the hydrolysis of RNA at positions where a pyrimidine nucleotide furnishes the 3′-hydroxyl for the phosphodiester linkage to yield oligonucleotides ending in 3′-pyrimidines. Ribonuclease is readily available, easily purified, and was one of the first proteins to be sequenced and the first enzyme to be chemically synthesized. It contains 124 amino acid residues, 4 disulfide bonds, and has a molecular weight of 13,680.

The protein is cleaved by a specific bacterial protease, subtilisin, between residue 20 and residue 21, to yield ribonuclease S. The peptide which has been cleaved off remains firmly bound to the rest of the molecule even though it is no longer covalently attached. This is an active form of the enzyme, but removal of the S peptide results in loss of activity. Various

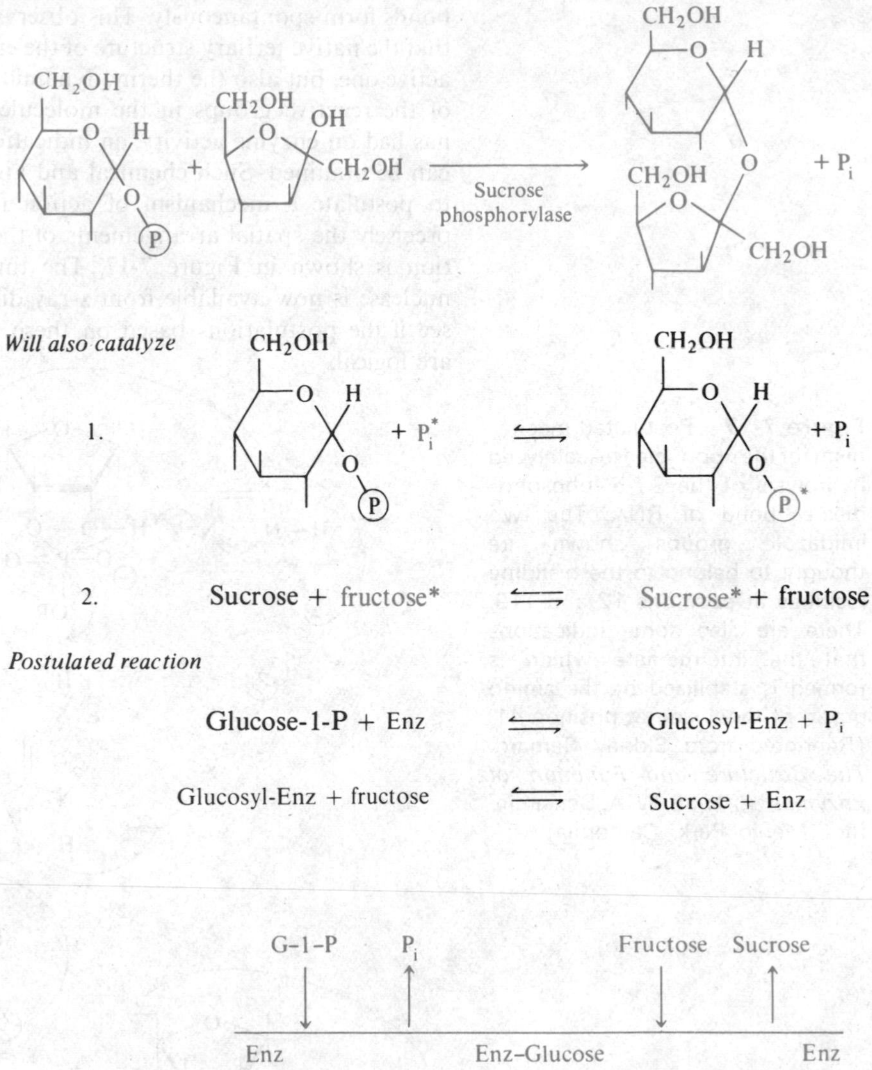

Will also catalyze

1.

2. Sucrose + fructose* ⟺ Sucrose* + fructose

Postulated reaction

Glucose-1-P + Enz ⟺ Glucosyl-Enz + P$_i$

Glucosyl-Enz + fructose ⟺ Sucrose + Enz

G–1–P   P$_i$ Fructose   Sucrose

Enz   Enz–Glucose   Enz

Figure 7-16 Action of sucrose phosphorylase. The enzyme will catalyze the synthesis of sucrose from glucose-1-phosphate and fructose. It will also catalyze the equilibration of radioactive phosphate into glucose-1-P or the equilibration of radioactive fructose with sucrose. The presence of these two partial reactions is consistent with the reaction mechanism indicated.

modifications of the amino acid sequence of the S peptide have been synthesized. It can be shown that a peptide containing the first 11 of the 20 residues will bind to the protein but will not restore activity, whereas a peptide containing the first 13 residues will bind and restore 70 percent of the activity of the native enzyme. Data such as these have been taken as evidence that residues 12 and 13, which are histidine and methionine, are essential for the catalytic mechanism or for formation of the correct tertiary structure.

The disulfide bonds in ribonuclease can be reduced with mercaptoethanol to form an enzymatically inactive protein, which will regain full activity if allowed to reoxidize to the disulfide form. The correct disulfide bonds form spontaneously. This observation has been interpreted to mean that the native tertiary structure of the enzyme is not only the enzymatically active one, but also the thermodynamically stable one. By blocking certain of the reactive groups in the molecule and determining what effect this has had on enzyme activity, an indication of which residues are important can be obtained. Such chemical and kinetic studies have made it possible to postulate a mechanism of action for ribonuclease without knowing precisely the spatial arrangements of the groups involved. Such a postulation is shown in Figure 7-17. The three-dimensional structure of ribonuclease is now available from x-ray diffraction data, and it is possible to see if the postulations based on these very indirect types of experiments are logical.

Figure 7-17 Postulated mechanism for the ribonuclease-catalyzed hydrolysis of the 3′, 5′-phospho-diester bond of RNA. The two imidazole groups shown are thought to belong to the histidine residues at positions 12 and 119. There are also some indications that the intermediate which is formed is stabilized by the amino group of the lysine at position 41. (Reprinted from Sidney Bernard, *The Structure and Function of Enzymes*, © 1968, W. A. Benjamin, Inc., Menlo Park, California)

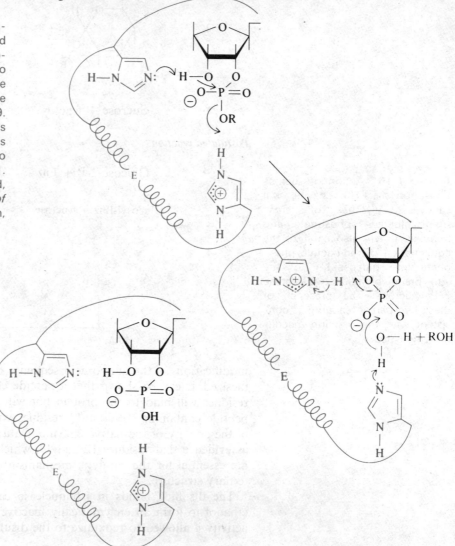

Chymotrypsin   Chymotrypsin is one of a class of proteolytic enzymes collectively called *serine proteases*. Other enzymes with a similar mechanism of action and many structural similarities are trypsin, elastase, thrombin, and the bacterial subtilisins. Because the pancreatic enzymes, trypsin and chymotrypsin, could be readily obtained and purified in high yields, they were among the first enzymes to be carefully studied. It was shown that organophosphate esters such as diisopropylfluorophosphate (see Figure 7-15) would irreversibly inactivate these enzymes and that this inactivation was due to the formation of a phosphate ester on a specific serine residue. These enzymes will also hydrolyze low molecular weight esters, and when the enzyme is mixed with a substrate such as *p*-nitrophenyl acetate a rapid burst of release of the *p*-nitrophenyl group can be seen, followed by a slower turnover. These data are consistent with the formation of an acyl enzyme intermediate and release of the alcohol, followed by a rate-limiting hydrolysis of the intermediate to release the acid portion of the ester. This type of action, illustrated in Figure 7-18, has been further verified by direct isolation of the acyl enzyme intermediate of some substrates.

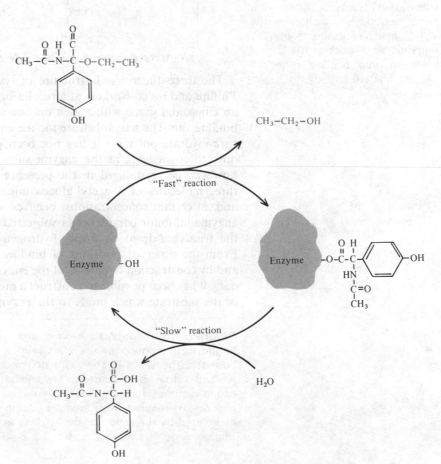

Figure 7-18 Mechanism of action of chymotrypsin. The enzyme will catalyze the hydrolysis of low molecular weight peptides and esters as well as proteins. The hydrolysis of *N*-acetyltyrosine ethyl ester shown here demonstrates the formation of the acyl enzyme intermediate.

177

Lysozyme One of the best examples of what can be determined about the detailed mechanism of an enzymatic reaction if the complete structure is known comes from a study of the enzyme lysozyme. This enzyme, which is widely distributed in nature, was discovered by Fleming as a substance in nasal mucus which had the ability to lyse or dissolve certain bacteria by hydrolyzing their cell walls. The sensitive linkage in the cell wall is a β-(1,4)-glycosidic bond in a polymer containing alternating *N*-acetyl-glucosamine (NAG) and *N*-acetyl-muramic acid (NAM) residues (Figure 7-19). The enzyme has now been obtained from a number of sources. The egg-white enzyme has a molecular weight of 14,500 and consists of a single polypeptide chain of 129 amino acids.

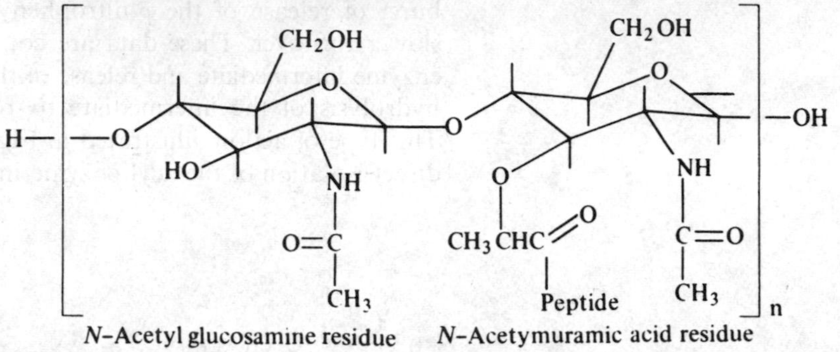

Figure 7-19 Basic cell-wall polysaccharide polymer which is the substrate for lysozyme. A tetra or a penta peptide is attached to the lactic acid substituent of the muramic acid residue. The amino acids in this peptide differ in various bacteria.

N–Acetyl glucosamine residue   *N*–Acetymuramic acid residue

The three-dimensional structure of lysozyme has been determined by Phillips and his co-workers at a resolution of 2 Å. The molecule forms into an ellipsoidal shape with a cleft on one side which contains the substrate binding site. The true substrate for the enzyme is a high molecular weight carbohydrate polymer; it has not been possible to look by direct x-ray diffraction methods at the enzyme-substrate complex. The enzyme can, however, be crystallized in the presence of a trisaccharide composed of three molecules of *N*-acetyl-glucosamine, which binds to the active site and, at certain concentrations, behaves as an inhibitor of lysozyme. The enzyme inhibitor complex was subjected to x-ray analysis which showed the trisaccharide to be firmly hydrogen-bonded to the enzyme surface. From the observed positions of binding of the tri-*N*-acetyl-glucosamine and by construction of models of the enzyme based on the x-ray diffraction data, it has been possible to construct a model showing the six-unit segment of the substrate which binds to the enzyme (Figure 7-20).

Figure 7-20 *(opposite)* Atomic arrangement in the lysozyme molecule in the neighborhood of the cleft with a hexa-*N*-acetylchitohexose shown bound to the enzyme. The main polypeptide chain is shown speckled and NH and O atoms are indicated by line and full shading, respectively. The positions of the sugar residues are those determined from the binding of known trisaccharides or inferred from model building. It is postulated that the linkage hydrolyzed by the action of the enzyme is between residues D and E. (D. C. Phillips, *Proc. Nat. Acad. Sci.*, Vol. 57, p. 493, 1967)

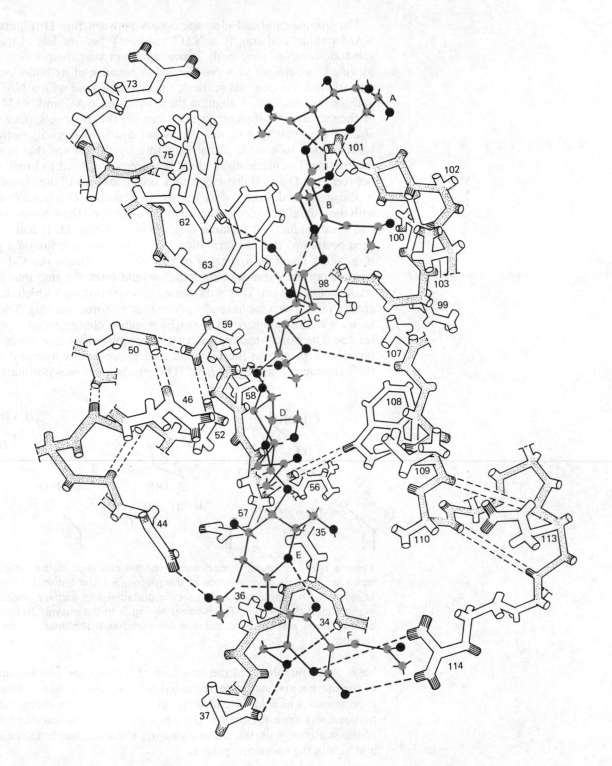

The enzyme-catalyzed cleavage occurs between ring D (Figure 7-20), a NAM residue, and ring E, a NAG residue. When models of the enzyme substrate complex were built, it was apparent that the position occupied by ring C would not fit a NAM residue because of its bulky constituent on carbon 3. Forcing this particular position to bind with a NAG residue had the overall effect of aligning the alternating NAG and NAM residues of the substrate to the correct positions on the enzyme surface. It could also be seen that ring D was somewhat distorted from its normal chair configuration to a more planar structure by the groups that bound it to the enzyme. The mechanism which has been postulated to break the bond between ring D and E involves attack on the oxygen of the glycoside bond by the un-ionized proton of a glutamic acid residue. This bond then breaks, with the glutamate hydrogen going to the oxygen on the 4 position of ring E, and a carbonium ion forming on carbon 1 of ring D. It had previously been postulated that the formation of a C-1 carbonium ion on a pyranose ring would result in a sharing of this charge between the C-1 and O-6 positions of the pyranose ring, which would force the ring into the "half-chair" configuration. This is the same conformation into which the binding groups on the enzyme have already tended to force this ring. The carbonium ion is also stabilized by a nearby negative charge on an aspartic acid residue. The bond-breaking reaction is completed when a water molecule furnishes a proton to the glutamic acid group and a hydroxyl to attack the carbonium ion (Figure 7-21). This mechanism was postulated on the

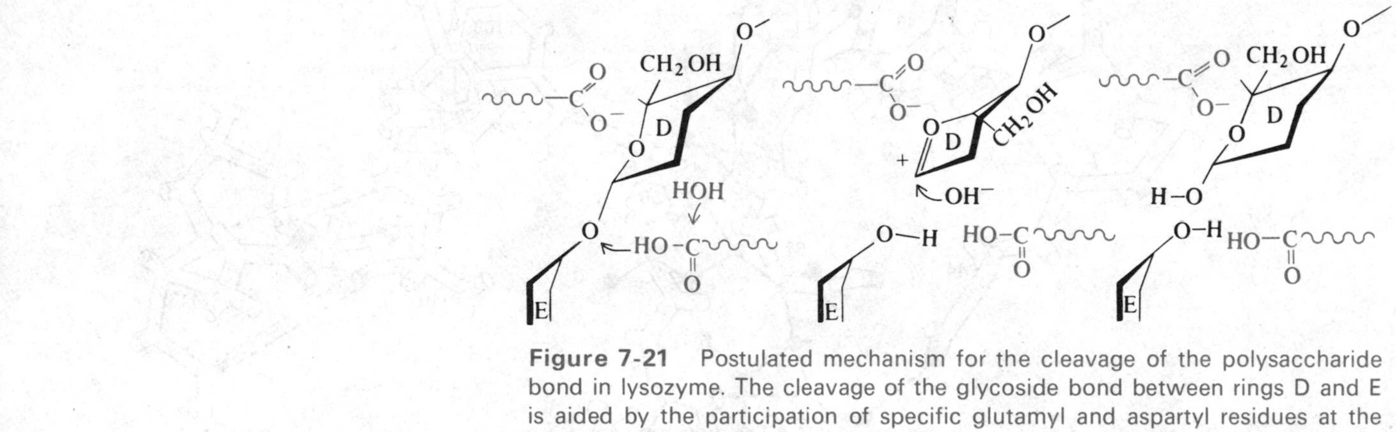

Figure 7-21 Postulated mechanism for the cleavage of the polysaccharide bond in lysozyme. The cleavage of the glycoside bond between rings D and E is aided by the participation of specific glutamyl and aspartyl residues at the active site of the enzyme. The binding of ring D to the enzyme has distorted its conformation and tends to stabilize the carbonium ion intermediate which is formed.

basis of the fine detail of the structure of the enzyme and the interaction of the enzyme and substrate. It has led to a number of chemical and kinetic experiments which have tended to substantiate it. This mechanism is the first real evidence to support the general hypothesis that the interaction of the substrate with the enzyme distorts chemical bonds in the substrate and aids in the enzymatic process.

PROBLEMS

1. The initial velocity (v) for the enzyme-catalyzed reaction $A \rightarrow B$ was determined at a number of concentrations of A, and v measured.

[A]	v
2×10^{-4}	0.16 units/min
1×10^{-3}	0.25 units/min
2×10^{-2}	0.5 units/min
8×10^{-2}	0.5 units/min

What would the initial velocity be when A is 5 mM [$5 \times 10^{-3}\ M$]?

2. For another enzyme, v was determined at three different substrate concentrations to give the following data:

[S]	v
1×10^{-3}	0.33 units/min
4×10^{-4}	0.222
2×10^{-3}	0.4

On graph paper construct a Lineweaver-Burk plot and find K_m and V_{max}.

3. An enzyme with a K_m of $2.4 \times 10^{-4} M$ was assayed at the following substrate concentrations: (a) $6.3 \times 10^{-5}\ M$, (b) $10^{-4}\ M$, and (c) $0.05\ M$. The velocity observed at $0.05\ M$ was 128 μmoles/liter-min. Calculate the initial velocities at the other substrate concentrations.

4. If the enzyme concentrations are increased fivefold in Problem 3, what will be the effect on the initial velocities at each substrate concentration?

5. An enzyme with a K_m of $1.2 \times 10^{-4}\ M$ was assayed at an initial substrate concentration of $0.02\ M$. By thirty seconds, 2.7 μmoles/liter of product had been produced. How much product will be present: (a) at one minute, (b) at three minutes; (c) what percent of the original substrate will be utilized by three minutes?

6. The initial rate of an enzyme reaction was determined at various substrate concentrations in the presence and absence of an inhibitor.

[S]	v	v (with inhibitor)
1.0×10^{-4}	56 μmoles/min	36 μmoles/min
2.0×10^{-4}	86	60
5.0×10^{-4}	126	102
7.5×10^{-4}	148	126

Construct a Lineweaver-Burk plot and determine if the inhibitor is competitive or noncompetitive.

Suggested Readings

Books

Bernhard, S. *The Structure and Function of Enzymes.* W. A. Benjamin, New York (1968).

Segel, I. H. *Enzyme Kinetics.* Wiley, New York (1975).

Articles and Reviews

Cleland, W. W. Steady State Kinetics, in *The Enzymes,* Boyer, P. D. (ed.), 3rd ed., vol. 2, pp. 1–65, Academic Press, New York (1970).

Koshland, D. E. Correlation of Structure and Function in Enzyme Action. *Science* **142**:1533–1541 (1963).

Monod, J., J. P. Changeux, and F. Jacob. Allosteric Proteins and Cellular Control Systems. *J. Mol. Biol.* **6**:306–329 (1963).

Phillips, D. C. The Three-Dimensional Structure of an Enzyme Molecule. *Sci. Amer.* **215**:78–90 (1966).

Stroud, R. M. A Family of Protein-Cutting Proteins. *Sci. Amer.* **231**:24–88 (1974).

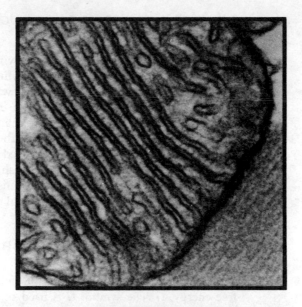

biochemical energetics

chapter eight

The energy available from a metabolic reaction can be measured as the change in free energy, ΔG, of the system. A limited number of catabolic, degradative, biochemical reactions produce energy which is then used to drive the anabolic, biosynthetic reactions. Oxidation in biochemical systems is accomplished by the reduction of coenzymes which are subsequently used to drive synthetic reactions or are reoxidized by the mitochondrial electron transport chain. The energy made available from the reoxidation of these coenzymes is conserved by coupling the oxidation to the formation of ATP by the process of oxidative phosphorylation. The formation of ATP or other nucleoside triphosphates is a way that cells can store energy, and the energy available from the hydrolysis of ATP is subsequently used to drive other energy-requiring reactions. Metabolic energy is also used to transport charged compounds across lipophylic biological membranes and to maintain thermodynamically unfavorable concentration gradients across these membranes. A number of different mechanisms are utilized to accomplish this important biological process.

The metabolic reactions that occur in living tissues can be divided into two general types: catabolic reactions, which degrade large molecules to smaller ones, and anabolic reactions, which synthesize larger molecules from various cellular precursors. The energy needed to drive the biosynthetic anabolic reactions must be made available from energy-yielding catabolic reactions. It will be seen in the discussions of metabolism that relatively few compounds are produced in catabolic reaction sequences which can subsequently be used to furnish the energy for synthetic reactions or to do other useful work for the organism. The relationships involved in these reactions fall in the area of chemical thermodynamics.

FREE ENERGY AND EQUILIBRIUM CONSTANTS

The thermodynamic concept most useful to the biochemist is that of the free energy, G. The symbol G is used to designate the Gibbs free energy, that energy available for useful work from chemical changes occurring at a constant pressure. The same thermodynamic parameter is referred to as F in the older literature, and many biochemists still use the term F for free energy. The free-energy content of a compound cannot be measured, but the change in free energy as one compound is transformed to another can be. This change, ΔG, is the amount of energy made available during this transformation. Reactions which go spontaneously in the direction written are those which have a negative ΔG and are called *exergonic reactions*. Those reactions which require an input of energy to drive them in the direction written have a positive ΔG and are called *endergonic reactions*. In a reaction system which has reached equilibrium, there is no tendency for a net movement of reactants in either direction, and the ΔG is equal to 0.

The free-energy change of a reaction can be related to the other thermodynamic parameters by the equation

$$\Delta G = \Delta H - T\Delta S$$

where ΔH, the enthalpy change, is a measure of the change in heat content of a reaction occurring at constant pressure and can be measured in a bomb calorimeter; T is the absolute temperature of the reaction; and ΔS is the change in entropy or degree of disorder in the system. Although the ΔH and ΔS of a reaction can, in some specific cases, be of interest to biochemists, free-energy changes are more generally used; they will be the only ones considered in this text.

It should always be remembered that free energy is a thermodynamic, not a kinetic, parameter; that is, the fact that a reaction has a large negative ΔG does not necessarily mean that the reaction will proceed at a measureable rate. The cellulose on a book page, for example, would burn to CO_2 and water, releasing a great deal of heat and generating a large negative ΔG, yet it will not do so unless the activation energy for initiating the combustion is first overcome by putting heat into the system.

The most useful property associated with the free-energy change of a reaction is, for the biochemist, its relationship to the equilibrium constant of a reaction. The generalized relationship between these two constants is

$$\text{if} \quad A + B \rightleftharpoons C + D$$

$$\text{then} \quad \Delta G = \Delta G^0 + RT \ln \frac{[C][D]}{[A][B]}$$

Substituting into this equation the value for the universal gas constant, R, of 1.987 cal mole^{-1} deg^{-1}, and a temperature of 25 °C or 298 °K, and converting natural logs to common logs, this expression becomes

$$\Delta G = \Delta G^0 + 1363 \log_{10} \frac{[C][D]}{[A][B]}$$

At equilibrium, the ratio of products to reactants becomes the equilibrium constant, K_{eq}, and because there is no net change in the concentrations of reactants and products, the ΔG becomes 0. Therefore,

$$0 = \Delta G^0 + 1363 \log_{10} K_{eq}$$

or

$$\Delta G^0 = -1363 \log_{10} K_{eq}$$

If the equilibrium constant for a reaction can be measured, the ΔG^0, or standard free-energy change, can therefore be calculated.

The standard free-energy change, ΔG^0, which can be calculated from the equilibrium constant, is defined as the free energy change of a reaction when all reactants and products are maintained at their standard state—that is, one molar for solutes, unit activity for solvents, and one atmosphere for gases. It is obvious from this equation that reactions which have equilibrium constants of greater than 1 will have a negative ΔG^0, those whose K_{eq} is equal to 1 will have a ΔG^0 of 0, and reactions where K_{eq} is less than 1 will have a positive ΔG^0. Each change in the K_{eq} by a factor of 10 will increase the log K_{eq} value by one unit and result in a change in the ΔG^0 of 1363 cal. Both the K_{eq} and ΔG^0 are, therefore, interchangeable indications of how far, and in which direction, a chemical reaction will have proceeded at equilibrium. These relationships are presented in Table 8-1.

The ΔG^0 of a reaction is calculated for standard-state conditions; this means that if a hydrogen ion is involved in a reaction, its concentration is taken to be 1 molar, and the pH would be 0. For this reason, tables of ΔG^0 values used by biochemists are often recalculated in terms of the equilibrium at some other pH, usually 7.0, and are then designated as $\Delta G^{0'}$ values. The pH should be indicated in the table, and if it is not, it was most likely pH 7.0. This distinction is important only if a proton is involved in a reaction; otherwise, the ΔG^0 is equal to $\Delta G^{0'}$.

An indication of the usefulness of these calculations can be gained from the following example of the reaction catalyzed by the enzyme phosphoglyceromutase. This enzyme catalyzes an interconversion of 2-phospho-

Table 8-1   Relationship between K_{eq} and ΔG^0

K_{eq}	ΔG^0	K_{eq}	ΔG^0
1000	−4089	1.0	0
500	−3678	0.1	1363
100	−2726	0.01	2726
20	−1640	0.001	4089
10	−1363		

glycerate and 3-phosphoglycerate in the glycolytic metabolism of glucose. At 25 °C, the K_{eq} for the reaction written in the direction of 3-phosphoglycerate formation is 6.0. The ΔG^0 can then be calculated.

$$2\text{-phosphoglycerate} \rightleftharpoons 3\text{-phosphoglycerate}$$

$$
\begin{aligned}
\Delta G^0 &= -RT \ln K_{eq} \\
&= -1363 \log_{10} K_{eq} \\
&= -1363 \log_{10} 6.0 \\
&= (-1363)(.778) \\
&= -1060 \text{ cal mole}^{-1}
\end{aligned}
$$

This, then, is the change in free energy which would be involved when 1 mole of 2PGA is converted to 1 mole of 3PGA at 25 °C, under conditions such that the concentration of both compounds was at all times 1 molar. The measured concentrations of these compounds in rat liver are on the order of 1×10^{-4} M 2PGA and 5×10^{-4} M 3PGA. The free-energy change associated with the conversion of 1 mole of phosphoglycerate if these concentrations are maintained can also be calculated

$$
\begin{aligned}
\Delta G &= \Delta G^0 + RT \ln \frac{[3\text{PGA}]}{[2\text{PGA}]} \\
&= -1060 + 1363 \log_{10} \frac{5 \times 10^{-4}}{1 \times 10^{-4}} \\
&= -1060 + 1363 \log_{10} 5 \\
&= -1060 + (1363)(0.70) \\
&= -1060 + 954 \\
&= -106 \text{ cal mole}^{-1}
\end{aligned}
$$

This is the change in free energy associated with the conversion of a mole of 2PGA to a mole of 3PGA at the steady-state concentrations maintained in rat liver. It can be seen from this calculation that this reaction is one which is functioning very close to equilibrium and which should not be expected to impair the flux of metabolites in either direction. Although the application of these thermodynamic equations to biological conditions, which are often far removed from the ideal conditions for which they were derived, results in potentially large errors, the concept is still very useful. It has provided a means of comparing various possible competing reactions and of interpreting the significance of various metabolic pathways.

FREE-ENERGY CHANGE AND REDOX POTENTIALS

Just as the ΔG^0 of any reaction can be calculated from a knowledge of the equilibrium constant, the ΔG^0 for a redox reaction can be calculated from the difference between the oxidation-reduction potentials of the reactants and the potentials of the products involved. This is particularly important to the biochemist because so many of the reactions of interest do involve an oxidation or a reduction of metabolic intermediates.

The oxidation or reduction of a biological compound is basically no different from the oxidation or reduction of any other chemical compound. In all cases a transfer of electrons from the reducing agent to the oxidizing agent is involved. For example, in the reduction of Cu^{2+} by Zn, the following two half-reactions occur.

Half-reaction	$Zn \longrightarrow Zn^{2+} + 2e^-$
Half-reaction	$Cu^{2+} + 2e^- \longrightarrow Cu$
Net reaction	$Zn + Cu^{2+} \longrightarrow Zn^{2+} + Cu$

The net reaction is an oxidation of Zn and a reduction of Cu^{2+}, with a transfer of electrons between them. Most biological oxidations differ only in that they involve a loss of hydrogen atoms as well as electrons, and are similar to the oxidation of hydroquinone by ferric ion.

Hydroquinone $+ 2Fe^{3+} \longrightarrow$ Quinone $+ 2Fe^{2+} + 2H^+$

The standard reduction potentials of some half-reactions which are of interest to biochemists are listed in Table 8-2. Tables of reduction potentials provide an indication of the relative potential of each of the compounds to accept electrons in relation to the standard hydrogen electrode, which has arbitrarily been given the oxidation-reduction potential, E_0, of 0.000 volt under standard conditions, that is, one atmosphere of H_2 and pH 0. The reaction is

$$H^+ + 1e^- \longrightarrow 1/2\ H_2$$

This reaction will be pH-dependent, as a proton is consumed, and the values in Table 8-2 are for half-reactions at pH 7.0 where the reduction potential, E_0', of the hydrogen electrode is -0.420 V. Tables of oxidation or reduction potentials found in different texts often seem confusing, because they are set up in many different ways and often do not show the complete half-reaction, but only indicate the pair of compounds undergoing the oxidation-reduction. Tables of redox potentials found in chemistry texts will usually be for oxidation potentials; that is, the signs will be reversed from those in Table 8-2, and will be E_0 values at pH 0 and not E_0' values at pH 7, as will be found in most biochemical references.

A complete oxidation-reduction reaction requires that two of these half-reactions be coupled. Table 8-2 contains reduction potentials, and when any two of these are coupled, the half-reaction with the more positive reduction potential, that is, the greatest tendency to gain electrons and become reduced, will go as written and the other half-reaction will be

Table 8-2 Standard Reduction Potentials of Some Oxidation-Reduction Half-Reactions[a]

Half-reaction (written as a reduction)	E'_0 at pH 7.0 (volts)
$\frac{1}{2}O_2 + 2H^+ + 2e^- \rightarrow H_2O$	0.81
Cytochrome-a-Fe^{+3} + 1e^- → cytochrome-a-Fe^{+2}	0.29
Cytochrome-c-Fe^{+3} + 1e^- → cytochrome-c-Fe^{+2}	0.25
Ubiquinone + 2H$^+$ + 2e^- → ubiquinone-H$_2$	0.10
Dehydroascorbate + 2H$^+$ + 2e^- → ascorbate	0.06
Fumarate + 2H$^+$ + 2e^- → succinate	0.03
FAD + 2H$^+$ + 2e^- → FADH$_2$	−0.06
Oxalacetate + 2H$^+$ + 2e^- → malate	−0.10
α-Ketoglutarate + NH$_3$ + 2H$^+$ + 2e^- → glutamate + H$_2$O	−0.14
Acetaldehyde + 2H$^+$ + 2e^- → ethanol	−0.16
Pyruvate + 2H$^+$ + 2e^- → lactate	−0.19
Riboflavin + 2H$^+$ + 2e^- → riboflavin H$_2$	−0.20
Acetoacetate + 2H$^+$ + 2e^- → β-hydroxybutyrate	−0.29
NAD$^+$(DPN$^+$) + 2H$^+$ + 2e^- → NADH(DPNH) + H$^+$	−0.32
NADP$^+$(TPN$^+$) + 2H$^+$ + 2e^- → NADPH(TPNH) + H$^+$	−0.32
Pyruvate + CO$_2$ + 2H$^+$ + 2e^- → malate	−0.33
Acetyl-CoA + 2H$^+$ + 2e^- → acetaldehyde + CoA	−0.41
CO$_2$ + 2H$^+$ + 2e^- → formate	−0.42
H$^+$ + 1e^- → $\frac{1}{2}$H$_2$	−0.42
Ferredoxin-Fe^{+3} + 1e^- → ferredoxin-Fe^{+2}	−0.43
Acetate + 2H$^+$ + 2e^- → acetaldehyde	−0.60

[a] Standard conditions: Unit activity of all components except H$^+$ which is maintained at 10^{-7} M. Gases are at 1 atm pressure.

reversed and be driven backwards. Stated another way, the oxidized form of compounds in the table with more positive potentials are good oxidizing agents (as they are readily reduced), and the reduced form of those with more negative potentials are good reducing agents.

The drive for an overall reaction to occur when two half-reactions are coupled is measured by the difference in reduction potential of the two half-reactions, or the $\Delta E'_0$.

$$\Delta E'_0 = \left[\begin{array}{c} E'_0 \text{ of half-reaction} \\ \text{containing the oxidizing agent} \end{array} \right] - \left[\begin{array}{c} E'_0 \text{ of half-reaction} \\ \text{containing the reducing agent} \end{array} \right]$$

As an example, consider the reaction between NADH and pyruvate:

$$\text{NADH} + \text{H}^+ + \text{pyruvate} \rightleftharpoons \text{NAD}^+ + \text{lactate}$$

The half-reactions are

$$NAD^+ + 2H^+ + 2e^- \longrightarrow NADH + H^+ \qquad E_0' = -0.32 \text{ V}$$

and

$$Pyruvate + 2H^+ + 2e^- \longrightarrow lactate \qquad E_0' = -0.19 \text{ V}$$

Therefore, as the equation is written, pyruvate is the oxidizing agent and NADH the reducing agent. The $\Delta E_0'$ can then be calculated as

$$\Delta E_0' = (-0.19) - (-0.32)$$
$$= -0.19 + 0.32$$
$$\doteqdot +0.13 \text{ V}$$

In order to combine half-reactions involving the production of different numbers of electrons, it is necessary to multiply one of the half reactions by the appropriate coefficient in order to establish stoichiometry of electrons. It should be noted that the numerical value of the reduction potential for that half-reaction remains unchanged. As the $\Delta E_0'$ of a reaction is a measure of the tendency of the reaction to go in one direction or the other, it can be related to the equilibrium constant or to the $\Delta F^{0'}$ of the reaction. The equation relating $\Delta G^{0'}$ and $\Delta E_0'$ is

$$\Delta G^{0'} = -n\mathscr{F} \, \Delta E_0'$$

where n is the number of electrons involved per mole of reaction, and $\mathscr{F}$ is Faraday's constant, the factor converting volts to calories, which is 23,063 cal V^{-1}. Returning to the example of the reduction of pyruvate to lactate, the $\Delta G^{0'}$ for the reaction can be calculated as

$$\Delta G^{0'} = -2(23,063)(\Delta E_0')$$
$$= -2(23,063 \ (+0.13)$$
$$= -6,000 \text{ cal}$$

Note that for the $\Delta G^{0'}$ of a reaction to be negative—that is, to be a reaction which proceeds spontaneously in the direction written—the $\Delta E_0'$ value must be positive. Even though the $\Delta G^{0'}$ indicates that the equilibrium for this particular reaction is greatly in favor of lactate, the reaction is one which is readily reversible in tissues. This can be accomplished by increasing the concentration of lactate or NAD^+, or by removing pyruvate by other metabolic reactions to keep its concentration low.

OXIDATION-REDUCTION COENZYMES

Three classes of coenzymes are usually associated with biological oxidations and reductions: pyridine nucleotides, flavin coenzymes, and heme compounds. An enzyme that catalyzes an oxidation or a reduction of a substrate cannot do so without some compound either to donate or to accept the electrons involved, and the redox coenzymes serve this function. In these types of reactions, the active enzyme, or the *holoenzyme*, is a combination of the protein or *apoenzyme* and its *prosthetic* group. The distinction between what is commonly called a coenzyme and what would normally be called a prosthetic group of a protein is purely operational. If the factor is tightly bound, it is called the prosthetic group of the protein; if it is more easily lost and is also a factor active in other systems, it is called a coenzyme. Of the three types of electron-transferring groups, the pyridine nucleotides are usually readily dissociable, flavins are sometimes dissociable but are usually not, and heme groups are generally tightly bound.

Pyridine Nucleotides

Two important coenzymes are collectively called the pyridine nucleotides, nicotinamide adenine dinucleotide (NAD$^+$) and nicotinamide adenine dinucleotide phosphate (NADP$^+$).

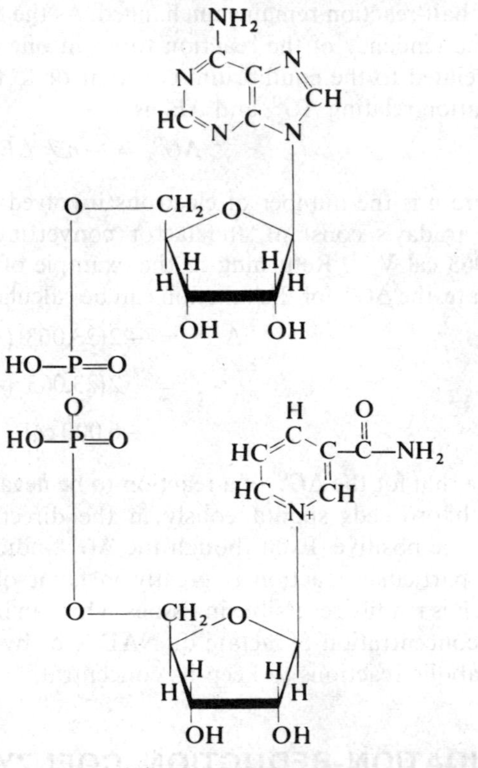

Nicotinamide adenine
dinucleotide (NAD$^+$)
or
Diphosphopyridine nucleotide (DPN$^+$)

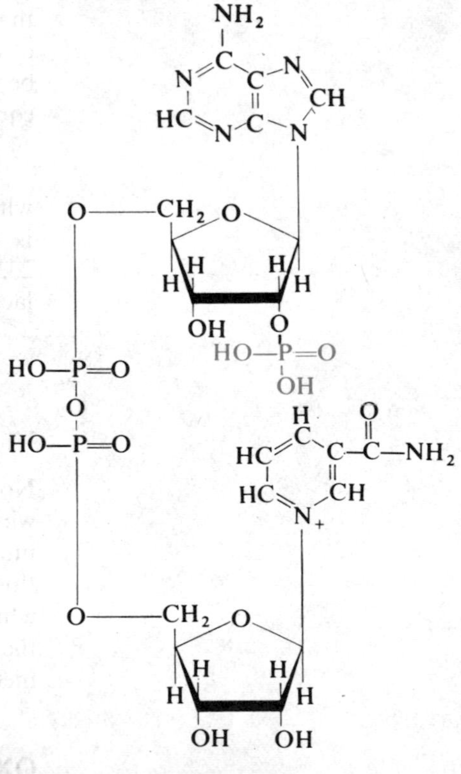

Nicotinamide adenine
dinucleotide phosphate (NADP$^+$)
or
Triphosphopyridine nucleotide (TPN$^+$)

These compounds were the first coenzymes to be discovered. When they were first discovered, they were called coenzyme I and coenzyme II; later, and until rather recently, they were called diphosphopyridine nucleotide (DPN^+) and triphosphopyridine nucleotide (TPN^+). Because these names are not generically correct, an international commission has proposed that the NAD^+ and $NADP^+$ nomenclature be used. Although NAD^+ is certainly much more descriptive, it is still chemically incorrect, and even though the newer terminology will be used in this text, many practicing biochemists still use the DPN^+ and TPN^+ terminology. The structures consist of a molecule of adenosine monophosphate joined through a phosphoanhydride bond to a second ribose-5-phosphate. The second ribose is linked through an *N*-glycoside at carbon 1 to a substituted pyridine ring. This portion of the molecule is the amide of nicotinic acid, or nicotinamide. Both nicotinic acid and nicotinamide are biologically active forms of the vitamin niacin. The discovery of these compounds provided one of the first examples of the coenzyme function of the water-soluble B-complex vitamins.

Niacin   Since 1873, niacin had been known as an organic compound, but it was not until 1937 that Elvehjem isolated it from a liver preparation and demonstrated that it would cure black tongue in dogs. Niacin was then shown to be of value in treating pellagra, a common human nutritional disease caused by a deficiency of the vitamin and characterized by oral lesions, diarrhea, dermatitis, and neurological lesions. It functions in metabolism as part of the coenzymes NAD^+ and $NADP^+$. Some of the best dietary sources of the vitamin are meat products and enriched grains. A portion of the requirement can be met by synthesis from the amino acid tryptophan, and a high tryptophan diet will decrease the dietary requirement for niacin. The adult human requirement of 18 mg per day is much higher than that of the other B-complex vitamins. Both nicotinic acid and nicotinamide are active forms of the vitamin.

Nicotinic acid   Nicotinamide

The function of the coenzyme is to oxidize or reduce a substrate by a transfer of a pair of electrons between the coenzyme and the substrate. This action, which is identical for both of the coenzymes, is illustrated in Figure 8-1. It can be viewed as though the substrate being oxidized were transferring a hydrogen atom with two electrons, a hydride ion (H:), to the coenzyme and leaving behind a proton to dissociate. Because of this, the reduced form of the coenzyme should always be written to indicate that a pair of electrons has been transferred and that a proton is associated

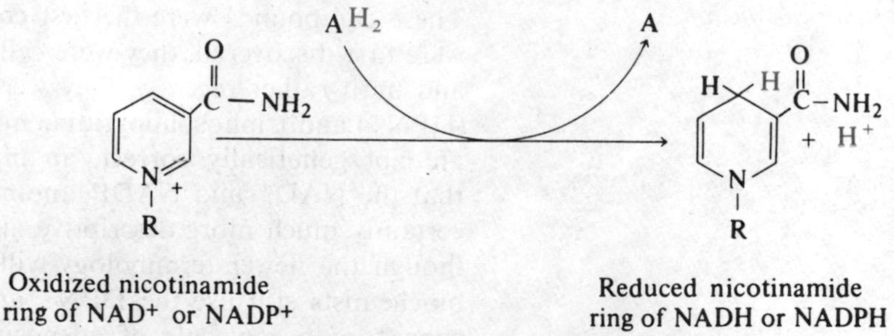

Figure 8-1 Reduction of the pyridine nucleotides. Both of the electrons and one of the hydrogens go to the nicotinamide portion of the molecule, while the other hydrogen is left as a proton in the medium.

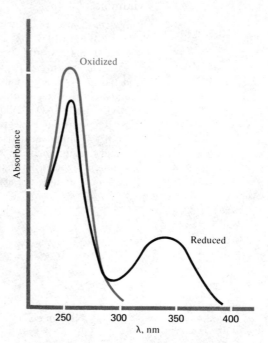

Oxidized nicotinamide ring of NAD^+ or $NADP^+$

Reduced nicotinamide ring of NADH or NADPH

with the reaction, that is, $NADH + H^+$, not simply NADH. The reduction of the pyridine ring results in the appearance of a new absorption peak at 340 nm (Figure 8-2), and because of this, enzyme reactions which use NAD^+ or $NADP^+$ as a cofactor can be very easily followed spectrophotometrically. The extent of appearance of the 340 nm absorption will be proportional to the amount of NAD^+ or $NADP^+$ reduced and of substrate

Figure 8-2 Absorption spectra of the oxidized and reduced form of the pyridine nucleotides. The peak near 260 nm is due to the absorption of the adenine ring, and the 340 nm absorption is from the reduced pyridine ring of NADH or NADPH.

oxidized. The reduced form of the nicotinamide ring is a planar structure, and the two hydrogens at the tetrahedral carbon extend above and below this ring. Reactions which utilize NAD^+ or $NADP^+$ are therefore stereospecific, and enzymes that use the reduced form of NAD^+ can be divided into an α and a β series depending on which of the two hydrogens of the ring is transferred to the substrate. This specificity of reaction cannot be determined unless the reduced coenzyme has previously been labeled in only one of the positions with a deuterium or tritium atom.

Flavin Coenzymes

The second of the important classes of redox coenzymes are the *flavins*. As with the pyridine nucleotides, two forms are known, flavin mononucleotide (FMN) and flavin adenine dinucleotide (FAD). Chemically, they are derived from the isoalloxazine derivative, riboflavin, another water-soluble vitamin.

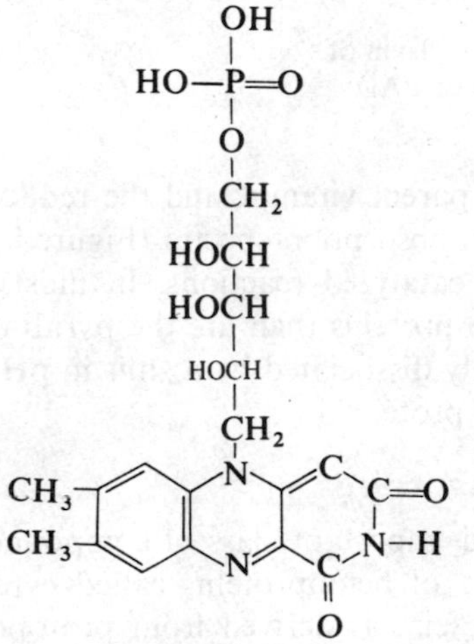

Flavin mononucleotide (FMN)

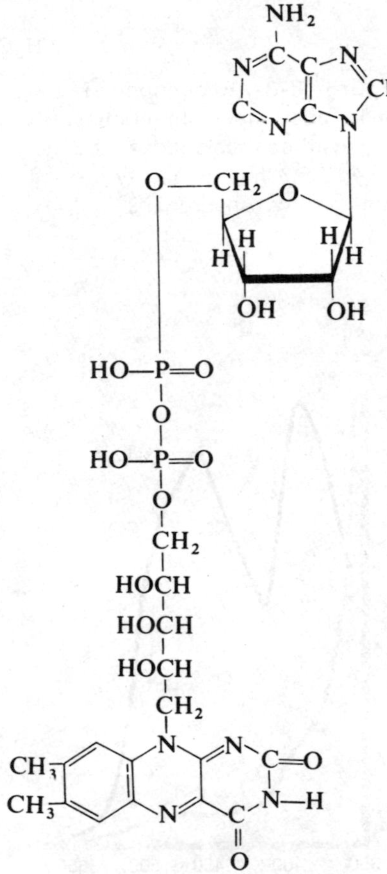

Flavin adenine dinucleotide (FAD)

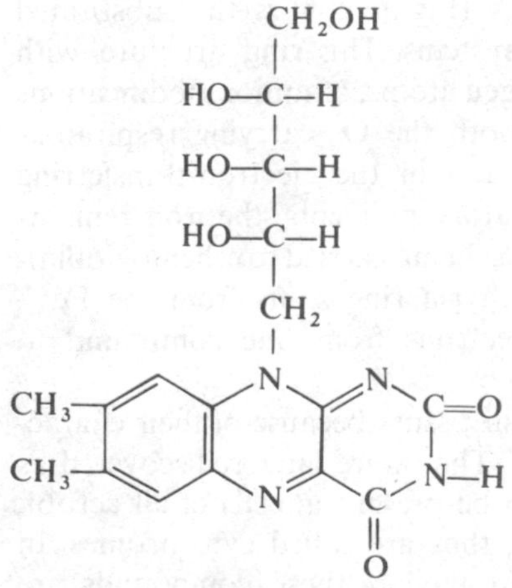

Riboflavin

Riboflavin   For many years, yellow-fluorescent pigments had been identified in biological material, and in 1933 a number of groups demonstrated that one of these pigments was a growth-promoting substance for the rat. The factor was identified and synthesized by 1935. The vitamin functions in metabolism as part of the coenzymes FMN and FAD. A deficiency of the vitamin causes numerous types of eye disorders, general atrophy of epidermis tissue, and specific lesions of mouth and tongue. A degeneration of nervous tissue is also seen in many species. Good dietary sources of the vitamin are milk, animal protein in general, and organ meats. The adult human requirement is 1.8 mg per day.

193

The flavin coenzyme can exist in either an oxidized form or a reduced form, in which a pair of electrons and two hydrogens have been added to the isoalloxazine ring (Figure 8-3). The oxidized form is bright yellow,

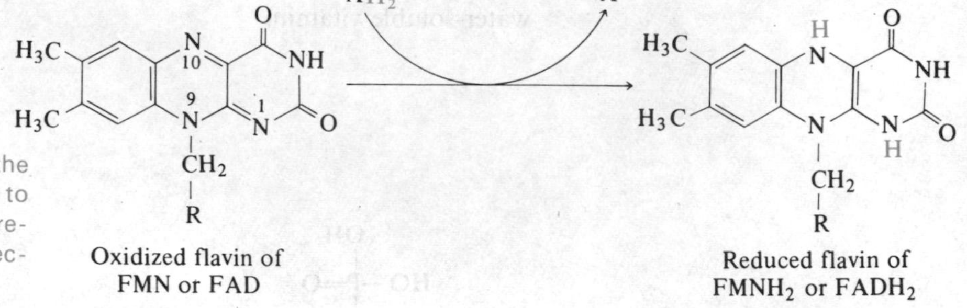

Figure 8-3 Reduction of the flavin coenzymes. In contrast to the pyridine nucleotides, the reduced form has gained two electrons and two hydrogens.

Oxidized flavin of
FMN or FAD

Reduced flavin of
FMNH$_2$ or FADH$_2$

as is the parent vitamin, and the reduced, or leuco, form is colorless. This change in absorption spectra (Figure 8-4) can be used to follow the extent of flavin-catalyzed reactions. In most cases the flavins are more tightly bound to proteins than are the pyridine coenzymes, but they can often be sufficiently dissociated by a shift in pH so that they can be dialyzed away from the protein.

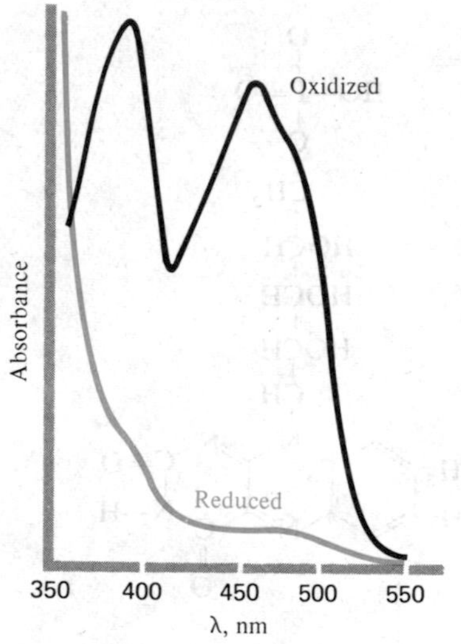

Hemoproteins

The third important class of compounds functioning as redox coenzymes is a class of hemoproteins called cytochromes. The prosthetic group of these proteins is derived from protoporphyrin IX, which is called heme or haeme when it has an Fe^{2+} ion chelated to it. Porphyrins are modifications of a parent tetrapyrrole compound, porphin, which has been substituted at the corners of the fused pyrrole rings by various side chains. The compound called protoporphyrin IX (Figure 8-5) is the substituted porphin most often found in biological systems. This ring structure, with an iron atom chelated to the pyrrole nitrogen atoms, or minor modifications of it, is the same compound found in both the O_2-carrying respiratory pigments, hemoglobin, and myoglobin, and in the electron-transferring pigments, the cytochromes. In the respiratory pigments, the iron remains in the ferrous state even when an O_2 atom is being carried (oxyhemoglobin); in the cytochromes the iron in the porphyrin ring shifts from the Fe^{2+} to the Fe^{3+} state while transferring electrons from one compound to another.

The cytochromes were first discovered in tissues because of their characteristic absorption spectrum (Figure 8-6). They were later rediscovered as a respiratory pigment which appeared to be present in cells of all aerobic organisms, and because of that property, they are called cytochromes. In contrast to the pyridine nucleotides and flavins, these compounds are

Oxidized

Reduced

350 400 450 500 550
λ, nm

Figure 8-4 Absorption spectra of the oxidized and reduced form of a flavoprotein. The wavelength of maximum absorption varies with different flavoproteins, and the oxidation state of these enzymes is followed by the change in absorption of the peak that is usually between 450–460 nm.

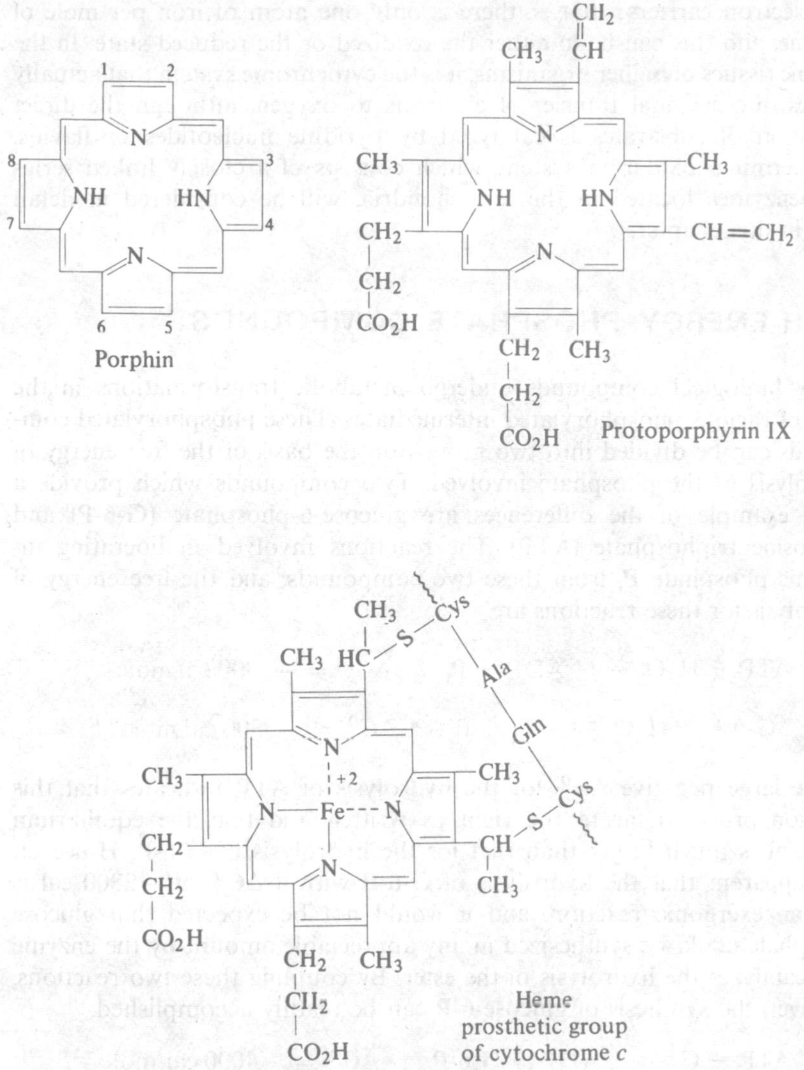

Porphin

Protoporphyrin IX

Heme
prosthetic group
of cytochrome c

Figure 8-5   Heme pigments. The tetrapyrrole, prophin, is the parent compound of a series of compounds called prophyrins. Prophyrins are classified on the basis of the number and types of constituents on the pyrrole rings. The protoporphyrins contain four methyl, two vinyl, and four propionic acid groups as side chains. Various isomers are possible, and protoporphyrin IX is the biologically important one. When an iron atom is chelated into this ring, the compound is called *heme* if it is Fe^{2+} and *hematin* if Fe^{3+}. In cytochrome *c* two of the vinyl groups have been modified to form thioether linkages to cysteine residues in the protein.

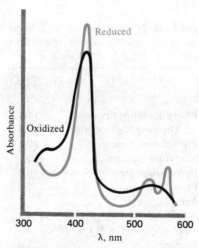

Figure 8-6   Absorption spectra of the oxidized and reduced form of cytochrome *c*.

one-electron carriers; that is, there is only one atom of iron per mole of enzyme, and this can be in either the oxidized or the reduced state. In the aerobic tissues of higher organisms, it is the cytochrome system that actually carries out the final transfer of electrons to oxygen, although the direct oxidation of substrates is catalyzed by pyridine nucleotides or flavins. This terminal oxidation system, which consists of a closely linked series of coenzymes located in the mitochondria, will be considered in detail later in this chapter.

HIGH-ENERGY PHOSPHATE COMPOUNDS

Many biological compounds undergo metabolic transformations in the form of various phosphorylated intermediates. These phosphorylated compounds can be divided into two groups on the basis of the free energy of hydrolysis of the phosphate involved. Two compounds which provide a good example of the differences are glucose-6-phosphate (G-6-P) and adenosine triphosphate (ATP). The reactions involved in liberating inorganic phosphate P_i from these two compounds, and the free energy of hydrolysis for these reactions are

$$ATP + H_2O \longrightarrow ADP + P_i \qquad \Delta G^{0'} = -7300 \text{ cal mole}^{-1}$$

$$G\text{-}6\text{-}P + H_2O \longrightarrow G + P_i \qquad \Delta G^{0'} = -3300 \text{ cal mole}^{-1}$$

The large negative $\Delta G^{0'}$ for the hydrolysis of ATP indicates that this reaction proceeds far to the right as written and that the equilibrium constant is much larger than that for the hydrolysis of G-6-P. However, it is apparent that the hydrolysis of G-6-P with a $\Delta G^{0'}$ of -3300 cal is also an exergonic reaction, and it would not be expected that glucose phosphate could be synthesized in any appreciable amount by the enzyme that catalyzes the hydrolysis of the ester. By coupling these two reactions, however, the synthesis of glucose-6-P can be readily accomplished.

$$ATP + G \longrightarrow ADP + G\text{-}6\text{-}P \qquad \Delta G^{0'} = -4000 \text{ cal/mole}^{-1}$$

This example illustrates the value in metabolic reactions of the class of compounds called high-energy phosphates or sometimes simply high-energy compounds. These compounds, which have a negative $\Delta G^{0'}$ of hydrolysis of >6500 cal/mole^{-1}, are for the most part acid anhydrides or other compounds which are chemically unstable. Figure 8-7 shows the chemical structures of the major types of high-energy compounds and examples of each that are important in metabolism. The currently acceptable values for the free energy of hydrolysis of these and some other compounds are listed in Table 8-3. It should be clear from these examples that a single compound may contain more than one type of phosphate ester bond. In 1,3-diphosphoglycerate, there is both a high-energy phosphate bond in

Table 8-3 Standard Free Energy of Hydrolysis of Some Important Bio-Chemical Compounds

	$\Delta G^{0'}$, kcal
Phosphoenolpyruvate	−14.8
1,3-Diphosphoglycerate	−11.8
Phosphocreatine	−10.3
Acetyl phosphate	−10.1
Pyrophosphate	−8.0
Phosphoarginine	−7.7
Acetyl-CoA	−7.5
ATP (to ADP + P_i)	−7.3
ADP (to AMP + P_i)	−7.3
Sucrose	−7.0
Glucose 1-phosphate	−5.0
Maltose	−4.0
Glucose 6-phosphate	−3.3

the form of a mixed anhydride, and a simple phosphate ester. The stepwise hydrolysis of ATP to ADP, AMP, and adenosine would involve two high-energy phosphate bond cleavages due to hydrolysis of the phospho-anhydride or pyrophosphate bonds, but the hydrolysis of AMP to phosphate and adenosine would be no different from the hydrolysis of a simple phosphate ester.

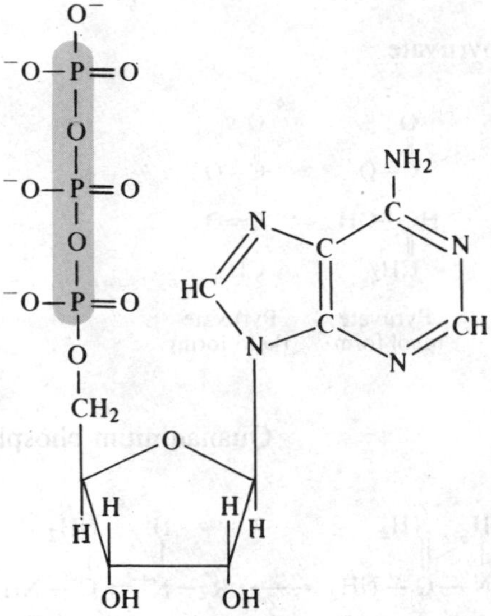

Adenosine triphosphate (ATP)
phosphoanhydride

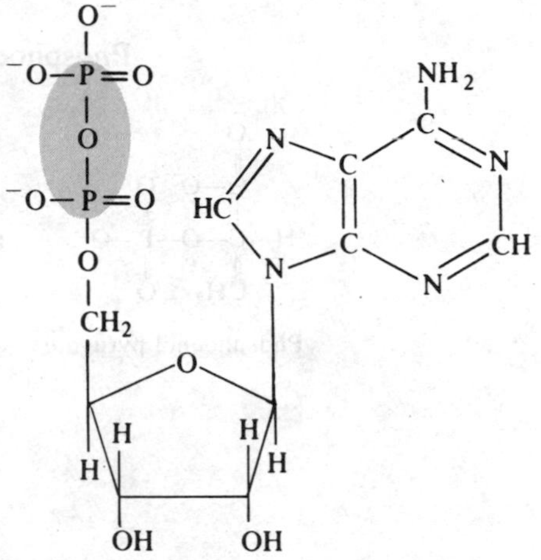

Adenosine diphosphate (ADP)
phosphoanhydride

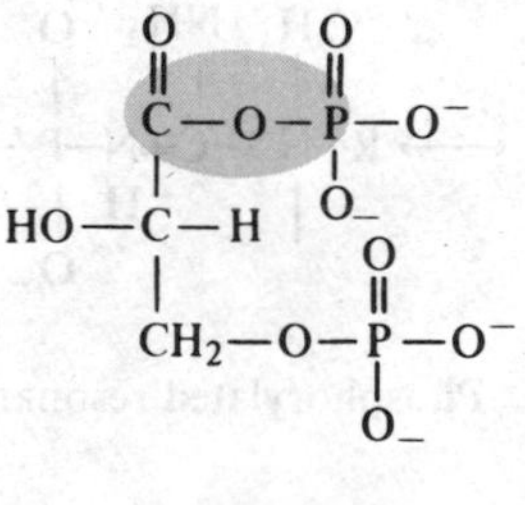

1,3–Diphospho-glyceric acid
mixed anhydride

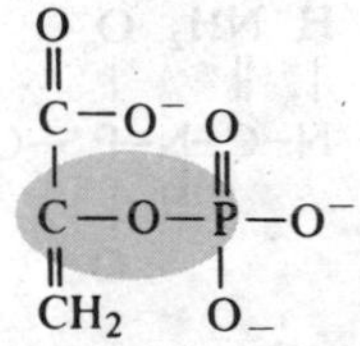

Phosphoenol-pyruvate
enolphosphate

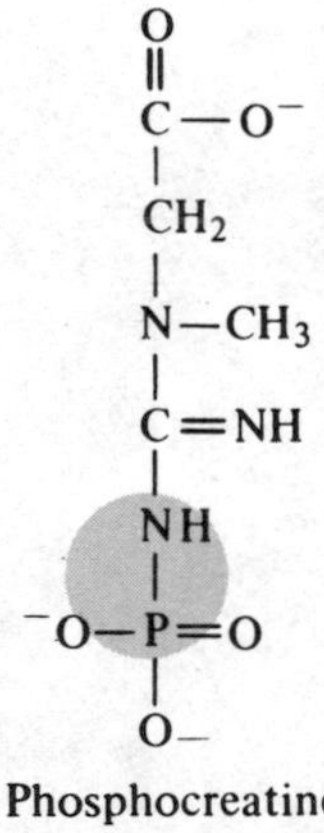

Phosphocreatine
guanidinium phosphate

Figure 8-7 Examples of compounds containing high-energy phosphate bonds.

The basis for the large negative $\Delta G^{0'}$ of hydrolysis of these compounds differs somewhat for the different types of compounds. In general, it is associated with either the release of some intramolecular strain in the compound by the hydrolysis or with a stabilization of the products of the hydrolysis by isomerization, resonance, or ionization. In most cases a number of factors contribute to the energy-rich nature of the molecule, and some of these are illustrated in Figure 8-8.

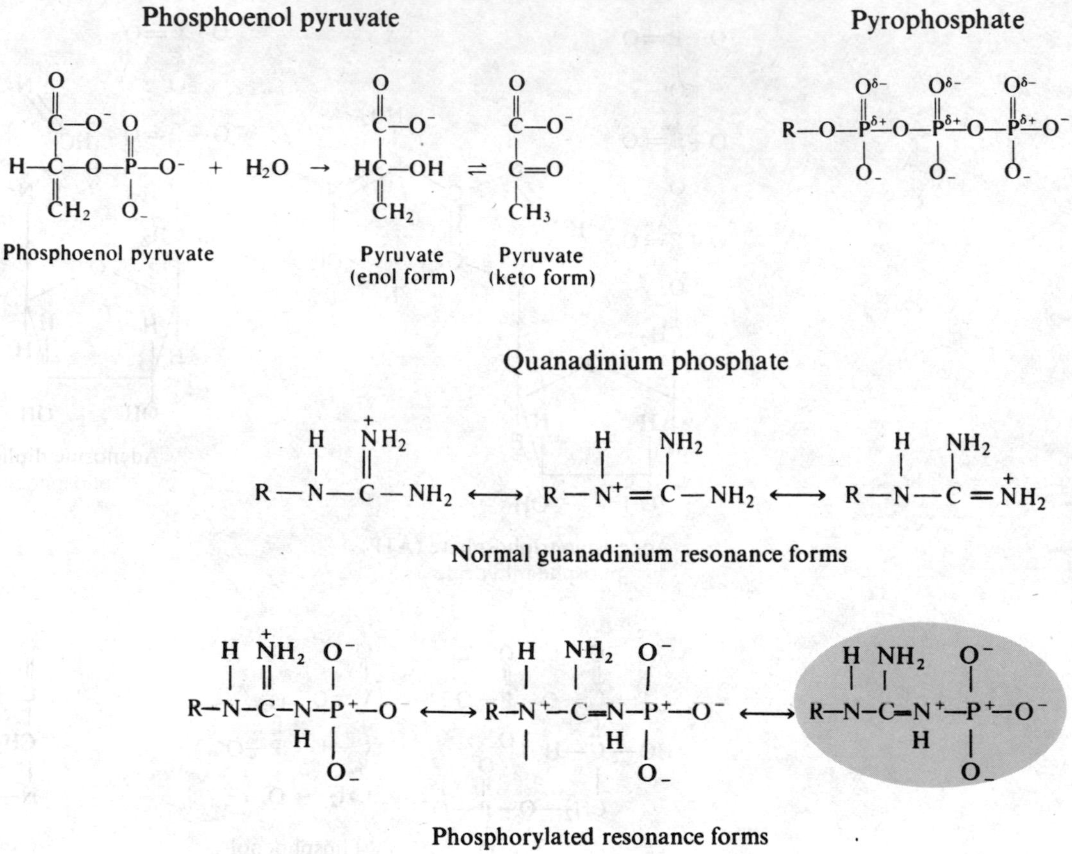

Phosphoenol pyruvate

Pyrophosphate

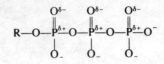

Phosphoenol pyruvate   Pyruvate (enol form)   Pyruvate (keto form)

Quanadinium phosphate

Normal guanadinium resonance forms

Phosphorylated resonance forms

Figure 8-8 Chemical basis for the large $-\Delta G^{\circ}$ of hydrolysis of some typical "high energy phosphate" compounds. In the case of phosphoenol pyruvate, the instability of the enol form leads to its complete conversion to the keto form, and removal of the product of hydrolysis forces the equilibrium to the right. The partial positive charges on the phosphorus atoms in pyrophosphate contribute to an instability of the oxygen-phosphorus bond and favor its hydrolysis. The phosphorylation of the quanidinium group of creatinine and arginine would place a positive charge on an adjacent phosphorus and nitrogen atom of one possible resonance form and drastically decrease its contribution to the resonance hybrid. Therefore, the unphosphorylated form has a much higher resonance energy and would be strongly favored.

It should be clear that these compounds do not possess, as is often stated, high-energy phosphate bonds in the usual chemical sense. The use of the term implies nothing about chemical bond energy; rather, the designation refers to the properties of the entire molecule. Because of this it has often been proposed that the term group transfer potentials, which are simply -1 times the $\Delta G^{0'}$ in kcal, or some similar designation, be used. These terms have never come into common usage.

Most tables of free energy of hydrolysis of high-energy compounds which are found in biochemical references are $\Delta G^{0'}$ values at pH 7, but the values in different tables seldom agree completely. This is because they have been calculated with somewhat different assumptions of temperature and Mg^{2+} concentration. The value of -7.3 for the free energy of hydrolysis of ATP used here is for pH 7.0, an excess of Mg^{2+}, and a temperature of $37°C$. The actual free energy available in the cell from the hydrolysis of ATP may be considerably different. A compound such as ATP has a pK_a near pH 7 on the phosphate group, and the free energy change associated with the hydrolysis will, therefore, be very pH-dependent. The actual concentration of the reacting species of ATP will change as the pH shifts slightly. The steady-state concentration of ATP and ADP in the cell are also important, as the free energy change of a reaction is a function of both the standard free energy change and the actual ratio of products to reactants. A shift in the ratio of product to reactant by a factor of 10 will result in a ΔG change of about 1400 cal, so that the energy available to the cell from the hydrolysis of a mole of ATP depends on both pH and concentration. These relationships are illustrated in Figure 8-9. When all these factors are considered, it can be calculated that the free energy of hydrolysis of ATP in the cell is probably about -12.5 kcal.

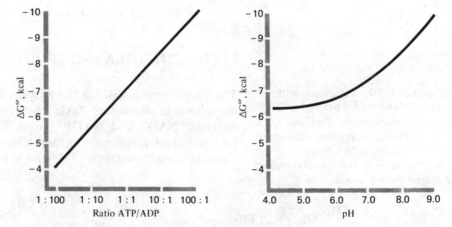

Figure 8-9   Effect of ATP/ADP ratio and of pH on the free energy of hydrolysis of ATP.

ENERGY CONVERSION IN BIOLOGICAL SYSTEMS

The close link between the reactions of energy metabolism and those of phosphate metabolism was recognized by some of the earliest workers in this field. In the early 1900s, Harden was able to show that phosphate was needed to support glycolysis in disrupted yeast cell preparations and that some phosphorylated sugars and other phosphorylated compounds were being accumulated. Later workers, studying the aerobic oxidation of compounds such as citrate, pyruvate, glutarate, and malate, noted that the oxidation of those compounds was stimulated by phosphate, and that if the correct type of preparation was obtained, the oxidation was dependent on phosphate. It was found that the phosphate must be involved in those steps which are responsible for the transfer of electrons from the oxidizable substrate to O_2 and that ATP was produced from ADP by a process which came to be called oxidative phosphorylation. It was shown that oxidative phosphorylation could be followed by determining the ratio of the moles of phosphate converted to organic forms to the atoms of oxygen taken up in the reaction. These measurements, called P:O ratios, were found to vary with the substrate used (Table 8-4). As the systems used were refined, it was shown that the reactions were localized exclusively in the mitochondrial fraction of the cell and that the variations in P:O ratio depended on which coenzyme was involved with the oxidation of the individual substrates.

Table 8-4 P:O Ratios for Mitochondrial Oxidations

Substrate	Product	Coenzyme	P:O
Succinate	Fumarate	FAD	2
Pyruvate	Acetyl CoA	NAD^+	3
Isocitrate	α-Ketoglutarate	NAD^+	3
Malate	Oxaloacetate	NAD^+	3
α-Ketoglutarate	Succinate	NAD^+	4^a

^{a} One of these is a substrate level phosphorylation.

ELECTRON TRANSPORT CHAIN

The system responsible for the reoxidation of reduced coenzymes by the mitochondria consists of flavin coenzymes; the pyridine nucleotide coenzymes, NAD^+ and $NADP^+$; iron in the form of the cytochromes; protein-bound nonheme iron and Cu; and an additional lipid-soluble electron carrier, coenzyme Q (Figure 8-10). The overall reaction mediated

Figure 8-10 Oxidized and reduced forms of coenzyme Q. In most mammalian tissues the coenzyme Q has ten isoprenoid units (CoQ_{10}), but can have less in other organisms. The compound is also called ubiquinone.

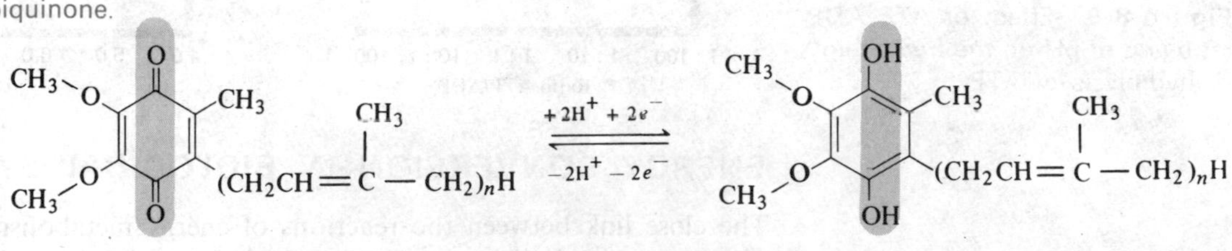

Oxidized coenzyme Q   Reduced coenzyme Q

by this series of mitochondrial enzymes is the transfer of electrons from NADH to O_2. Each component of this system can exist in an oxidized and reduced form, and each is arranged generally in order of increasing redox potential. The total energy that is potentially available from this reaction is indicated in Figure 8-11.

$$NADH + H^+ + 1/2O_2 \longrightarrow NAD^+ + H_2O$$

The half-reactions involved are

$$1/2O_2 + 2H^+ + 2e^- \longrightarrow H_2O \qquad E_0' = 0.816$$

$$NAD^+ + 2H^+ + 2e^- \longrightarrow NADH + H^+ \qquad E_0' = -0.320$$

The energy available is

$$\Delta G^{0'} = n \mathscr{I} \Delta E_0'$$
$$= (-2)(23,063)(1.136)$$
$$= -52,400 \text{ cal}$$

Figure 8-11 Energy available from the reoxidation of NADH by oxygen.

The general scheme of enzymes and electron carriers required for the transfer of electrons through the electron transport chain of the mitochondria is diagrammatically illustrated in Figure 8-12; the manner in

NAD-linked
dehydrogenases

Flavoprotein-linked
dehydrogenases

$$NAD^+ \rightarrow \underset{\text{(nh Fe)}}{\text{Flavoprotein}} \rightarrow CoQ \rightarrow \underset{\text{(nh Fe)}}{\text{Cyt}_b, \text{Cyt}_{c_1}} \rightarrow Cyt_c \rightarrow \underset{\text{(Cu Prot)}}{\text{Cyt}_{aa_3}} \rightarrow O_2$$

Figure 8-12 Respiratory chain. Electrons are transferred from one of these electron carriers to the other, alternatively oxidizing and reducing them until O_2 is reduced to H_2O. Some of these transfers are also associated with the oxidation or reduction of sulfur containing nonheme iron proteins (nhFe) or Cu containing proteins which are complexed to the cytochromes. Although this general order of the carriers is well established, there is still some question about the specific transfer of electrons in the vicinity of coenzyme Q. In addition to the flavoprotein which links NAD to coenzyme Q, there are also flavoproteins associated with fatty acid and succinate oxidation which feed electrons into the chain at the level of coenzyme Q. The membrane bound NAD can be reduced by NADH which can arise from a number of pyridine nucleotide-linked dehydrogenases in the mitochondrion.

which the components are alternatively oxidized and reduced and the redox potentials of the components involved are indicated in Figure 8-13. From these diagrams it can be seen that two types of flavoprotein systems are involved, both of which feed electrons into coenzyme Q. This coenzyme serves as the link between those coenzymes which require two electrons for their oxidation and reduction (the pyridine nucleotides and flavins) and the cytochrome system, which is a single electron carrier. The enzymes of the terminal portion of this chain, the cytochrome a and a_3 complexes, pass electrons on to O_2 and are also called the cytochrome oxidase system.

Figure 8-13   Electron carriers of the respiratory chain. The diagram on the left indicates the redox potential of the various electron carriers present in the chain. The cyclic manner in which each pair is oxidized and reduced by the adjacent members of the chain is illustrated at the right in the figure. Note that the oxidation and reduction of NAD, the flavins, and coenzyme Q involve both electrons and hydrogens, while the cytochromes transfer only electrons. The overall chain balances because the protons liberated at the interaction between coenzyme Q and cytochromes are balanced by the utilization of protons at the level of cytochrome oxidase.

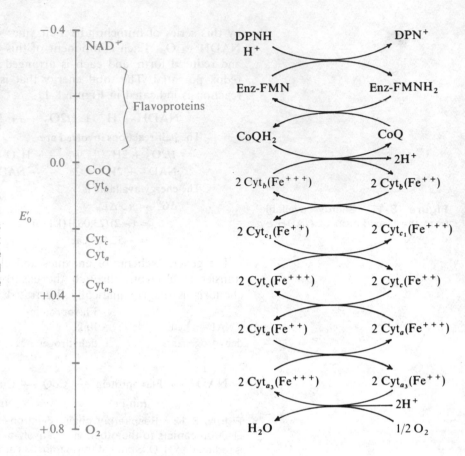

OXIDATIVE PHOSPHORYLATION

The total energy available from the reoxidation of NADH by the electron transport chain was calculated in Figure 8-11 to be a $-52,400$ cal/mole. The utilization of this energy to form ATP by the phosphorylation of ADP is the process called *oxidative phosphorylation*. The formation of 3 moles of ATP between NADH and O_2 is equivalent to a total $\Delta G^{0'}$ of three times 7.3 kcal or 21.9 kcal out of the potentially available 52.8 kcal for this oxidation. This amounts to conservation of about 40 percent of the energy which can be calculated to be potentially available under standard conditions. If the steady-state conditions which exist in the mitochondria are taken into consideration, the actual efficiency may be considerably higher than this.

Sites of Phosphorylation

The determination of the order of the different electron carriers in the chain and the elucidation of the sites of phosphorylation have not been a simple matter. One approach to the problem has been the determination of the redox potentials of different redox pairs so that their relative order in the electron-transport chain could be calculated. These values also give some indication of the possible phosphorylation sites. For a segment of the electron-transport chain to constitute a phosphorylation site, the energy available must be in excess of that associated with the hydrolysis of ATP. A second approach has been to use inhibitors that block electron transfer at specific points, or substrates that will feed in electrons to the chain at certain points. This approach has allowed investigators to isolate for study specific parts of this complex system.

A number of distinctly different classes of chemical compounds will interfere with the normal capacity of the electron chain to carry out oxidative phosphorylation. One class, called uncoupling agents, consists of compounds like 2,4-dinitrophenol (DNP). These compounds do not inhibit the oxidation of NADH by the electron-transport chain, but they do "uncouple" the process from ATP production. Other agents, including many antibiotics, are true inhibitors of specific segments of the chain. Antimycin, for example, specifically prevents the transfer of electrons from cytochrome b to cytochrome c. The use of this agent causes all those electron carriers in the chain prior to cytochrome c to become completely reduced, whereas cytochrome c and those electron carriers after it are all in the oxidized form. The barbiturate, amytal, and an insecticide, Rotenone, will block the flow of electrons from the NADH-linked flavoproteins to coenzyme Q, and compounds such as cyanide and carbon monoxide have long been recognized as inhibitors of the final transfer of electrons to oxygen by the cytochrome oxidase system. Another antibiotic, oligomycin, is of a different type; it appears to act directly at the phosphorylation-coupling sites rather than on the electron chain. By the use of these specific inhibitors and by feeding electrons in at different sites and through the use of the NAD or flavin-linked dehydrogenases, it has been possible to isolate segments of the chain and determine if phosphorylation still occurs. A combination of these approaches has led to an understanding of the position of the phosphorylation sites indicated in Figure 8-14. As the first of these sites involves the transfer of electrons from NADH to FAD, it is clear why a substrate such as succinate, which feeds electrons into the chain at the level of a flavin coenzyme, produces only two ATPs/atom of oxygen rather than the three produced by a NADH-linked substrate.

Figure 8-14 Electron transport and energy conservation. The sites in the electron transport chain where ADP is phosphorylated, as well as the sites at which the common inhibitors of electron transport have been shown to interact, are indicated. The energy available ($\Delta G°$) at each of the phosphorylation sites is also indicated.

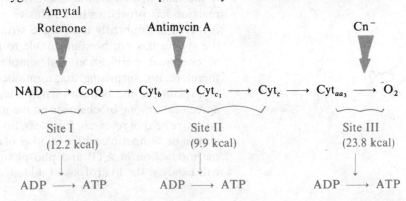

Respiratory Control

If mitochondria are carefully prepared and handled, the flow of electrons is closely coupled to the phosphorylation events and respiration is inhibited in the absence of ADP, which is required as an acceptor of the high-energy phosphate formed. This phenomenon can be most easily demonstrated if oxygen consumption in a mitochondrial suspension is continuously monitored by the use of an oxygen electrode. Under these conditions the addition of ADP to mitochondrial suspension (Figure 8-15) results in a

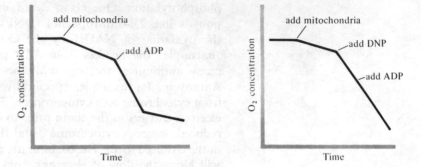

Figure 8-15   Effect of ADP and dinitrophenol on the rate of mito-chondrial respiration.

rapid increase in the rate of oxygen consumption, which continues until the added ADP has been converted to ATP. A value called the *respiratory control ratio* can be calculated as the ratio of the rates of oxygen consumption in the presence and absence of ADP. As both ADP added and O_2 consumed can be measured in this type of experiment, P:O ratios can also be calculated from the same data. The examples in Figure 8-15 also illustrate that in the presence of an uncoupling agent such as 2,4-dinitrophenol, the addition of ADP has little if any effect on the rate of respiration, because electron flow is no longer controlled by the phosphorylation events.

MECHANISM OF OXIDATIVE PHOSPHORYLATION

A mechanism that will adequately explain the oxidative phosphorylation reaction has proved to be an elusive one for the biochemist. The enzymes involved undoubtedly need some structural integrity for their action, and the system has not been amenable to the standard biochemical technique of continual purification and simplification of complex systems. It is, therefore, not surprising that a number of hypotheses have been put forth to explain the mechanism of oxidative phosphorylation. There is still no agreement among biochemists on the nature of the reactions, and it remains an active field of research. It is clear, however, that whatever the mechanism, it must be compatible with a number of experimental observations regarding the interaction of ATP and phosphate with mitochondria. Mitochondria will catalyze the hydrolysis of added ATP, ATPase activity, and will also catalyze a number of isotope exchange reactions, which are illustrated in Figure 8-16. All the reactions shown in Figure 8-16 are inhibited by oligomycin, and the exchange reactions are also inhibited by 2,4-dinitrophenol.

Figure 8-16 Three isotope exchange reactions and the ATPase activity that are catalyzed by mitochondria. Any mechanism postulated to explain oxidative phosphorylation must be consistent with these reactions.

1. ATPase activity

$$ATP + H_2O \longrightarrow ADP + P_i$$

2. ATP-^{32}P exchange

$$AMPPP + P^* \rightleftharpoons AMPP\,P^* + P$$

3. ATP-ADP exchange

$$ATP + ADP^* \rightleftharpoons ATP^* + ADP$$

4. HPO$_4^-$-^{18}O exchange

$$HPO_4^- + H_2{}^{18}O \rightleftharpoons HP^{18}O_4^- + H_2O$$

The ATPase activity is strongly stimulated by DNP, and is very active in damaged mitochondria which have lost their ability to demonstrate respiratory control, or in mitochondrial fragments. When mitochondria have been treated so that they have lost the ability to catalyze these exchange reactions, they have also lost the ability to carry out oxidative phosphorylation, and these reactions are often considered to represent partial reactions of the enzymes responsible for the phosphorylation events.

Chemical Coupling Hypothesis

The oldest hypothesis regarding the mechanism of oxidative phosphorylation is one which postulates a series of chemical intermediates between the point at which electrons are transferred between components of the electron transport chain and the formation of ATP. A general mechanism

Reaction 1.	$AH_2 + B + X \rightleftharpoons A + BH_2 \sim X$
Reaction 2.	$BH_2 \sim X + Y \rightleftharpoons X \sim Y + BH_2$
Reaction 3.	$X \sim Y + P_i \rightleftharpoons Y \sim P + X$
Reaction 4.	$Y \sim P + ADP \rightleftharpoons Y + ATP$
Net reaction	$AH_2 + B + ADP + P_i \longleftarrow A + BH_2 + ATP$

Figure 8-17 Postulated mechanism for oxidative phosphorylation.

of this type is shown in Figure 8-17. In this scheme, A and B could be considered to be any of the electron carriers, and $BH_2 \sim X$ would be assumed to contain a high-energy bond at about the same energy level as ATP. Compounds X and Y are purely hypothetical intermediates whose chemical nature has not been determined. In a hypothetical scheme like this, an uncoupling agent such as dinitrophenol is assumed to act by reacting with and splitting $X \sim Y$, while the antibiotic oligomycin would act to prevent the formation of $Y \sim P$ but would not split the $X \sim Y$ bond. There is considerable evidence that the mitochondrion can use the hypothetical high-energy intermediate, $X \sim Y$, for purposes other than ATP formation, and it is apparently the source of energy for ion transport across the cell membrane. This type of mechanism is also supported by the observation that under some conditions the electron transport chain can be driven backward. If succinate is added to oligomycin-blocked mitochondria, NAD^+ can be reduced. This would be consistent with the formation of $X \sim Y$ (reactions 1 and 2) by succinate oxidation, and because oligomycin is blocking the formation of $Y \sim P$ (reaction 3), the energy from the breakdown of $X \sim Y$ can be used to drive the reduction of NAD^+.

Although this theory is accepted by a large number of biochemists, there is little direct experimental evidence to support it. It has not yet been possible to isolate any of the hypothetical intermediates in these reactions. A number of compounds have been proposed, but have all eventually been ruled out as true intermediates in the pathway. It can, however, be argued that these high-energy intermediates would be very unstable and difficult to isolate, and this hypothesis does explain the various isotope-exchange reactions which are observed.

Chemiosmotic Hypothesis

The second postulated mechanism of oxidative phosphorylation which has received serious consideration is the chemiosmotic theory. This theory, originally proposed by Mitchell, assumes that the ATPase activity of the mitochondria is due to an enzyme located in the mitochondrial membrane, and that this enzyme is isolated from the aqueous phase on both sides of the membrane by the lipid components of the membrane. The basic ATPase reaction

$$ATP + H_2O \longrightarrow ADP + P_i$$

can be written as two half-reactions

$$ATP + H^+ + OH^- \longrightarrow ADP + P_i$$

and

$$H_2O \longrightarrow H^+ + OH^-$$

If it is assumed that the ATPase is accessible to H^+ from one side of the membrane and OH^- from the other side, then ATP hydrolysis will be accompanied by a movement of protons from one side of the membrane to the other. If splitting of ATP is caused by a movement of protons in one direction, synthesis of ATP would result from movement in the other; and if it is possible to form a gradient of protons across the membrane, synthesis of ATP could result. It is an experimental observation that respiring mitochondria can be made to extrude protons from the inner membrane and it is assumed that the components of the electron transport chain are physically located in the mitochondrial membrane such that the passage of electrons through the components accomplishes this trans-

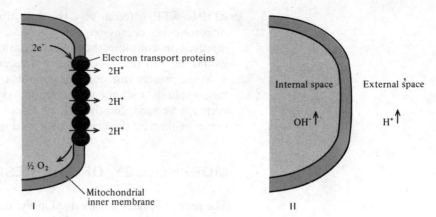

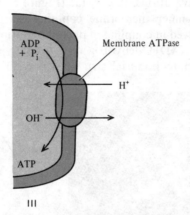

Figure 8-18 Basic components of a chemiosmotic theory of oxidative phosphorylation. The passage of electrons through the electron transport chain extrudes protons (I) to the outer surface of the inner mitochondrial membrane which results in an increased concentration of H^+ on the outside and in an increased concentration of OH^- on the inside of this membrane (II). The movement of protons back across the membrane (III) is then coupled to a reversal of the mitochondrial ATPase reaction to form ATP by the phosphorylation of ADP.

location. This hypothesis is diagrammatically illustrated in Figure 8-18. In this theory the action of the various chemical uncoupling agents is postulated to be that of allowing protons to diffuse back across the membrane without interacting with the ATPase. It is now felt that the hydrogen ion gradient generated in the mitochondria is not sufficient to account for the amount of ATP produced, and theory has been expanded to include the possibility that gradients of other ions also contribute to chemical potential needed to drive ATP formation.

Mitochondrial Structural Hypothesis

It is possible to show that the ultrastructural arrangement of the mitochondrial inner membrane is dependent on this respiration state. In rapidly phosphorylating mitochondria the inner membranes are in a highly condensed state compared to the more orthodox form generally seen. It has been proposed that the process of electron transport produces an energized state owing to conformational changes in the inner membrane, and that the transition to the alternate conformation results in the conversion of

ADP to ATP. Alternatively, the transition could result in ion accumulation or some other energy-requiring process. It is not clear what the mechanism involved in coupling these conformational changes to chemical changes would be, and this theory has not received the support of the other two.

At the present time it is not possible to determine which of these theories best explains the phenomenon of oxidative phosphorylation, and it is certainly conceivable that a new hypothesis will be developed which will better explain all the present observations.

MORPHOLOGY OF THE RESPIRATORY CHAIN

The series of reactions involved in the transfer of electrons between NADH and O_2 and the coupling of this transfer to ATP formation occur in the mitochondria. Specifically, the inner membrane of the mitochondria and its involutions, the cristae (Figure 8-19) have been shown to be involved. This inner membrane contains the cytochrome system, those enzymes involved in coupling of phosphorylation to electron transport, the succinate and NADH dehydrogenase activity, as well as a large amount of lipid and mitochondrial structural protein. The mitochondrial matrix contains

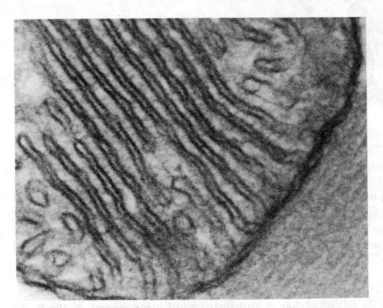

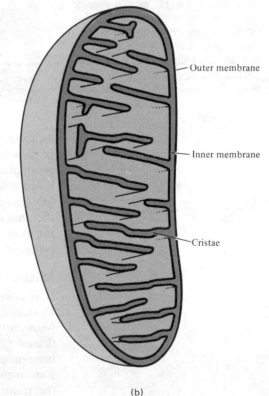

Outer membrane

Inner membrane

Cristae

(a)

(b)

Figure 8-19 (*a*) Electron micrograph of a thin section of a beef heart mitochondrion. (Courtesy of D. E. Green) (*b*) Schematic drawing of a heart mitochondrion showing the three-dimensional arrangement of the membranes.

those enzymes of the Krebs cycle (Chapter 13) which generate the intra-mitochondrial NADH for ATP production. Although mitochondria will oxidize exogenous NADH, the mitochondrial membrane is relatively impermeable to it, and they must be specially treated to fully demonstrate this activity.

Submitochondrial Particles

An understanding of the nature of oxidative phosphorylation has been hampered by the difficulty in obtaining submitochondrial preparations which will carry out the individual reactions involved. It is now possible to isolate small fragments of the mitochondria which will carry out electron transport. These can be subfractionated by detergent treatment and sonication into a series of four lipoprotein complexes containing only portions of the electron-transport chain. These complexes, together with coenzyme Q and cytochrome *c*, which is the one enzyme in the chain that is easily extracted as a soluble protein, can be recombined to form a fully functional electron transport chain. These particular preparations of particles do not contain the enzymes needed for coupling electron flow to ATP formation. It has also been possible to remove from the mitochondria specific proteins needed for phosphorylation but not for electron transport. In some cases these *coupling factors* can be added back to submitochondrial particles to restore their ability to phosphorylate ADP. One of these factors, called f_1, is a 200,000 molecular weight protein which has ATPase activity, and another, called f_0, is a protein which confers oligomycin sensitivity to the system.

High-resolution electron microscopy has revealed that the inner membrane of the mitochondria is not smooth surfaced, but that it is lined with spherical particles, about 80 Å in diameter, which appeared to be connected to the inner membrane by a short stalk (Figure 8-20). At one time it was believed that these contained the entire complex of respiratory enzymes. It would now seem that these enzymes are contained in the membrane itself and that the spherical particles which are seen contain coupling factor f_1, the enzyme responsible for the coupling of electron transport to phosphorylation.

MICROSOMAL ELECTRON TRANSPORT

Not all the cellular cytochromes and flavoproteins are located in the mitochondria, and there are short electron transport chains in the microsomes. These respiratory chains are involved in complex hydroxylations catalyzed by enzyme systems called *mixed function* oxidases. In these systems, NADPH is oxidized through a flavoprotein cytochrome chain, and oxygen is used as an electron acceptor. However, only one of the atoms of oxygen is reduced to water, and the other is used to hydroxylate some acceptor molecule (Figure 8-21). A particular heme protein called

Figure 8-20 Electron micrograph (*a*) of a preparation of cristae from heart mitochondria showing the spheres attached to the membrane (low magnification); and (*b*) of a high magnification of such a preparation showing more detail of the spherical projections. (Courtesy of D. E. Green)

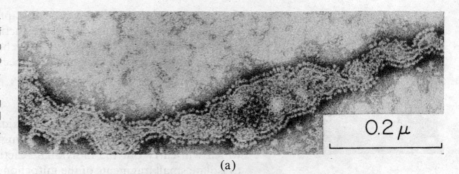

0.2 μ

(a)

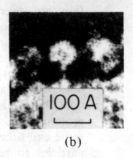

100 A

(b)

cytochrome P-450 is involved in most of these systems, which are often also characterized by their sensitivity to carbon monoxide. These enzymes are responsible for the metabolism of a large number of drugs by the liver, and are also involved in steroid hormone biosynthesis and formation of unsaturated fatty acids.

MEMBRANE TRANSPORT—ACTIVE TRANSPORT

Biological membranes are semipermeable; only certain molecular species can cross such membranes. In general, highly polar or charged compounds do not diffuse through a lipid bilayer, presumably because they are insoluble in the highly nonpolar region of the lipid side chain. One exception to this is water, which apparently is freely diffusible across either model lipid membranes or real biological membranes. There are several means by which a solute can cross a biological membrane. In *simple diffusion*, a compound which is relatively soluble in the lipid region of the bilayer dissolves in it and diffuses across it. In contrast with simple diffusion, *mediated transport* is a process which is both specific and saturable. Only specific stereoisomers of a given compound are transported, and the rate of transport is a saturable function of substrate concentration in the same manner that an enzymatic reaction is saturable. Mediated transport may be divided in turn into two types of processes: facilitated (passive) transport

Figure 8-21 Postulated mechanism of action of a mixed function oxidase. Electrons are passed from NADPH through flavoproteins and nonheme iron containing proteins (nh(Fe^{++})prot) to specialized cytochromes. When the electrons are transferred to O$_2$, one of the atoms is reduced to an OH$^-$ and the other is used to hydroxylate some acceptor molecule.

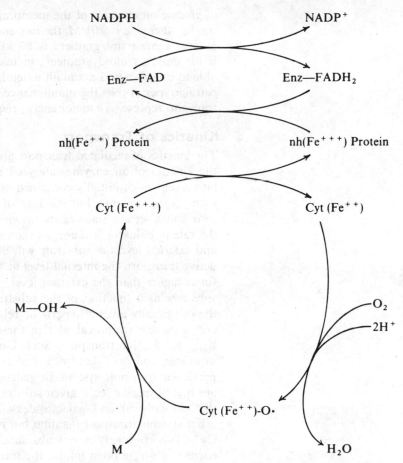

and active transport. *Facilitated transport* is catalyzed flux across the membrane from the side with higher concentration to the side with lower; the flux continues until the concentration of solute is equal on both sides of the membrane. *Active transport* represents the "uphill" carrying of solutes across the membrane against a concentration gradient, resulting in an increased trans-membrane concentration difference. The distinction between these two processes is best seen in a consideration of the energetics of transport.

Energetics of Transport

Any process in which a solute moves from a region of low concentration to one of higher concentration occurs with a decrease in entropy; and the process will occur only if energy is supplied to the system. The amount of free energy involved in the transport of a mole of uncharged solute from one side of a membrane to another is given by the expression $\Delta G = RT \ln C_2/C_1$, where C_2 is the higher concentration. For example, if glucose is to be actively transported against the gradient where the concentration

211

of glucose on one side of the membrane is 0.001 M and the concentration on the other side is 0.01 M, the free-energy change involved in transporting glucose against this gradient is 1.3 kilocalories per mole of glucose. This is for only a tenfold gradient; in many cases biological membranes are able to establish and maintain much larger gradients than this. In fact, for certain types of cells the maintenance of concentration gradients by active transport represents a major energy requiring reaction in the cell's economy.

Kinetics of Transport

The kinetics of mediated transport are in many respects comparable with the kinetics of an enzyme-catalyzed reaction. The uptake of a substrate into a cell which initially contained none of that substrate is at first linear with respect to time, but the rate of net uptake eventually decreases to zero and a steady state exists in which the rate of influx exactly equals the rate of efflux or leakage. In the case of passive transport, the internal and external levels of substrate will be equal in the steady state, but with active transport, the internal level at steady state may be several or many times higher than the external level. A plot of the initial rate of entry of substrate as a function of the substrate concentration on the outside of the cell usually gives a hyperbolic relationship such as that in Figure 7-7, and a double reciprocal plot of these same data (see Figure 7-8) yields V_{max} and K_m for transport. Such kinetic studies show that transport is saturable, and provide a means of evaluating the relative specificity of a particular transport system. In general, it is found that transport systems are highly specific for a given substrate. For example, a transport system that normally carries β-galactosides will not transport an α-galactoside, and other systems transport L-amino but not D-amino acids, or Mg^{++} but not Ca^+. This specificity is not absolute, however, and close analogs of the correct substrate often inhibit the transport of the real substrate. Another type of inhibition seen in catalyzed transport systems is that which is caused by reagents which react covalently with proteins. Inhibitions by substrate analogs, or by protein-derivatizing reagents, would not be expected if transport occurred by simple diffusion, but are expected for a protein-mediated transport.

These kinetic and energetic considerations provide the criteria for distinguishing biological transport from simple diffusion. Unlike simple diffusion, mediated transport is (1) stereospecific, (2) saturable, and (3) inhibitable by substrate analogs or by reagents which react with proteins. In the case of active transport, (4) energy is required for the establishment of a concentration gradient.

SPECIFIC TRANSPORT SYSTEMS

Biological transport systems for many different substrates in a wide variety of organisms and cell types have been characterized to some degree, and three particularly well-studied systems can be used to illustrate several general features of transport.

Lactose Transport in *E. coli*

In the early 1950s, Cohen and Rickenberg isolated mutants of the bacterium *Escherichia coli*, which possessed all the enzymes required for lactose metabolism, but which would not use lactose provided in their growth medium. Such "cryptic" mutants proved to be unable to take up lactose—they were defective in the *lactose permease*. All such mutants mapped in a single gene (the *y* gene of the lactose operon) (Figure 17-10) which is responsible for lactose permease production.

The lactose transport system can be assayed by measuring the uptake of radioactive lactose added to the suspending medium, but such assays are complicated by the fact that, in normal cells, soon after lactose enters, it is metabolized by conversion into glucose and many of its catabolites. To circumvent this difficulty, it is possible to use an analog of lactose, thio-digalactose, or TDG, which is recognized and taken up by the transport system but cannot be further metabolized and thus accumulates inside the cell as unaltered TDG. The time course of uptake of TDG is shown in Figure 8-22. Uptake and accumulation are rapid, and finally a steady-state level of internal TDG is reached which is approximately 100 times higher than the external level of the sugar. That is, a 100-fold gradient can be established by this pump. In the presence of an energy poison such as cyanide, which prevents the cells from producing energy from electron transport, very slow TDG uptake occurs, and no net accumulation can be seen. Only equilibration of internal and external TDG occurs under "poisoned" conditions. Similarly, if unpoisoned cells are first allowed to accumulate TDG to their final extent and then poisoned with cyanide, the accumulated TDG is quickly given up to the medium. Thus, the cells deprived of energy-yielding metabolism can neither establish nor maintain a concentration gradient of the transported substrate. The product of the *y* gene in the lactose operon, the lactose permease protein which has been partially purified and characterized by Kennedy and his co-workers, is a

Figure 8-22 Uptake of the lactose analog, thiodigalactoside (TDG) by the bacterium *E. coli*. The amount of TDG inside the cell can be measured as a function of time in the presence or absence of metabolic poisons such as NaN_3 or iodoacetate. When the cells are poisoned after they have accumulated a steady-state level of TDG, the concentration rapidly falls to the level present in the media.

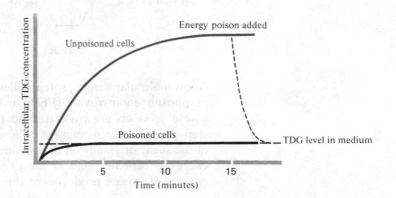

relatively small protein of molecular weight about 35,000, and it is apparently the only protein required for lactose transport. This single, relatively small protein, therefore, carries out three important functions: it *recognizes* lactose, the specific substrate for this permease system; it catalyzes the *mediated transport* of lactose across the membranes; and it somehow *couples the accumulation* of lactose or lactose analogs to *energy-yielding metabolism* of a cell, allowing "uphill" transport. In the absence of metabolic energy, the lactose permease is able still to catalyze the facilitated diffusion of lactose across the membrane, but in this case no net accumulation of lactose can occur. Once lactose has entered a cell through this system, it can be further metabolized, but its metabolism is not essential to its transport; the very first thing that accumulates within cells is free lactose, not a derivative. In this respect the lactose transport system differs significantly from the phosphotransferase system.

Phosphotransferase System in *E. coli*

When cells of *E. coli* are placed in a medium containing radioactive glucose, the very first product of glucose uptake which appears inside cells is not free glucose, but is instead glucose-6-phosphate. Careful studies by Roseman and Kundig and their collaborators have established that the transport and the phosphorylation of glucose are inseparable, both being catalyzed by the phosphotransferase system, which is also responsible for the uptake of several other sugars (but not lactose) in this bacterium. The phosphorylation and uptake of glucose are catalyzed in two steps.

$$\begin{array}{c} \text{COO}^- \\ | \\ \text{C}-\text{O}-\text{PO}_3^{-2} + \text{HPr} \xrightarrow{\text{Enzyme I}} \text{HPr}-\text{PO}_3^{-2} + \\ \| \\ \text{CH}_2 \end{array} \qquad \begin{array}{c} \text{COO}^- \\ | \\ \text{C}=\text{O} \\ | \\ \text{CH}_3 \end{array}$$

$$\text{HPr}-\text{PO}_3^{-2} + \underset{\text{(outside)}}{\text{[sugar]}-\text{OH} + \text{HOH}} \xrightarrow{\text{Enzyme II}} \text{HPr} + \underset{\text{(inside)}}{\text{[sugar]}-\text{O}-\text{PO}_3^{-2} + \text{HOH}}$$

A low molecular weight protein called HPr is phosphorylated at the expense of phosphoenolpyruvate (PEP) in the reaction catalyzed by enzyme I, whose products are pyruvate and the phosphorylated HPr. The second step occurs at the membrane of *E. coli* and involves the phosphorylation of the sugar originally on the outside of the cell at the expense of the high-energy phosphate bond in the phosphoHPr in a reaction catalyzed by enzyme II, whose products are the sugar phosphate *now on the inside of the cell*, with the regeneration of unphosphorylated HPr. Enzyme I is a soluble protein, but enzyme II is very tightly membrane-associated. Mutation in either enzyme I or enzyme II results in the inability to take up the sugar; thus both are essential parts of the transport system. Because glucose is immediately phosphorylated during its transport, little or no free glucose builds up within cells. Glucose-6-phosphate does accumulate; being highly charged and thus membrane impermeant, it cannot leak back out and is trapped in the cell. Since phosphorylation is the first step in the oxidative metabolism of glucose, the cell has accomplished both the uphill accumulation of glucose and its phosphorylation at the cost of only one high-energy bond, that of PEP.

Primary Sodium Pump

The uptake of a number of sugars and amino acids by animal cells occurs through a third mechanism in which the substrate is cotransported with sodium ion. Cells from a large variety of animal tissue have an internal sodium concentration appreciably lower than the external level of sodium. This concentration gradient is maintained by a membrane-localized sodium pump, which constantly pumps sodium out of the cell cytoplasm at the expense of metabolic energy. The uptake of amino acids, or of sugars such as glucose, by animal cells requires this sodium gradient; in the absence of external sodium, neither amino acids nor sugars are accumulated to any appreciable extent. This observation led to the hypothesis that the sodium gradient provided the driving force for the transport of a variety of substrates, and subsequent experimental work has established the essential correctness of this hypothesis. The sodium pump in a variety of animal tissues is inhibited strongly and very specifically by ouabain, which also blocks specifically the uptake of amino acids and sugars into such cells. In the late 1950s, Skou discovered and characterized an enzymatic activity in the membranes of nervous tissue which split ATP only in the presence of Na^+ and K^+ ions. This membrane ATPase was strongly and specifically inhibited by ouabain. Na^+, K^+-ATPases were subsequently discovered in the membranes of a variety of animal tissues. The outward pumping of Na^+ would establish an electrical gradient across the membrane unless balanced by an equal inward flux of some other cation, and in animal cells K^+ is the cation normally used to maintain electrical neutrality across the membrane. The Na^+, K^+-ATPase catalyzes the outward flux of Na^+ across the membrane and the inward flux of K^+, and is thus responsible for the establishment of the Na^+ gradient which drives amino acid and sugar transport. As Na^+ is allowed to fall from its high concentration outside to its low concentration inside, the free energy thereby gained (see equation above) is used to drive the uphill transport of some sugar or amino acid. Figure 8-23 is a schematic representation of such a transport system driven by a downhill flow of Na^+. From this diagram it should be apparent that even in the absence of a Na^+ gradient the carrier might transport its substrate from one side of the membrane to the other, but

Figure 8-23 Schematic model for the transport of glucose into animal cells against a concentration gradient. Energy must be expended to pump sodium out of the cell, and the free energy gained as sodium ions diffuse back with the concentration gradient is used to move glucose into the cell. In the presence of a high concentration of Na^+ (I), glucose is bound to the glucose carrier protein. A translocation of this protein within the membrane then occurs (II) and glucose and Na^+ are released inside of the cell.

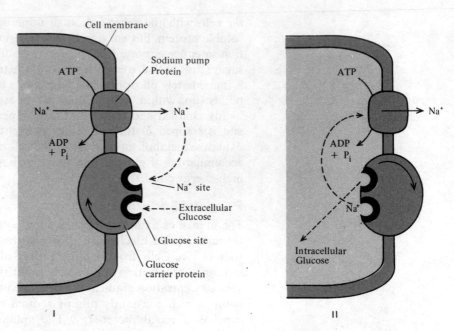

only facilitated diffusion could be catalyzed under these circumstances; with no driving force there can be no uphill transport. In fact, as noted above, in the presence of the poison ouabain where no Na^+ gradient exists, only facilitated diffusion occurs. Although this model is consistent with the known fact, it is purely schematic; the precise molecular details of this process remain to be determined.

PROBLEMS

1. Given the reaction

$$uridine + P_i \rightleftharpoons uracil + ribose\text{-}1\text{-}P$$

the equilibrium constant is 0.91 at pH 7.0 and 38 °C. Calculate the standard free energy and the ratio of products to reactants required to produce 1000 cal mole^{-1} of energy. (Note, at 38 °C, $RT \ln K_{eq} = 1420 \log K_{eq}$.)

2. For the reaction $A \rightarrow B$ catalyzed by enzyme E, it was found that if the reaction is started with 22 μmoles of A, at equilibrium there are 20 μmoles of B. What is the ΔG^0 for this reaction at 25 C?

3. An enzyme catalyzes this reaction: X-CHO + DPN$^+$ → X-COOH + DPNH + H$^+$ at 25° and pH 7.0, the $\Delta G^{o\prime}$ is 9,200 cal/mole, the E'_0 for DPNH/DPN$^+$ is $-0.32v$. What is K_{eq} and $\Delta E'_0$ for the reaction? What is the E'_0 for the X-CHO/X-COOH pair?

216

4. Using the E'_0 values from the text, calculate the free energy change, $\Delta G'$ associated with the following reactions: (a) Riboflavin H_2 + DPN$^+$ → DPNH + Riboflavin + H$^+$ (b) Ethanol + DPN$^+$ → acetaldehyde + DPNH + H$^+$.

5. Assume that you can determine if the oxidized or reduced form of four different electron carriers is present in an electron transport system. In the presence of excess substrate and oxygen and a number of inhibitors of respiration, you observe the following

	Electron Carrier			
Inhibitor	A	B	C	D
I	ox	red	red	ox
II	ox	ox	red	ox
III	red	red	red	ox

What must the order of these carriers be in this electron transport chain?

6. Assume the following compounds were found to pass through a biological membrane by simple diffusion: propionic acid, 1,3-propandiol, propionamide, propionic acid, 1-propanol. List them in the order of their probable diffusion rate (fastest to slowest).

Suggested Readings

Books

Christensen, H. N. *Biological Transport*. Benjamin, Menlo Park, Calif. (1974).

Klotz, I. M. *Energy Changes in Biochemical Reactions*. Academic Press, New York (1967).

Lehninger, A. L. *Bioenergetics,* 2nd ed. Benjamin, Menlo Park, Calif. (1972).

Sato, S. (ed.). *Mitochondria*. University Park Press, Baltimore (1972).

Articles and Reviews

Chance, B. The Nature of Electron Transfer and Energy-Coupling Reactions. *FEBS Lett.* **23**:3–19 (1972).

Dickerson, R. E. The Structure and History of an Ancient Protein. *Sci. Amer.* **226**:58–72 (1972).

Griffiths, D. E. Oxidative Phosphorylation, in *Essays in Biochemistry,* Campbell, P. N., and Greville, G. D. (eds.), vol. 1, pp. 57–90. Academic Press, New York (1965).

Kaback, H. R. Transport Across Isolated Bacterial Cytoplasmic Membranes. *Biochim. Biophys. Acta* **265**:367–416 (1972).

Racker, E. The Membrane of the Mitochondrion. *Sci. Amer.* **218**:32–39 (1968).

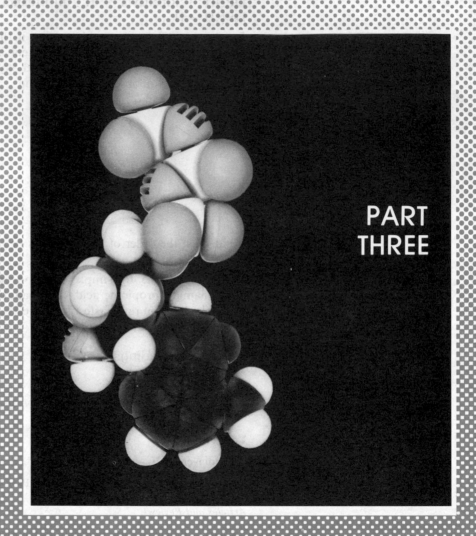

PART
THREE

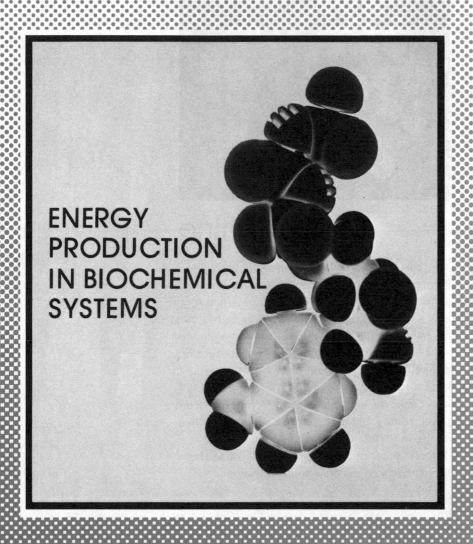

ENERGY
PRODUCTION
IN BIOCHEMICAL
SYSTEMS

Space-filled
model of ATP.

methods for studying intermediary metabolism

chapter nine

Our current knowledge of metabolic pathways has been gained by many different types of investigations. Some information has been obtained by the use of intact animals or isolated organs, but most investigators have utilized tissue homogenates or subcellular organelles. A standard procedure of differential centrifugation has made it possible to obtain relatively pure preparations of nuclei, mitochondria, and microsomal membranes from a tissue homogenate. These preparations can then be used to study the metabolic capability of different parts of the cell and to determine the location of various metabolic pathways.

An extensive body of information is now available describing the metabolic fate of the hundreds of compounds which can be biochemically transformed by living systems. This knowledge has been accumulated through the use of many different types of experiments. Studies of the metabolism of compounds in living tissue have been carried out at many levels of cellular organization, and there are advantages and disadvantages to each approach.

WHOLE ANIMAL

Although studies utilizing an intact animal can yield little information about detailed biochemical reactions, they have been used to gain a great deal of information about important metabolic processes. Intact animals have been used to establish the nutritional requirement for dietary factors, such as the amino acids and vitamins (Figure 9-1) and their quantitative requirements. Whole animals can be used to establish ingestion-excretion balances of certain body constituents, such a mineral elements or nitrogen. For example, if the administration of some compound caused an animal to excrete more nitrogen per day than it was ingesting, the animal would be in "negative nitrogen balance," an indication that the synthesis of tissue protein was proceeding less rapidly than degradation.

The administration of a compound to an animal, followed by an identification of the urinary excretion products of the compound, will give some indication of how the compound was metabolized. The identification of radioactive intermediates in the tissues or of radioactive excretion products following the administration of a radioactive substrate to an animal is a widely used method for obtaining a preliminary idea of the metabolic fate of that compound. In some cases the level of particular enzymes rises or falls in response to dietary or hormonal treatment, and preliminary studies of these responses are often carried out in intact animals.

ISOLATED ORGANS

A number of organs (most commonly the liver) have been extensively used as isolated systems in which the metabolic fate of a compound may be observed. If the isolated organ is supplied with an oxygenated blood supply and provisions are made for circulating this through the organ, it will carry out many of the reactions that the same tissue would in the intact animal (Figure 9-2). The advantages of this technique are that the concentrations of the compound being studied can be closely regulated, and the metabolic reactions of the tissue are not subjected to nervous or hormonal control from other tissues. Perfused organs are often used to determine which tissues in the body are able to carry out a particular metabolic function.

Figure 9-1 Effect of dietary zinc on the young chick. The chick at the top has received adequate zinc in its ration and weighs 430 g at five weeks. The chick at the bottom has been fed a zinc-deficient diet. It is also five weeks old but weighs only 125 g and has the weak legs and feather abnormalities characteristic of a zinc deficiency in this species. (Courtesy of W. G. Hoekstra)

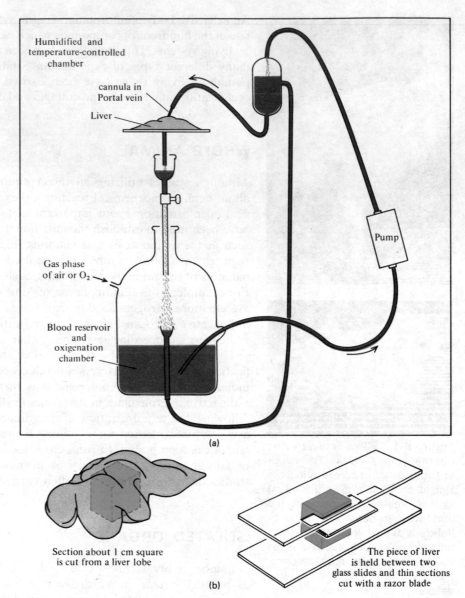

Humidified and
temperature-controlled
chamber

cannula in
Portal vein

Liver

Gas phase
of air or O$_2$

Blood reservoir
and
oxigenation
chamber

Pump

(a)

Section about 1 cm square
is cut from a liver lobe

(b)

The piece of liver
is held between two
glass slides and thin sections
cut with a razor blade

Figure 9-2 (*a*) Apparatus for liver perfusion; (*b*) preparation of slices
from liver.

TISSUE SLICES

Thin slices of liver, kidney cortex, or muscle tissue which are incubated in a
buffered salt solution have also been used to study metabolic transformations. In the slices, most of the cells are still intact, and it is easy to change

the concentration of metabolites in the incubation medium or to assay for products produced. The slices must be carefully prepared if they are to be thin enough to allow gaseous exchange of O_2 and CO_2 in and out of the cells. The ability of the compounds being studied to diffuse across the cell wall must also be determined before the results can be interpreted. The ability of these preparations, as well as of tissue homogenates and isolated mitochondria, to oxidize various substrates has been extensively studied through the use of the Warburg apparatus for measuring O_2 uptake (Figure 9-3). The rate of O_2 uptake is the most common measure of overall metabolic activity of aerobic tissues.

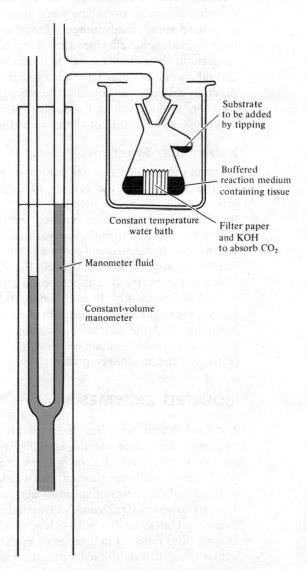

Substrate
to be added
by tipping

Buffered
reaction medium
containing tissue

Constant temperature
water bath

Filter paper
and KOH
to absorb CO_2

Manometer fluid

Constant-volume
manometer

Figure 9-3 Warburg-Barcroft apparatus for measurement of oxygen consumption by tissue slices or homogenates.

HOMOGENATES

If tissues are subjected to mechanical and shearing forces such as those generated by a Waring blender, or are obtained by forcing cells between the stationary wall of a tube and a rapidly rotating pestle (a Potter-Elvehjem homogenizer), most of the external cell membranes will be disrupted. The cellular contents and subcellular organelles will be mixed with whatever medium the tissue is suspended in. This suspension of broken cells, the *homogenate*, will carry out many of the reactions of the intact tissue. This technique has the advantage that it is easy to work with and there is no diffusion barrier to overcome when substrates are added. The preparation does have some disadvantages: Essential cofactors in the cell may have been diluted, some enzymes may have been inactivated, many hydrolytic enzymes in the cell may have been activated by the process of disruption, and inhibitors that may have been localized in one part of the cell are now generally available throughout the media. Some of the mechanical devices used for breaking cells can also disrupt cellular organelles, and problems may arise with heating of the system during cellular disruption.

Subcellular Fractions

Much of the detailed study of metabolic pathways has been carried out in cell-free systems. By the process of differential centrifugation of homogenates relatively clean preparations of different cellular organelles can be obtained. These can be assayed for their enzymatic activity, or their ability to utilize or metabolize various substrates can be determined. It will be seen in subsequent chapters that many biochemical reactions are localized in specific parts of the cell. By separating the organelles, reactions which would normally compete for the same substrate in the cell can be studied independently. In addition to the disadvantages inherent in any system where the cell has been broken open, these studies have the disadvantage that the normal interaction of cellular components has been lost and controlling influences from metabolites produced in other parts of the cell are no longer present.

ISOLATED ENZYMES

The final details regarding the individual steps in metabolic pathways, and particularly their control, is often gained from a study of isolated enzymes. At this level of investigation the influence of other metabolites which may modify the reaction can be determined, and some knowledge of the details of the mechanism of action of the enzyme can be obtained. The physiological significance of isolated enzyme reactions is often difficult to assess. Unless studies with a less purified system have proved that an enzyme does function in a series of metabolic transformations in the particular tissue it was isolated from, the reaction studied may not have any physiological importance.

TISSUE METABOLITE CONCENTRATIONS

Useful information regarding metabolic pathways can be obtained at almost all levels of organization if the concentration of various tissue metabolites can be measured. The concentration of most tissue metabolites is less than millimolar, and is often only a few micromolar. Because of this, standard chemical methods of analysis are usually not sufficiently sensitive, and enzymatic techniques are used. If a tissue such as perfused liver is frozen very quickly, very few changes in the concentrations of the metabolic intermediates occur. These intermediates can be extracted from the tissue with perchloric acid, and the amounts of each of them determined after neutralization of the acid. This is most commonly done by using the intermediates as the substrate for an enzyme which has NAD^+ or $NADP^+$ as a cofactor. The amount of the compound present can then be determined by measuring the amount of the pyridine nucleotide oxidized or reduced in this reaction (Figure 9-4).

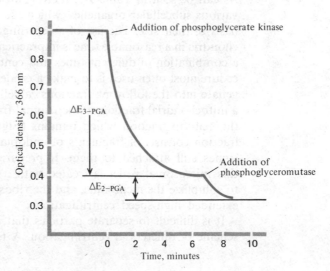

Figure 9-4 Assay of glycolytic intermediates by an enzymatic method. The reaction is started with glyceraldehyde-3-phosphate dehydrogenase, ATP, and NADH, and after the system has stabilized, phosphoglycerate kinase is added: This enzyme will convert all the 3-phosphoglycerate to 1,3-diphosphoglycerate, which will be reduced by glyceraldehyde-3-phosphate dehydrogenase. The amount of NADH oxidized is, therefore, equal to the amount of 3-phosphoglycerate which was present. After all the 3-phosphoglycerate has been converted to glyceraldehyde-3-phosphate, phosphoglyceromutase is added and the amount of 2-phosphoglycerate determined in the same way. The amount of each of the metabolites can be calculated from the determination of the change in optical density and the extinction coefficient for NADH.

DIFFERENTIAL CENTRIFUGATION

The study of the metabolism of subcellular particles is so widespread and so important to the biochemist that it might well be considered an essential technique in any biochemical laboratory. Once a cell has been broken and the various subcellular particles liberated into the medium, usually a buffered isotonic or somewhat hypertonic sucrose solution, the problem of separating these particles remains. The most commonly used procedure is differential centrifugation.

Figure 9-5 Equation for the sedimentation of particles in a centrifuge.

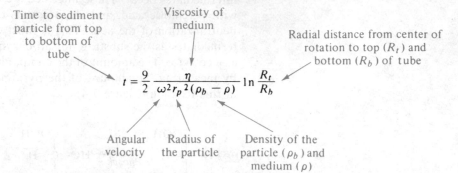

The time required for a roughly spherical particle to sediment from the top to the bottom of a tube is given by the equation in Figure 9-5. If all the particles in a mixture are in the same medium and are subjected to the same gravitational force, the only variables in the equation are the relative size of the particle $r_p{}^2$, and the density of the particle, ρ_b. At any given speed of a centrifuge rotor, the time required to separate a population of particles will be inversely proportional to the square of the radius of the particle and to the difference in density between the particle and the medium. As can be seen in Table 9-1, there is little difference in the density of the various subcellular organelles, which are usually separated mainly on the basis of their size. The use of a centrifugal force that will sediment mitochondria in a reasonable time is impractical for the isolation of microsomes; a combination of different times and centrifugal forces are used. The procedure most often used is to make a crude separation of the cellular homogenate into the following fractions: a cellular debris and nuclear fraction, a mitochondrial fraction, a microsomal fraction, and the soluble protein—the cell-sap fraction which remains (Figure 9-6). The crude microsomal fraction consists of fragments of endoplasmic reticulum with many ribosomes still attached to them. If preparations of ribosomes are desired, this microsomal pellet is treated with a detergent such as deoxycholate to solubilize the membrane, and the ribosomes are then sedimented by an extended high-speed centrifugation.

It is difficult to separate particles that are similar in size by this simple scheme of differential centrifugation. A technique called density gradient

Table 9-1 Properties of Rat Liver Subcellular Fractions

Fraction	Diameter (μ)	Density (g/ml)	Fraction of cellular protein (percent)
Nuclei	5–10	1.30	15
Mitochondria	0.7–1.2	1.20	25
Lysosomes	0.4–0.6	1.21	2
Microsomes	0.05–0.25	1.20	20

velocity centrifugation (Figure 5-13) is often used for this more difficult separation. This method takes advantage of the small differences in density of various subcellular particles and is usually used to separate lysosomes from mitochondria or to obtain clean preparations of nuclei. The heavier particles sediment faster, therefore, after a given time, they are distributed farther down the tube. The concentration gradient which is used enhances the separation by stabilizing the separated bands.

Figure 9-6 Differential centrifugation. The specific times and gravitational forces used will depend on the sucrose concentration of the homogenizing medium and the equipment available. All the crude fractions obtained are usually washed by resuspending them in the homogenizing medium and recentrifuging the suspension. Ribosomes are obtained from the crude microsome fraction by treating it with a detergent to break up the lipid-rich membranes and then centrifuging the ribosomes out of the detergent solution.

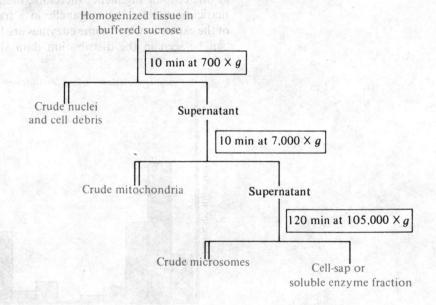

Homogenized tissue in buffered sucrose

10 min at 700 × *g*

Crude nuclei and cell debris

Supernatant

10 min at 7,000 × *g*

Crude mitochondria

Supernatant

120 min at 105,000 × *g*

Crude microsomes

Cell-sap or soluble enzyme fraction

SUBCELLULAR DISTRIBUTION OF ENZYMES

A great deal of useful information has been gained by studying the distribution of enzyme activities in these various subcellular fractions. The data in Table 9-2 indicate some activities known to be associated with the

Table 9-2 Some of the Enzyme Activities in Different Cell Fractions

Nuclei	Mitochondria	Lysosomes	Microsomes	Soluble Fraction
DNA-dependent, RNA-polymerase Biosynthesis of histones Biosynthesis of NAD$^+$	Pyruvate dehydrogenase Electron transport chain Citric acid cycle enzymes Oxidative phosphorylation system Most of urea cycle enzymes Fatty acid oxidizing enzymes	Various hydrolases with a low pH optimum	Cholesterol biosynthesis Triglycerides biosynthesis Phospholipid biosynthesis Gluose-6-phosphatase "Mixed function" oxidases	Glycolytic enzymes Pentose cycle enzymes Fatty acid biosynthesis Amino acid activating enzymes

various components of the cell. It should be remembered that the localization of these activities requires that the cell be broken and separated into fractions before the analyses can be made and a number of artifacts can be introduced. In spite of this, information on the localization of enzyme activity can be useful in interpreting and understanding metabolic reactions and their control. Some activities are thought to be located almost exclusively in one cellular organelle; therefore, their activity is often used as an enzymatic marker for that organelle in a fractionation scheme. An indication of the extent to which some enzymes are located in one subcellular organelle can be seen in the distribution data shown in Figure 9-7. Although its

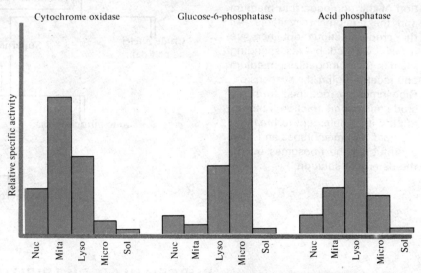

Figure 9-7 Distribution of marker enzymes. The enzymes cytochrome oxidase, glucose-6-phosphatase, and acid phosphatase are often used as markers for mitochondria, microsomes, and lysosomes. The data in the figure indicate the extent to which these enzymes are localized in the subcellular fractions prepared by standard methods.

enzymatic makeup seems to be a fairly general property of an isolated organelle, there are a number of cases known where there are differences between species or between different tissues of the same species. A particular enzyme may be mainly mitochondrial in one tissue and yet be found largely in the soluble portion of the cell in another.

GLUCONEOGENESIS—EXAMPLE OF METABOLIC RESEARCH

Although most organs and tissues of the body can metabolize fatty acids for energy, nervous tissue utilizes them poorly and has a requirement for glucose. The metabolic process by which the body provides glucose for nervous tissue metabolism during periods of low carbohydrate intake is therefore of fundamental importance to the animal. The overall process, the synthesis of glucose from noncarbohydrate precursors, is called gluconeogenesis and is considered in detail in Chapter 15. Because this metabolic process has been so extensively studied, it serves as a good example of the way in which information from various levels of cellular organization and different types of experiments has been used to arrive at a detailed understanding of a metabolic pathway.

Early observations, using intact animals, identified many of the compounds capable of serving as a source of carbon for glucose synthesis. Rats were starved for thirty-six to forty-eight hours, at which time the liver glycogen content was reduced to very low values, and various compounds were administered to the animals by stomach tube. The rats were killed a few hours after they were given the compounds. Those compounds which could be converted to glucose caused an increase in liver glycogen during this period, while others did not. These studies indicated to early investigators that lactic acid, pyruvic acid, glycerol, and many of the amino acids could be utilized as precursors for glucose formation, but that fatty acids could not.

A similar experiment can be carried out with dogs which have been treated with phlorizin to prevent reabsorption of glucose in the kidney. These animals are able to utilize very little of the glucose they synthesize and most of it will be excreted in the urine. The excretion of glucose in the urine can therefore be measured after giving the animals a test dose of some compound to see if it is a good precursor of glucose.

Tissue slices were exclusively used to demonstrate that the capacity for gluconeogenesis was present in the liver and kidney cortex, but not in other tissues. Although liver slices have not been very useful for studying gluconeogenesis, kidney cortex slices have been. The data in Table 9-3 indicate that this type of preparation can be used to determine which of the common amino acids are the most effective precursors of glucose, and also to demonstrate that tissues from animals fed a carbohydrate-free diet have a greater capacity to synthesize glucose.

Table 9-3 Effect of Substrate and Diet on Glucose Formation from Various Substrates in Rat Kidney Cortex Slices[a]

Substrate Added	μ moles of Glucose/g of Dry wt./hr	
	Standard Mixed Diet	Carbohydrate-Low Diet
None	11	24
Pyruvate	250	404
Aspartate	33	73
Glutamate	137	269
L-Arginine	22	37

[a] Adult rates were used, and the slices were incubated for one hour at a substrate concentration of 0.01 M. The carbohydrate-low diet consisted of three parts of casein and one part of margarine and was given for three to five days. [Adapted from *Advan. Enz. Reg.*, **1**, 385 (1963).]

Table 9-4 Gluconeogenosis in the Perfused Rat Liver[a]

Substrate Added	Glucose Formed μ moles/min/g of Liver
None	0.14
Pyruvate	1.02
L-Alanine	0.66
L-Arginine	0.27
L-Glutamine	0.45
L-Aspartate	0.23

[a] Female rats were starved for 48 hours before livers were removed and perfused. The different substrates were all added as the neutral sodium salts at a concentration of 10 mM. (Adapted from *Biochem. J.*, vol. 102, p. 943, 1967.)

The isolated perfused liver has been very useful in studies of gluconeogenesis, and the data in Table 9-4 show that this system can furnish much the same kind of information on utilizable substrates as that for slices. The advantage of this system, however, is that it will respond to hormonal regulation. The data in Figure 9-8 illustrate the effect of the hormone glucagon on the rate of production of radioactive glucose from ^{14}C labeled alanine, which is a good substrate for gluconeogenesis.

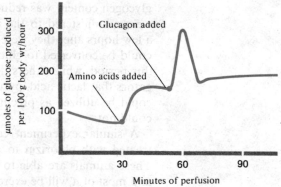

Figure 9-8 Effect of glucagon on glucose production from amino acids in a perfused liver. Note that the curve is for the rate of glucose production, not accumulative production. (L. E. Mallette, J. H. Exton, and C. R. Park, *J. Biol. Chem.*, Vol. 244, p. 5713, 1969)

The isolated perfused liver has also been a good system in which to observe changes in the cellular content of metabolic intermediates and to gain information on the significance of metabolic pathways from studies of these changes. The data in Figure 9-8 merely indicate that the hormone glucagon has had a profound effect on the formation of glucose, but give no indication as to which steps in the metabolic pathway (see Chapter 15) have been influenced by the hormone. The data in Figure 9-9 in which the concentration of the various intermediates in the pathway for the conversion of alanine to glucose have been plotted, show that the hormone has influenced the conversion of pyruvate to phosphoenopyruvate. It can be concluded from these data that the hormone must be exerting its effect at this point.

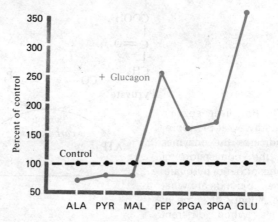

Figure 9-9 Crossover point plot. Some of the metabolic intermediates in the conversion of alanine to glucose are indicated at the bottom of the graph in the order they occur in the pathway. The concentration of each of these in a control perfused liver and in a perfused liver which has had glucagon added to the perfusate is indicated. The point where the glucagon curve goes from below the control value to above it is called a crossover point. It indicates that the glucagon-sensitive step is somewhere in the conversion of pyruvate to phosphoenolpyruvate. (L. E. Mallette, J. H. Exton, and C. R. Park, *J. Biol. Chem.*, Vol. 244, p. 5713, 1969)

Analyses of enzyme activities have also been useful in determining the importance of metabolic pathways. In the process of gluconeogenesis it is necessary to carboxylate pyruvate to a 4-carbon acid before it can be converted to phosphoenolpyruvate. This can be accomplished by the two different sets of metabolic reactions shown in Figure 9-10. Conditions that are known to favor gluconeogenesis, such as fasting or glucagon administration, or to decrease it, such as insulin administration, can be imposed on animals, and the activity of some of the key enzymes in these separate pathways determined. When this is done, it can be seen (Figure 9-10) that the PEP-carboxykinase reaction must be an important one in regulating gluconeogenic activity, but that the malic enzyme does not constitute a key control point.

These few examples serve to illustrate some kinds of information used to arrive at a detailed explanation of the pathways of intermediary metabolism. In the subsequent chapters, very little effort will be made, except

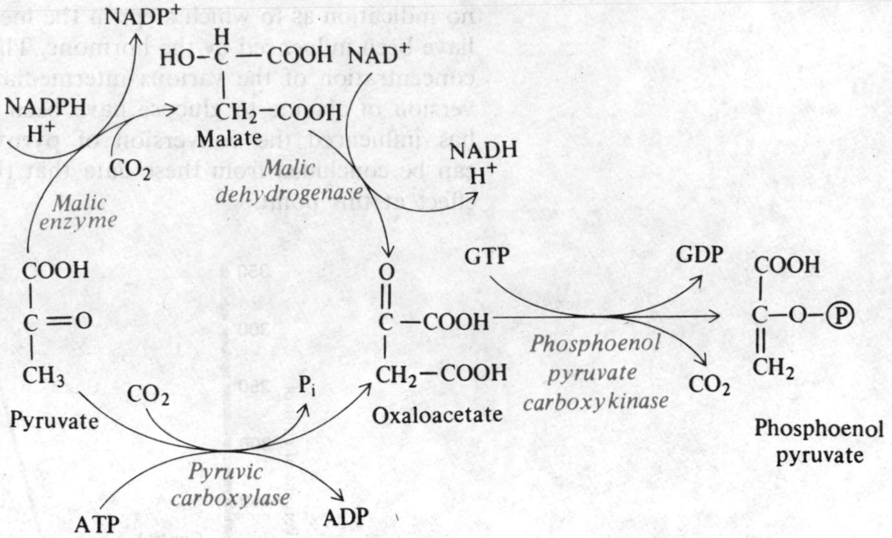

Figure 9-10 Rat liver enzyme activity. The pathway at the top of the figure indicates the enzymes involved in the conversion of pyruvate to phosphoenol pyruvate. The data in the table indicate what occurs to the activity of some of these enzymes when different treatments are imposed on the animal. Hydrocortisone, fasting, and glucagon all increase the rate of gluconeogenesis and also increase phosphoenolpyruvate carboxykinase activity without significantly changing the activity of malic enzyme. These data have been taken to mean that malic enzyme is not an important enzyme in the overall conversion and that the pyruvate carboxylase pathway is the important one. Insulin administration decreases gluconeogenesis and also decreases the activity of the enzymes involved. (H. A. Lardy, *Harvey Lectures,* Series 60, Academic Press, 1964–1965)

| | Units of Enzyme Activity | |
| | Phosphoenol Pyruvate | |
Experimental treatment	Carboxykinase	Malic Enzyme
Control	3.9	0.68
Hydrocortisone treatment	8.8	0.77
Fasted for 48 hours	11.7	0.53
Glucagon treated	8.4	0.34
Insulin treated	2.8	1.11

in cases of historical interest, to indicate how the knowledge about a particular metabolic conversion was gained. It should be remembered, however, that information from a number of different sources probably contributed to present knowledge, and, as with all biochemical knowledge, much of the detail still remains unclarified, and indeed many concepts which seem well established at the present time might require extensve revision in the future.

PROBLEMS

1. If you were interested in studying the mitochondrial respiration of hamster kidney cortex, could you assume that published methods for the isolation of rat liver mitochondria would yield a good clean preparation of this subcellular particle? If not, what would you have to do to verify this?

2. If you were following a microsomal isolation procedure that called for a 60-min centrifugation at 135,000 $\times$ g to pellet the microsomes, what could you do if you had available a centrifuge that would generate a force of only 80,000 $\times$ g?

3. Assume you had a multistep *in vitro* system that converted a substrate S to intermediates A, B, C, D, and then to a final product P. In an attempt to locate the point of inhibition of an inhibitor I, you measured the increase or decrease in the steady-state levels of A–D in the presence of I, and found they all decreased. What could you conclude?

4. What effect would each of the following have on the time required to form a firm pellet of a subcellular particle in a centrifuge? (a) using more sucrose in the suspension medium, (b) using a rotor with a larger radius, (c) increasing the rotor speed.

Suggested Readings

Books

Birnie, G. D. (ed.). *Subcellular Components: Preparation and Fractionation,* 2nd ed. University Park Press, Baltimore (1972).

Roodyn, D. B. (ed.). *Enzyme Cytology.* Academic Press, New York (1967).

Umbreit, W. W. R., R. H. Burris, and J. F. Stauffer. *Manometric and Biochemical Techniques.* 5th ed. Burgess Pub., Minneapolis (1972).

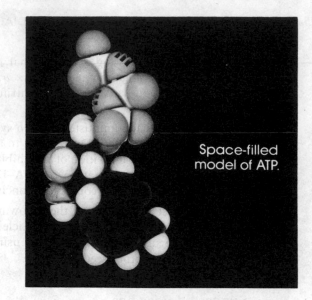

Space-filled
model of ATP.

carbohydrate metabolism

chapter ten

Carbohydrates are consumed mainly as starch and
sucrose, and are hydroxylized in the gastrointestinal
tract to monosaccharides by digestive enzymes.
Following absorption from the small intestine and
transport into tissues, these sugars are converted
to sugar phosphates by a series of kinases. The
enzyme phosphorylase also breaks down liver and
muscle glycogen to form sugar phosphates. These
sugar phosphates can serve as substrates for the
enzymes of the glycolytic pathway which results in
the conversion of all hexoses to pyruvate. In
anaerobic muscle and in some bacteria, pyruvate is
reduced to lactic acid as an end product of
carbohydrate metabolism. In anaerobic yeast,
pyruvate is converted to CO_2 and ethanol, while
in aerobic tissues and cells, pyruvate is completely
oxidized to CO_2. Although glycolysis is the major
pathway of carbohydrate metabolism, there are a
number of alternative pathways, and many cells
and tissues can utilize the pentose phosphate
pathway. The enzymes of this system decarboxyl-
ate glucose-6-P and then subject the remaining
pentose phosphate to a series of interconversions
that regenerate glucose-6-P for further oxidation.

The series of biochemical reactions involved in the utilization of carbohydrates for energy constitute one of the most important metabolic pathways in living organisms. This series was the first segment of intermediary metabolism, as the total enzymological system is called, to be studied extensively by biochemists. The pathways involved in carbohydrate degradation differ only slightly in a wide range of organisms, and have been most thoroughly studied in yeast, liver, and muscle. A wide variety of soluble carbohydrates can be metabolized by various microorganisms; a limited number of different carbohydrates are found in the diet of higher animals.

DIGESTION AND ABSORPTION OF CARBOHYDRATES

In the average American diet, carbohydrates, in the form of relatively few chemical compounds, may contribute about 40 percent of the caloric value. The mixture of amylose and amylopectin (starch) present in potatoes and in the cereal grain products is the most prevalent source of carbohydrate in the human diet. Sucrose, used for sweetening many food products, and lactose from dairy products are the only other significant sources. The average diet contains only a small amount of monosaccharides. The ingested disaccharides and polysaccharides must be digested before they can be absorbed by the intestinal tract. Some understanding of the relationships between the various internal digestive organs of the body is essential to an understanding of digestion and absorption. These relationships are illustrated in Figure 10-1.

Digestive Enzymes

Digestion of carbohydrates begins in the mouth, where the ingested food is masticated and mixed with saliva. The combined production of saliva (an alkaline solution with a very low solid content) from the three oral salivary glands may amount to about 1.5 liters per day. The main protein present in saliva is a glycoprotein, mucin, which because of its lubricating properties aids in swallowing food, but saliva also contains the enzyme *ptyalin* or *salivary amylase*. This hydrolytic enzyme is activated by chloride ion, to catalyze a random cleavage of the α-1,4-glycosidic bonds of starch to yield a limited amount of smaller oligosaccharides. The action of salivary amylase would be halted by the low pH of the stomach when food is swallowed, but starch digestion by this enzyme may persist until there is complete mixing of the ingested food mass with the gastric fluids.

The majority of the digestion of carbohydrates occurs as they pass into the small intestine and are mixed with *pancreatic amylase*, or *amylopsin*, which enters the small intestine through the pancreatic duct. The alkaline nature of the pancreatic secretion itself, and the flow of bile from the liver into the small intestine, raises the pH of the intestinal contents to the point where this amylase is active. The enzyme has the same specificity as

Figure 10-1 Diagrammatic representation of the digestive tract. Note that venous blood going from the intestine to the heart must first go through the portal vein to the liver. All the nutrients absorbed from the small intestine must therefore pass through the liver before being circulated to the rest of the tissues.

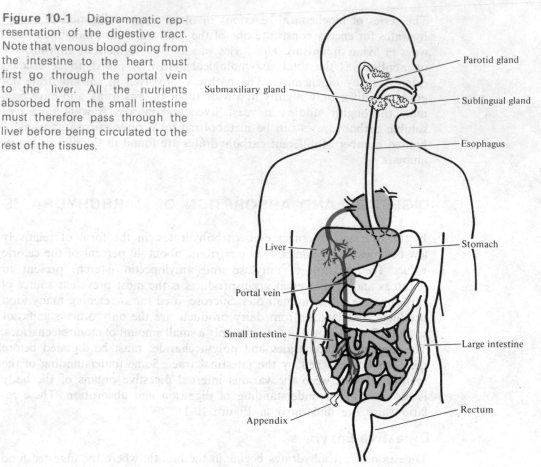

Parotid gland

Submaxiliary gland

Sublingual gland

Esophagus

Liver

Stomach

Portal vein

Small intestine

Large intestine

Rectum

Appendix

salivary amylase and converts amylose to a mixture of maltose and glucose and amylopectin to maltose, glucose, and small oligosaccharides called *limit dextrins*. The enzyme will split only 1,4-glycosidic bonds, and the remaining limit dextrins are the fragments that contain the 1,6-branch points of amylopectin.

There is little evidence that the mixture of disaccharides and small oligosaccharides which are formed by the action of amylase on the ingested carbohydrate can be absorbed. Rather, they are further hydrolyzed to monosaccharides. The action of the various enzymes which complete this process is shown in Table 10-1. These enzymes can be found in the lumen of the small intestine, which is not simply a cylindrical tube, but as illustrated in Figure 10-2, a tissue highly differentiated for efficient absorption. It is likely that the disaccharidases carry out most of their activity within the microvilli or "brush border" of the small intestine. These surface cells are continually being renewed, and as the old cells are sloughed off, they will contribute enzymatic activity to the lumen contents. Because of this, it is easy to measure maltase, sucrase, lactase, and α-1,6-glucosidase activity in samples of the lumen contents. There is considerable evidence

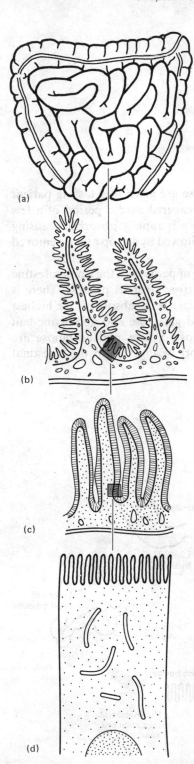

(a)

(b)

(c)

(d)

Figure 10-2 Morphological organization of the alimentary tract which tends to increase efficiency by increasing the area. (a) First the hollow organ is greatly elongated. (b) Its internal surface is further increased by the circumferential folds of the mucosa. (c) These in turn are covered with microscopic villi. (d) The individual cells on the villi are covered with myriad microvilli at the absorptive end. (D. W. Fawcett, *Circulation*, Vol. 26, p. 1105, 1962, by permission of the American Heart Association, Inc)

Table 10-1 Intestinal Oligosaccharidases and Disaccharidases

Enzyme	Substrate[a]	Products
Sucrase	Sucrose	Glucose and fructose
Maltase	Maltose	Glucose
Lactase	Lactose	Glucose and galactose
Isomaltase	1,6-α-glucosidic bonds	Glucose and 1,4-α-glucosides

[a] Most of these enzymes will have some activity toward a large number of other disaccharides.

that these enzymes are located on the brush border membranes and that by hydrolyzing the oligosaccharides at this position they aid in furnishing a high concentration of monosaccharides to the cellular transport sites.

Cellulose, as well as many pentosans, is not significantly digested by the enzymes available in the gut of monogastric animals. Because of this, only ruminants can use the potentially large amounts of energy present in the diet in the form of the β-linked glucose polymer, cellulose. In these animals, such as cattle and sheep, the rumen bacterial population metabolizes cellulose to short-chain fatty acids. These fermentation products are utilized for energy by the animal.

A number of hereditary diseases are known to be caused by a lack of one or more of the enzymes of intestinal carbohydrate degradation. Most common of these is a low level, or a lack, of lactase. If lactose is not degraded to glucose and galactose, it will not be absorbed in the small intestine and will pass into the large intestine. It will increase the osmotic pressure of the intestinal fluid in this part of the digestive tract and will also be subjected to some bacterial fermentation. Both these conditions can result in a severe diarrhea which can be alleviated only by removing lactose from the diet.

Intestinal Absorption

The monosaccharides released in the digestive process are absorbed into the mucosal cells and pass into the circulation through the portal vein. The intestinal absorption of sugars is a rapid process, and the concentration of glucose present in the blood is a balance between its rate of removal by body tissues, its rate of absorption from the intestine, and its rate of synthesis from other compounds. The influence of intestinal absorption can be seen in the results of the standard glucose tolerance test, which is used as an indication of the ability of the body tissues to remove glucose

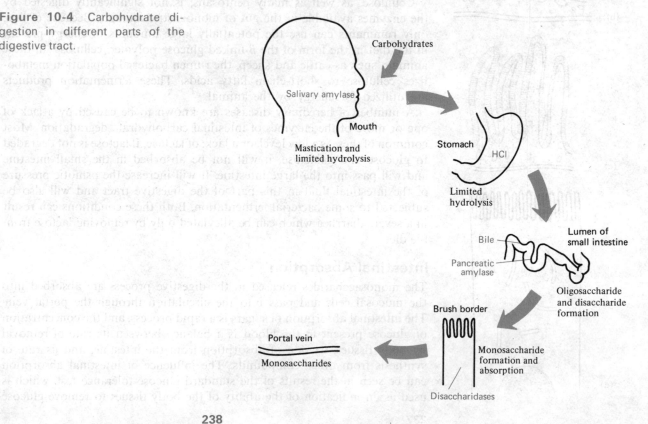

Figure 10-3 Changes in blood glucose concentration following the ingestion of 100 g of glucose by an adult human.

Blood glucose, mg/100 ml
160
140
120
100
80
60
1
2
3
4
Hours

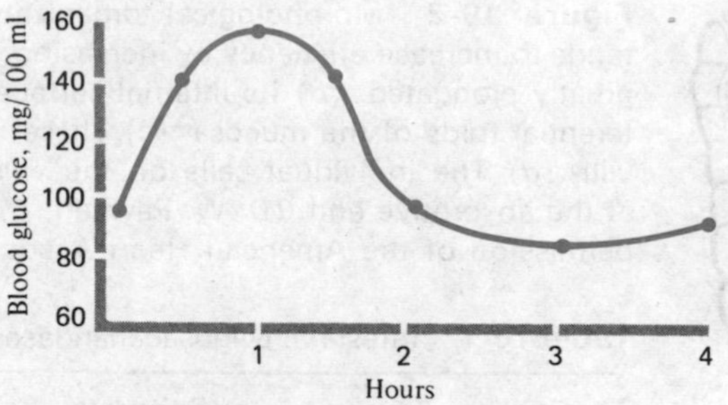

from the blood. In this test, 100 g of glucose are given to a fasting patient and the concentration of blood glucose measured over a period of a few hours. In the normal condition this glucose is rapidly absorbed, causing a sharp increase in glucose concentration followed by a drop as it is removed from the blood (Figure 10-3).

The absorption of sugars occurring in the upper part of the small intestine is not due to simple diffusion, but is an active transport process. There is considerable specificity with regard to the sugar being absorbed; the highest rate of absorption is seen with glucose and galactose and about one-half this rate with fructose; sugars like mannose, xylose, and arabinose are absorbed much more slowly. This transport of sugars into the intestinal

Figure 10-4 Carbohydrate digestion in different parts of the digestive tract.

Carbohydrates
Salivary amylase
Mouth
Mastication and limited hydrolysis
Stomach
HCl
Limited hydrolysis
Bile
Lumen of small intestine
Pancreatic amylase
Oligosaccharide and disaccharide formation
Brush border
Portal vein
Monosaccharides
Monosaccharide formation and absorption
Disaccharidases

cells is facilitated by mobile sugar carriers in the membrane and is linked to the extrusion of sodium from the cell. The overall process of carbohydrate digestion and absorption is summarized in Figure 10-4.

FORMATION OF SUGAR PHOSPHATES

The cells of higher animals are presented with a limited number of saccharides to metabolize: glycogen, which serves as an intracellular storage form of carbohydrate; glucose from intestinal absorption and cellular synthesis; and galactose and fructose from intestinal absorption. Yeast and other microorganisms have available whatever soluble carbohydrate is present in the growth media or various polysaccharides which they can synthesize for energy storage.

The metabolism of these carbohydrates can be conveniently divided into two phases: a series of reactions which convert sugars to lactic acid in cells of higher animals (anaerobic glycolysis) or to ethanol in yeast (alcoholic fermentation), and a second phase which can occur only under aerobic conditions where pyruvic acid is oxidized to CO_2 and H_2O rather than reduced to lactate or ethanol.

The first steps in this process, regardless of the end product, are a conversion of each of the monosaccharides to a phosphorylated intermediate and the interconversion of these phosphorylated sugars to a common intermediate, fructose-6-phosphate. The overall reactions involved are

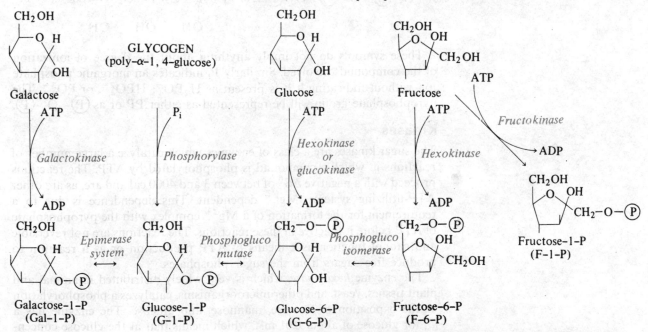

Figure 10-5   Reactions involved in the activation and interconversion of the common hexoses.

239

shown in Figure 10-5. These few reactions are all that are needed to activate any of the sugars commonly available and convert them to any of the other sugar phosphates.

Symbols for Phosphate Groups

The symbol Ⓟ will be used in this text to indicate

$$\begin{array}{c} O \\ \| \\ -P-OH \\ | \\ OH \end{array}$$

except where it occurs in the middle of a series of phosphates linked by ester bonds, where it indicates

$$\begin{array}{c} O \\ \| \\ -P- \\ | \\ OH \end{array}$$

Therefore

$$R-O-\text{Ⓟ} \equiv \begin{array}{c} O \\ \| \\ R-O-P-OH \\ | \\ OH \end{array}$$

and

$$R-O-\text{Ⓟ}-O-\text{Ⓟ}-O-\text{Ⓟ} \equiv \begin{array}{ccc} O & O & O \\ \| & \| & \| \\ R-O-P-O-P-O-P-OH \\ | & | & | \\ OH & OH & OH \end{array}$$

These symbols do not imply anything about the degree of ionization of the compounds involved. Similarly P_i indicates an inorganic phosphate ion without indicating if it is present as $H_2PO_4^-$, HPO_4^{2-}, or PO_4^{3-}. The pyrophosphate group will be represented as either PP or as Ⓟ—O—Ⓟ.

Kinases

The sugar kinases are a class of enzymes which catalyze a large number of reactions in which a compound is phosphorylated by ATP. The reactions proceed with a negative ΔG^0 of between 3 and 4000 cal and are, as are other ATP-utilizing systems, Mg^{2+} dependent. This dependence is due to a requirement for the formation of a Mg^{2+} complex with the pyrophosphate of ATP before it is active in these reactions. The reactions are not reversible and cannot function as a means of ATP production or as a reaction to produce free sugars from the sugar phosphates.

The enzyme *hexokinase*, which is very widely distributed in animal and plant tissues, yeast, and other microorganisms, catalyzes a phosphorylation at the 6 position of fructose, mannose, and glucose. The enzyme has a k_M for glucose of about 0.01 mM, which means that at the glucose concentration prevailing in most cells it is saturated and is reacting at its maximal

240

velocity. A second glucose activating enzyme, *glucokinase*, is much more specific, and has been found to be particularly important in mammalian liver. This enzyme has a k_M of about 20 mM, which is roughly four times the concentration of glucose normally found in blood. If the blood glucose concentration rises because of glucose absorption or inability of the periferal tissues to utilize it, the hexokinase system cannot increase its activity, as it is already saturated, but the activity of glucokinase will increase in response to the higher glucose concentrations.

In mammalian liver and muscle it is likely that much of the fructose utilized is activated by a specific *fructokinase* rather than by hexokinase, although both contribute. The fructokinase reaction leads to phosphorylation of fructose at the 1 position to form fructose-1-P, rather than the fructose-6-P formed by hexokinase. The enzyme which specifically phosphorylates galactose, *galactokinase*, also results in the formation of the 1 phosphate, galactose-1-P.

Phosphorylase

This enzyme, which is extremely important in the overall energy economy of the cell, does not need ATP to form an activated sugar, but rather catalyzes the phosphorolysis of an α-1,4-glycosidic bond. The reaction is analogous to the hydrolysis of starch by amylase except that it is a phosphorolysis, not a hydrolysis. *Phosphorylase* acts on glycogen by removing the terminal glucose residue from the nonreducing end and converting it to glucose-1-P (Figure 10-6). It cannot act at an α-1,6 bond, and its action results in the formation of limit dextrins when the enzyme reaches or nearly reaches a branch point. The branch point residues are removed by the combined action of an *oligo-1,4 → 1,4-glucan transferase*, which transfers an oligo-

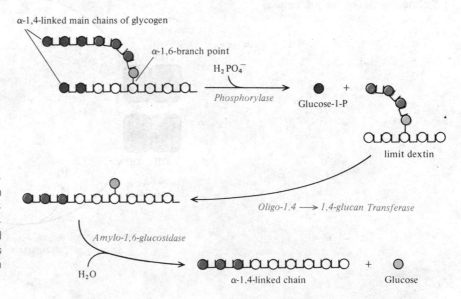

Figure 10-6 Enzymatic degradation of glycogen by the action of phosphorylase, oligo-1,4—1,4-glucan transferase, and amylo-1,6-glucosidase. The α-1,4-linked chain formed after these reactions is a substrate for further action of phosphorylase.

α-1,4-linked main chains of glycogen

α-1,6-branch point

$H_2PO_4^-$

Phosphorylase

Glucose-1-P

limit dextin

Oligo-1,4 → 1,4-glucan Transferase

Amylo-1,6-glucosidase

H_2O

α-1,4-linked chain

Glucose

saccharide from the shortened branch to the nonreducing end of the main chain. This action leaves only a single α-1,6-linked residue at the branch point, and it is removed as glucose by a specific *amylo-1,6-glucosidase*.

In contrast to the kinase reactions, which are driven to completion by the large negative $\Delta G°$ of hydrolysis of ATP, the phosphorylase reaction is readily reversible. Because each cleavage exposes a new nonreducing group, the effective concentration of the substrate does not change when the terminal glucose residue is cleaved off. The equilibrium constant is therefore determined by the ratio of glucose-1-P and free phosphate. As this ratio is only about 3, the reaction is readily reversible, and it was thought at one time that the synthesis of glycogen involved the same

$$K_{eq} = \frac{[\text{glucose-1-P}]\,[\cancel{\text{glycogen}}]}{[\text{H}_3\text{PO}_4]\,[\cancel{\text{glycogen}}]} = \frac{[\text{glucose-1-P}]}{[\text{H}_3\text{PO}_4]} = 3$$

enzymes. As will be seen in Chapter 15, it is now clear that this is not the case.

The phosphorylase enzyme found in muscle is much different from that found in liver, but in both cases both are key enzymes in controlling carbohydrate metabolism. Muscle phosphorylase can exist in two forms (Figure 10-7), an active "a" form and a relatively less active "b" form, which can, however, be activated by high concentrations of AMP. Varying the amount of enzyme in each form has a profound effect on the rate of glycogen de-

Figure 10-7 Activation of phosphorylase b to phosphorylase a. There is a serine residue in each of the monomeric units which must be phosphorylated for full activity.

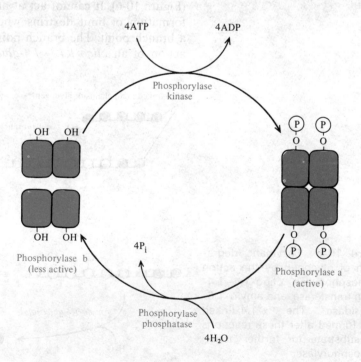

4ATP   4ADP

Phosphorylase
kinase

OH OH

P P
O   O

OH OH
Phosphorylase b
(less active)

4P$_i$

O   O
P   P

Phosphorylase a
(active)

Phosphorylase
phosphatase

4H$_2$O

gradation. The enzyme *phosphorylase kinase*, which converts the less active "b" form to the active "a" form, is in itself subject to an intense regulation. The manner in which hormonal-induced increases in cyclic AMP levels ultimately lead to increased phosphorylase activity is discussed in Chapter 17.

Glycogen metabolism is another area of carbohydrate metabolism affected by a number of known hereditary diseases. As a class, these defects are called glycogen storage diseases. There are examples which are due to a very low activity, or complete lack of activity, of muscle or liver phosphorylase, and also to the lack of the amylo-1,6-glucosidase. In the latter case, no branches are ever removed, but new ones are formed during synthesis. This results in the formation of an abnormal, highly branched, glycogen. The most common of these diseases is due to the lack of the enzyme *glucose-6-phosphatase*, which converts glucose-6-P to glucose. In this case the tissue concentration of glucose-6-P and ultimately glucose-1-P builds up and pushes the phosphorylase equilibrium toward glycogen.

Enzymes Catalyzing the Interconversion of Sugar Phosphates

The utilization of galactose proceeds by means of the transformation of galactose-1-P to glucose-1-P. The reaction requires the two enzymes shown in Figure 10-8, and involves the utilization of a new coenzyme,

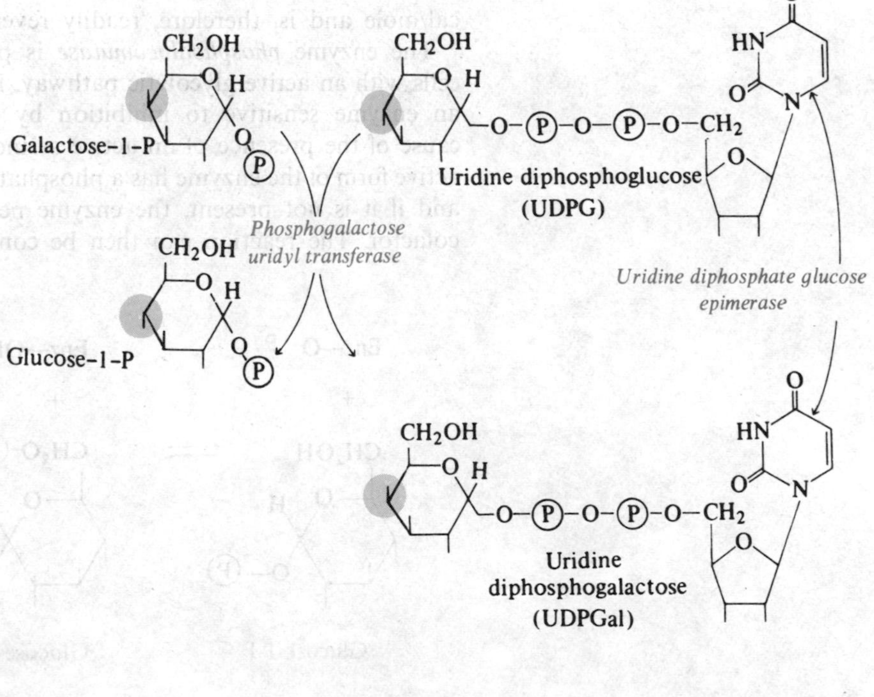

Figure 10-8 Conversion of galactose-1-P to glucose-1-P.

uridine diphosphoglucose, UDPG. The first enzyme, *phosphogalactose uridyl transferase*, catalyzes the conversion of UDP-glucose to UDP-galactose with the formation of glucose-1-P. The second enzyme, *uridine diphosphate glucose epimerase*, functions to regenerate, from UDP-galactose, the UDP-glucose which was used up. The enzyme requires bound NAD^+ as a cofactor, and present evidence would support the mechanism involving an oxidation of the hydroxyl at the 4 position of the hexose to a keto group, which can be reduced to give either of the isomers.

This system has been extensively studied because of the occurrence of a human hereditary disorder called galactosemia. The disease results from a lack of the transferase enzyme and causes the accumulation of galactose-1-P and eventually galactose in the tissues of infants with the disease. The high tissue galactose concentrations cause eye cataracts and mental disorders. Later in life, a second enzyme, a *uridine diphosphate galactose pyrophosphorylase* develops. This enzyme is able to convert Gal-1-P and UTP to UDPGal and PP, and to reduce the Gal-1-P concentration. The epimerase is then able to convert UDPGal to UDPG, which can be utilized in a number of metabolic reactions. If galactosimia is recognized early enough, milk and other galactose-containing foods are removed from the diet until adolescence, when the activity of the second enzyme is developed. There are also indications that some galactose can be metabolized by a completely different pathway which oxidizes it to xylulose and CO_2.

The two other reactions which cause interconversions of phosphorylated hexoses are also readily reversible. The *phosphoglucoisomerase* reaction has a ΔG^0 in the direction of fructose-6-P formation of only about 400 cal/mole and is, therefore, readily reversible.

The enzyme *phosphoglucomutase* is present in high concentrations in cells with an active glycolytic pathway. It was one of the first examples of an enzyme sensitive to inhibition by diisopropylphosphofluoridate because of the presence of an active serine residue at the catalytic site. The active form of the enzyme has a phosphate esterified to this hydroxyl group, and if it is not present, the enzyme needs glucose-1,6-diphosphate as a cofactor. The reaction can then be considered to be the following:

Glucose-1-P   Glucose-1,6-P   Glucose-6-P

As with many phosphate-transferring enzymes, Mg^{2+} is required, and at equilibrium the distribution of products is about 95 percent in the direction of glucose-6-P.

Conversion of Fructose-6-P to Pyruvate

The reactions just discussed are responsible for the conversion of all the common sources of carbohydrate to one common intermediate, fructose-6-P. The series of reactions, shown in Figure 10-9, then act on fructose-6-P to convert it to pyruvic acid.

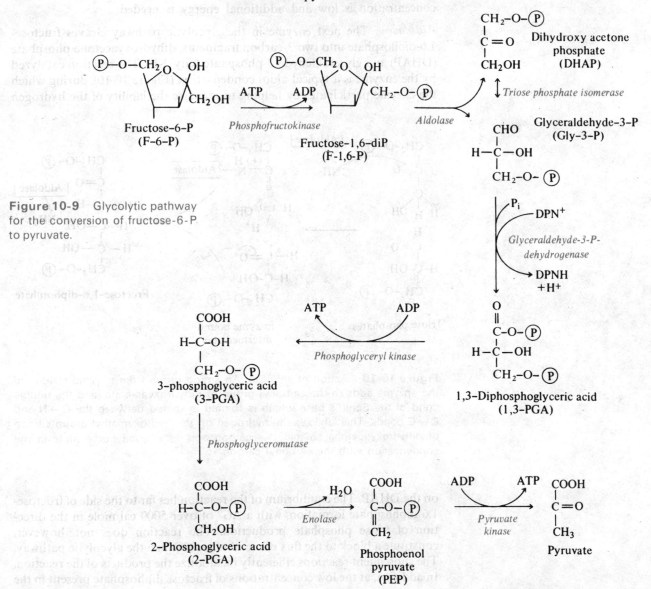

Figure 10-9 Glycolytic pathway for the conversion of fructose-6-P to pyruvate.

Phosphofructokinase This enzyme is one of the rate-limiting reactions in the entire series, and as such it is a key regulatory enzyme in the glycolytic pathway. It is a typical kinase, with a ΔG^0 of about -3500 cal/mole and is subject to an allosteric inhibition by an increase in cellular ATP or citrate. Both these metabolites will be present in high concentrations when the cell is carrying out an active aerobic metabolism. This inhibition will prevent the entry of unneeded substrates into the energy-yielding pathways. Conversely, the enzyme is stimulated by AMP and ADP, both of which will be present in high concentration when the cellular ATP concentration is low and additional energy is needed.

Aldolase The next enzyme in the glycolytic pathway cleaves fructose-1,6-diphosphate into two 3-carbon fragments, dihydroxyacetone phosphate (DHAP) and glyceraldehyde-3-phosphate (Gly-3-P). The reaction catalyzed by the enzyme is a typical aldol condensation (Figure 10-10), during which the enzyme participates by helping to increase the lability of the hydrogen

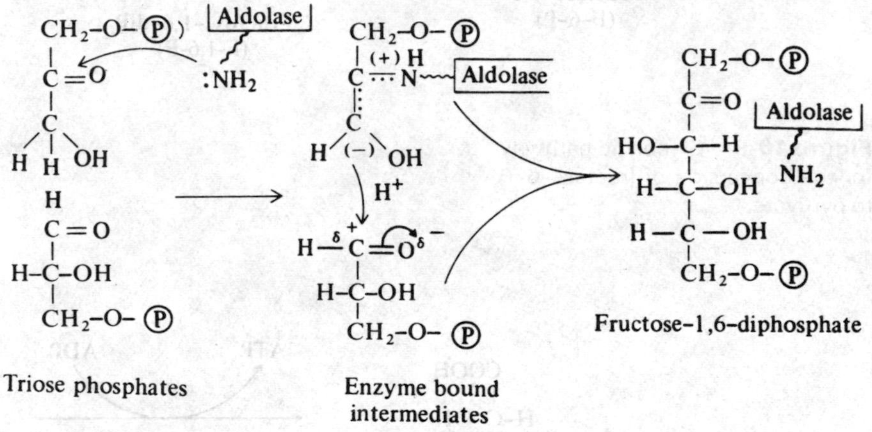

Triose phosphates   Enzyme bound intermediates   Fructose-1,6-diphosphate

Figure 10-10 Action of aldolase. An amino group from a lysine residue of the enzyme adds to the carbonyl group of dihydroxyacetone, and the double bond of the Schiff's base which is formed is shared between the C—N and C—C bonds. This labilizes the hydrogen on the hydroxymethyl group carbon of dihydroxyacetone so that this carbon acts as a good nucleophile in the condensation with the carbonyl carbon.

on the DHAP. The equilibrium of the reaction lies far to the side of fructose-1,6-diphosphate formation, with a ΔG^0 of over 5000 cal/mole in the direction of triose phosphate production. The reaction does not, however, constitute a block to the flux of metabolites through the glycolytic pathway. The subsequent reactions efficiently metabolize the products of the reaction. In addition, at the low concentrations of fructose diphosphate present in the cell, the equilibrium is much further in the direction of triose formation than the calculation based on standard-state free energy change would indicate.

Liver aldolase will also catalyze the conversion of fructose-1-P to DHAP and glyceraldehyde. The glyceraldehyde formed from this reaction can be phosphorylated to Gly-3-P by a kinase. Therefore, any fructose metabolized

Fructose-1-P   Dihydroxyacetone phosphate   Glyceraldehyde

in the liver to fructose-1-P by the specific fructokinase is ultimately metabolized in the same fashion as that activated by hexokinase.

Triose Phosphate Isomerase   Subsequent metabolism of the trioses formed by aldolase proceeds through Gly-3-P, and triose phosphate isomerase catalyzes an interconversion of the two sugar phosphates formed by aldolase. The equilibrium of this isomerase reaction is such that less than 10 percent is in the form of Gly-3-P. From a consideration of the ΔG^0 of both the aldolase and the triose phosphate isomerase reactions, it would appear that very little glyceraldehyde-3-P would be present in equilibrium between fructose-1,6-diphosphate, dihydroxyacetone phosphate, and glyceraldehyde-3-phosphate. This set of reactions is a good example of the fact that the ΔG^0 values for metabolic reactions are useful only in providing an indication of what equilibrium conditions would be, and in giving some indication of what reactions might be useful in driving other reactions. They do not, however, predict what the steady-state, nonequilibrium concentrations of metabolites will be in cells.

Glyceraldehyde-3-P Dehydrogenase   This enzyme catalyzes the first NAD^+-linked oxidation of a substrate in the glycolytic pathway. As the enzyme oxidizes the aldehyde group of the sugar, energy is conserved by the formation of a mixed anhydride between a phosphate and the newly formed carboxylic acid. The enzyme is fairly sensitive to the action of sulfhydryl poisons such as iodoacetic acid, and it has been shown that an intermediate in the enzymatic reaction is an acylated cysteine group. This thioester is then cleaved by phosphate to form the final product (Figure 10-11). The compound which is formed, 1,3-diphosphoglyceric acid (1,3-PGA), has a very high negative free energy of hydrolysis and can be used to drive a reaction which converts ADP to ATP. It should be noted that, in most tissues, dihydroxyacetone phosphate is not oxidized directly; rather, the equilibrium established between it and Gly-3-P accounts for its metabolism.

Figure 10-11 The mechanism of action of glyceraldehyde-3-P dehydrogenase. The enzyme is a tetramer containing four subunits of the type shown. Each of the subunits contains a tightly, but noncovalently bound molecule of NAD^+. It is this bound NAD^+ which oxidizes the added glyceraldehyde-3-P to a thioester, and the external NAD^+ is used to reoxidize the enzyme-bound reduced pyridine nucleotide.

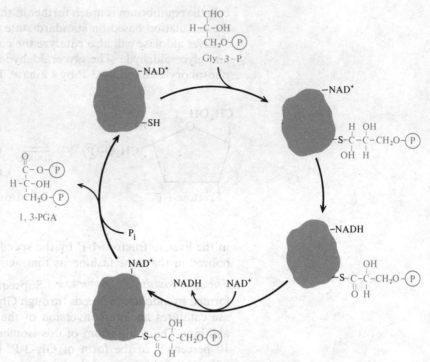

Phosphoglycerate Kinase The high-energy phosphate formed in the preceding reaction is transferred to ADP in a reaction catalyzed by phosphoglycerate kinase, which is named for the reaction that it catalyzes in the reverse direction. The transfer to ADP is strongly exergonic, with a ΔG^0 in the direction of ATP formation of -4500 cal/mole. This reaction is, therefore, one which is able to effectively lower the cellular concentration of diphosphoglyceric acid and to maintain the flow of metabolites in the direction of pyruvate in spite of the rather unfavorable equilibrium constant of the aldolase and isomerase reactions which preceded it. This reaction is also the first one of the pathway that has produced an energy-rich compound, in the form of ATP, which can be used to drive other cellular energy-requiring reactions.

Phosphoglyceromutase The product of the transfer of phosphate from diphosphoglyceric acid to ATP is 3-phosphoglyceric acid (3-PGA), which is converted to 2-phosphoglyceric acid (2-PGA) by phosphoglyceromutase. The mechanism of the enzymatic reaction is analogous to that seen for the conversion of glucose-6-P to glucose-1-P by phosphoglucomutase in that 2,3-diphosphoglyceric acid is an activator of the enzyme. The enzyme catalyzes a reaction that is readily reversible, and it requires Mg^{2+} ion.

Enolase This Mg^{2+}-requiring enzyme is the site of the classical inhibition of glycolysis by fluoride ion. In the presence of phosphate, a Mg^{2+}-fluoride-phosphate complex is formed which effectively prevents the Mg^{2+} ion from participating in the enzymatic reaction. The reaction is readily reversible, and the removal of a mole of water from 2-phosphoglyceric acid results in the formation of phosphoenol pyruvate (PEP), a compound with a high-energy phosphate bond.

Pyruvate Kinase The enol phosphate formed by the enolase reaction has a very large negative free energy of hydrolysis, and its reaction with ADP to form ATP is associated with a ΔG^0 of about -7500 cal/mole. The pyruvic kinase reaction is, for all practical purposes, irreversible in mammalian systems. It constitutes an effective block to the flow of metabolites from pyruvate back through the glycolytic reactions. Although the removal of the phosphate would be expected to result in the formation of enol pyruvic acid, this compound is not stable, and the more common keto tautomer of pyruvate is formed. This reaction and the phosphoglyceryl kinase reaction are the only two in the glycolytic series which directly result in the formation of ATP.

UTILIZATION OF PYRUVATE

The disposition of the pyruvate formed as the end product of the glycolytic pathway depends upon the system studied. In glycolyzing muscle, erythrocytes, and many anaerobic systems, the pyruvate is reduced to L-lactate by NADH. This reaction, catalyzed by *lactic dehydrogenase*, is highly exergonic, and the equilibrium is far in the direction of lactate formation. If the entire pathway is considered, this reaction can be looked at as one which reoxidizes the NADH previously formed by the oxidation of glyceraldehyde-3-P. In this sense, the oxidation and reduction of NAD^+ is merely serving as a mechanism for a transfer of electrons between these two redox pairs (Figure 10-12). Therefore, there is no net accumulation of reduced pyridine nucleotides during the reaction, and only catalytic amounts are required in the cell.

In yeast, where glucose is undergoing an alcoholic fermentation, the pyruvate which is formed is decarboxylated by *pyruvic decarboxylase* to acetaldehyde and CO_2, and the acetaldehyde formed is reduced to ethanol by *alcohol dehydrogenase*.

The decarboxylation reaction requires Mg^{2+} ion, and thiamine pyrophosphate functions as a tightly bound coenzyme.

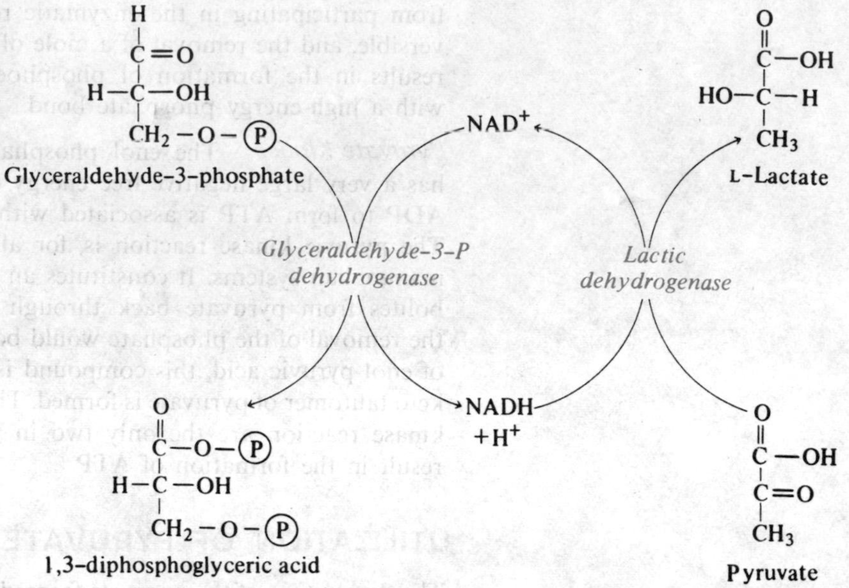

$$
\begin{array}{c}
\text{C}=\text{O} \\
\text{H}-\text{C}-\text{OH} \\
\text{CH}_2-\text{O}-\text{P}
\end{array}
$$

Glyceraldehyde-3-phosphate

NAD$^+$

$$
\begin{array}{c}
\text{O} \\
\text{C}-\text{OH} \\
\text{HO}-\text{C}-\text{H} \\
\text{CH}_3
\end{array}
$$

L-Lactate

Glyceraldehyde-3-P dehydrogenase

Lactic dehydrogenase

NADH +H$^+$

$$
\begin{array}{c}
\text{O} \\
\text{C}-\text{O}-\text{P} \\
\text{H}-\text{C}-\text{OH} \\
\text{CH}_2-\text{O}-\text{P}
\end{array}
$$

1,3-diphosphoglyceric acid

$$
\begin{array}{c}
\text{O} \\
\text{C}-\text{OH} \\
\text{C}=\text{O} \\
\text{CH}_3
\end{array}
$$

Pyruvate

Figure 10-12 Action of the two NAD$^+$-linked dehydrogenases of the glycolytic pathway. The net result is an oxidation of glyceraldehyde-3-P and a reduction of pyruvate.

Thiamine This was the first of the water-soluble vitamins to be discovered. As early as 1912, Funk produced a polyneuritis condition in pigeons fed polished rice and cured it by feeding an extract of rice bran. In 1915 McCollum demonstrated that there was a water-soluble factor B different from the earlier identified fat-soluble growth factor A. It was later found that there was more than one such factor, and thiamine became vitamin B$_1$, which was isolated in 1926 and synthesized by 1936. Vitamin B$_1$ functions in metabolism as the coenzyme, thiamine pyrophosphate. A deficiency of this vitamin leads to a loss of appetite, nervous disorders, and increased levels of blood and urine pyruvate. The human disease resulting from a lack of vitamin B$_1$ is called beriberi, which can take a number of clinical forms.

The vitamin is found in whole cereals, pork, kidney, yeast, and commercial vitamin-enriched bread. The requirement for the human is 1.4 mg per day.

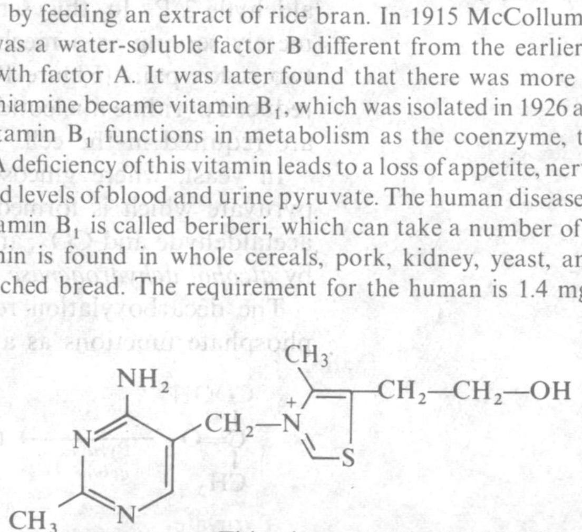

Thiamine

The mechanism of action of this coenzyme, which is found to be associated with a large number of different decarboxylation reactions, is shown in Figure 10-13. In yeast, the reduction of acetaldehyde to ethanol serves

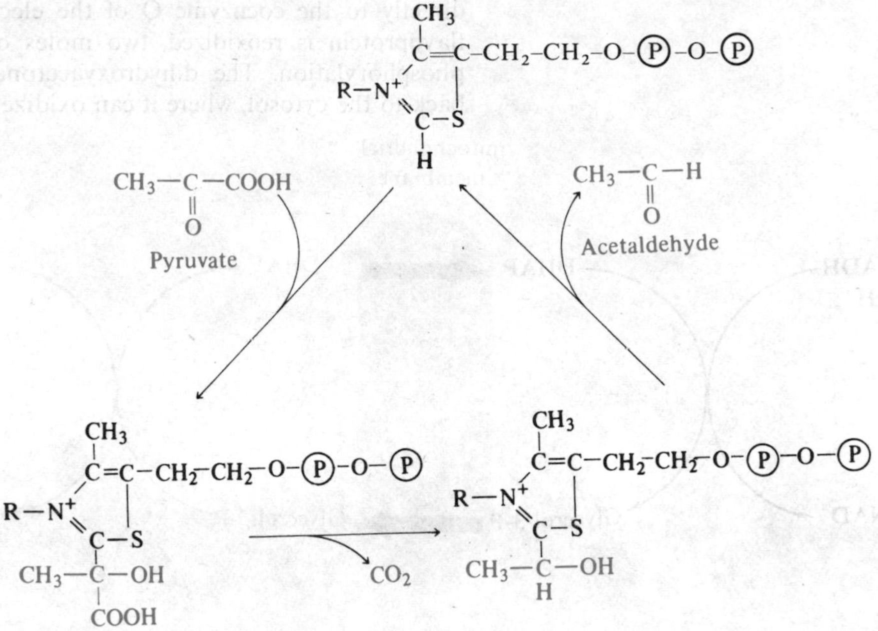

Figure 10-13 Decarboxylation of pyruvate. The pyruvate reacts with the thiazole ring portion of thiamine pyrophosphate to form an intermediate which is decarboxylated and then released as acetaldehyde to regenerate free thiamine pyrophosphate.

the same function of reoxidizing the reduced NAD^+ which has been previously produced, as does the lactic dehydrogenase system in muscle tissue.

In aerobic systems, the pyruvate which is formed is neither reduced to lactate nor converted to ethanol, but is further oxidized to CO_2 and water by a series of reactions which represent a large gain in energy for the organism. These reactions will be discussed in Chapter 13.

MITOCHONDRIAL NADH OXIDATION

An alternative to the reoxidation of NADH by pyruvate would be the reoxidation of the reduced coenzyme by the mitochondrial electron transport chain. This does occur to an appreciable extent in brain and skeletal muscles, but it requires a special mechanism to get the reducing equivalents into the mitochondria. The mitochondrial membrane is not permeable to the cytoplasmic NADH, and the mechanism employed is a reduction of cytoplasmic dihydroxyacetone phosphate to glycerol-3-phosphate by NADH. The glycerol phosphate can diffuse into the mitochondria, where it is reoxidized to dihydroxyacetone phosphate by a flavoprotein-linked glycerolphosphate dehydrogenase. This is a membrane-bound enzyme similar to succinic dehydrogenase in that the electrons are transferred directly to the coenzyme Q of the electron transport chain. When this flavoprotein is reoxidized, two moles of ATP are formed by oxidative phosphorylation. The dihydroxyacetone phosphate formed will diffuse back to the cytosol, where it can oxidize more NADH. In these reactions,

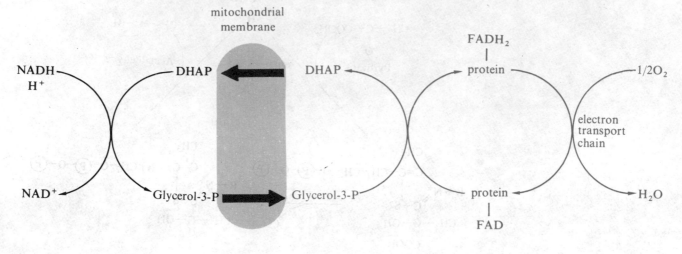

CYTOPLASM   MITOCHONDRIA

Figure 10-14 Mitochondrial oxidation of NADH. Free NADH does not pass the mitochondrial membrane, but it can be used to reduce DHAP to form Gly-3-P. The passage of Gly-3-P into the mitochondria and its subsequent oxidation serves as a shuttle to transfer reducing equivalents into the mitochondria.

which are illustrated in Figure 10-14, the triose phosphates oxidation-reduction has served to transfer electrons into the mitochondria, but there has been no net utilization of dihydroxyacetone phosphate.

OTHER PRODUCTS IN FERMENTATION OF SUGARS

A small amount of glycerol is usually produced in the alcoholic fermentation of carbohydrates by yeast. This occurs when the enzyme, *glycerolphosphate dehydrogenase*, reduces dihydroxyacetone phosphate to L-glycerol-3-phosphate, and couples the reaction to the oxidation of NADH. A specific *phosphatase* is also present in the cell which will hydrolytically remove the phosphate from the product to produce glycerol. The production of glycerol is usually enhanced early in a fermentation, before sufficient acetaldehyde has been formed to reduce the NADH produced by glyceraldehyde-3-P oxidation. The amount of glycerol which can be produced in the fermentation can be greatly enhanced if $NaHSO_3$ is added to the medium. Bisulfite will combine with acetaldehyde to form an aldehyde addition product which

252

Dihydroxyacetone phosphate — L-Glycerol-3-P — Glycerol

cannot be reduced. In this situation, the accumulated NADH will have to be reoxidized by the alternate substrate, dihydroxyacetone phosphate. This process, which has been used commercially, is one of the first examples of a regulated industrial fermentation.

Alterations of the basic glycolytic pathway occur in a wide variety of microorganisms. In some systems the NADH from glyceraldehyde-3-P oxidation is used to form glycerol from dihydroxyacetone phosphate, and acetaldehyde is oxidized to acetate by a second mole of NAD^+. This leaves an excess of NADH, which can be used to reduce a second mole of acetaldehyde to ethanol. A metabolic pathway of this type would be called a *hetero* fermentation, because there is more than one primary fermentation product.

ENERGY YIELD FROM GLYCOLYSIS

The gain or loss of ATP from the reactions of the glycolytic pathway are indicated in Table 10-2. In an alcoholic fermentation there is a net production of 2 moles of ATP for each mole of hexose degraded to ethanol and CO_2.

Table 10-2 Energy Production during Glycolysis

Reaction	ATP/Hexose
1,3-Diphosphoglyceric acid → 3-phosphoglyceric acid	+ 2
Phosphoenol pyruvate → pyruvate	+ 2
	+4
Glucose → glucose-6-P	− 1
Fructose-6-P → fructose-1,6-P	− 1
	Net = +2

The apparent efficiency of muscle glycolysis is higher, as there is no need to use an ATP to form the hexose phosphate. However, if it is considered that ATP is required to form the glycosidic bond of glycogen which is cleaved by phosphorylase, the overall yield of energy to the organism from glycogen utilization is no greater. The energy available from the hydrolysis of the two moles of ATP produced by glycolysis is only between 2 and 3

253

percent of the energy potentially available (686,000 cal/mole) from the complete oxidation of glucose to CO_2 and H_2O. Anaerobic systems such as an alcoholic fermentation are extremely inefficient for two reasons: the substrate has not been completely oxidized, and the NADH formed in the reaction has been used to reduce some other product, rather than being subjected to an efficient energy-yielding aerobic reoxidation. On the other hand, if the organism is being grown primarily for the production of one of its fermentation products, this apparent inefficiency is an asset because it insures that more of the substrate will be converted to the desired product.

REVERSIBILITY OF THE REACTIONS

The majority of the enzymatic reactions in the glycolytic pathway are readily reversible, but a few key steps are not. As is true for all such reactions, the two sugar kinases, hexokinase and the phosphofructokinase, are for all practical purposes irreversible. Formation of ATP catalyzed by pyruvate kinase is so highly exergonic that it also constitutes an essentially irreversible step in the pathway. In spite of this, lactate can be converted to glucose by much the same pathway used in the degradation of glucose. The irreversibility of the key steps involved is overcome by the use of separate enzymes at these points. Specific phosphatases cleave glucose-6-P and fructose-1,6-diphosphate. Note that in these reactions there is no regeneration of the high-energy phosphate bond used to phosphorylate the sugar, and that inorganic phosphate is the product of the reaction. In general, biosynthetic reaction pathways require energy, while degradative reactions are energy producing. The second of these two enzymes, fructose-1,6-diphosphatase, and the enzyme catalyzing the opposite reaction, phosphofructokinase, are key enzymes in controlling the direction of metabolite flow in the glycolytic pathway.

Bypassing the third reaction which limits the reversal of glycolysis, pyruvate kinase, is much more involved and is not accomplished by a single enzyme. There is more than one series of metabolic reactions which could conceivably function in this capacity. These reactions will be discussed in detail in the consideration of gluconeogenesis (Chapter 15).

HISTORY OF STUDIES OF GLYCOLYSIS PATHWAY

This pathway of glucose utilization was one of the first series of biochemical reactions to be studied and understood. It is well to keep in mind that it has been only slightly more than one hundred years since Pasteur demonstrated that the ability to ferment sugar was a property associated with living yeast cells, and that it was not until 1892 that Buchner prepared a cell-free yeast extract that could convert sugar into ethanol and CO_2.

This yeast extract system was extensively studied by Harden and Young in England. They demonstrated that inorganic phosphate was required for fermentation and that phosphate-containing sugars were accumulating in the medium. They were also the first to demonstrate a requirement for the two heat-stable organic factors in the process which were eventually shown to be the NAD^+ and the adenosine nucleotide phosphates. The identification of the various sugar phosphates, which are the first intermediates in the pathway, was carried out largely through the efforts of Harden, Young, Robison, and Neuberg.

At much the same time that alcoholic fermentation was being studied in yeast extracts, the relationship between the contraction of an isolated frog muscle and its lactic acid content was being studied by Hopkins. It was shown that the stimulation of muscle contraction in an anaerobic atmosphere resulted in an accumulation of lactic acid up to the point where the muscle became fatigued and would no longer respond to stimulation. If the system was maintained under aerobic conditions, less lactate accumulated, and if the fatigued muscle was allowed to rest under aerobic conditions, the lactate would disappear. Meyerhof was able to show that the substrate being converted to lactate was glycogen, and later, using muscle extracts, he and Embden studied many of the individual reactions of the sequence. Through the use of specific inhibitors of individual reactions and through the isolation of the individual enzymes involved, the general series of glycolytic reactions was established. Although much current work is devoted to mechanisms of control of glycolysis, the essential pathway as we know it today was established by the early 1940s.

PENTOSE PHOSPHATE PATHWAY—ALTERNATIVE TO GLYCOLYSIS

Although the Embden-Meyerhof glycolytic pathway is the main route by which glucose is metabolized in most tissues, it is not the exclusive pathway

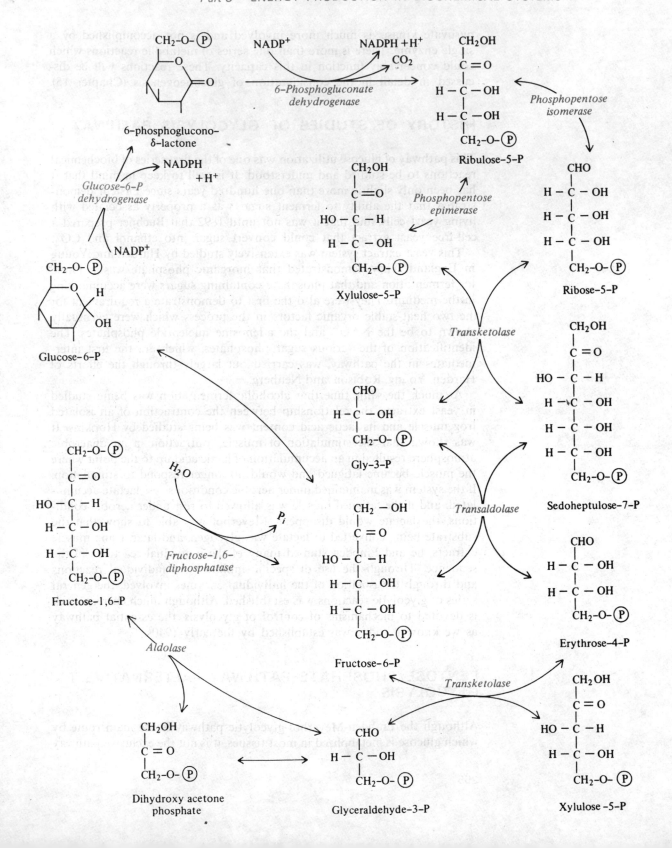

for the metabolic degradation of glucose. A second pathway, which is of fairly general distribution in biological tissues, is the pentose phosphate pathway. This series of reactions is also called the hexose monophosphate shunt, or sometimes the Warburg-Dickens pathway, after two of the investigators who contributed extensively to its elucidation. The reactions involved in the pathway are shown in Figure 10-15.

Oxidation Enzymes

There are two oxidative enzymes in the pathway. The first of these, *glucose-6-phosphate dehydrogenase*, catalyzes the oxidation of glucose-6-phosphate to the corresponding δ-lactone coupled to the reduction of $NADP^+$. It was the study of this reaction that led to the discovery of $NADP^+$ as a coenzyme. The lactone formed as a product of this reaction is chemically unstable and hydrolyzes spontaneously. There is, however, a specific *lactonase* which speeds up the rate of hydrolysis and insures the formation of the free acid.

The second oxidative enzyme, *6-phosphogluconate dehydrogenase*, catalyzes an oxidative decarboxylation of the gluconic acid by effecting the oxidation of the 3-hydroxyl of the sugar acid to a keto group with the loss of carbon-1 as CO_2. The 5-carbon sugar which is formed is ribulose-5-P. Although it seems a likely intermediate, it has not been possible to show the presence of the 3-keto-sugar phosphate during the reaction, and the oxidation and decarboxylation apparently go as a single step. This reaction is again coupled to the reduction of $NADP^+$. These three enzymes have the overall effect of oxidizing a hexose phosphate to a pentose phosphate, with the concurrent formation of two moles of NADPH. The subsequent reactions in the pathway merely provide a way of redistributing the carbon atoms of the pentose phosphate to reform a hexose phosphate. The hexose phosphate which is formed can then be reoxidized by the first enzymes in the pathway.

Pentose Metabolizing Enzymes

Two separate enzymes can act on the ribulose-5-P generated by glucose oxidation and convert it into substrates which can undergo further metabolic transformation. *Phosphopentose epimerase* catalyzes an equilibration of ribulose-5-P with its carbon atom number 3 epimer, xylulose-5-P, and *phosphopentose isomerase* catalyzes an equilibration of ribulose-5-P and and ribose-5-P. Both these reactions are readily reversible, and the result of these activities is a mixture of the three pentose phosphates.

Transketolase and Transaldolase

The two pentose phosphates formed from ribulose-5-P are converted by the action of two separate enzymes to a hexose phosphate, fructose-6-P, and a 4-carbon sugar phosphate. The first enzyme involved is *transketolase*. The enzyme contains tightly bound thiamine pyrophosphate as a cofactor, and catalyzes the transfer of a 2-carbon fragment, a glycoaldehyde group,

Figure 10-15 *(opposite)* Reactions of the pentose phosphate pathway.

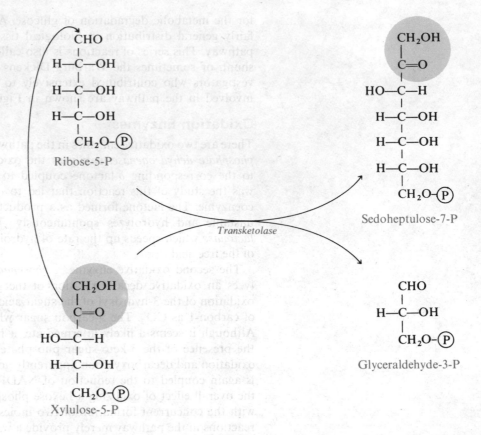

Ribose-5-P

Xylulose-5-P

Transketolase

Sedoheptulose-7-P

Glyceraldehyde-3-P

from xylulose-5-P to ribose-5-P. This action of the enzyme forms a 7-carbon sugar, sedoheptulose-7-P, and leaves the 3-carbon fragment, glyceraldehyde-3-P. The 2-carbon fragment is transferred first to the coenzyme, thiamine pyrophosphate, in the same manner as the acetaldehyde group is carried in the pyruvate decarboxylase reaction (Figure 10-13), and then to the acceptor, ribose-5-P. Although the major action of the enzyme in the pentose phosphate pathway is on xylulose-5-P, it will catalyze the removal of a 2-carbon fragment from a number of ketoses as long as they have an L-hydroxyl at carbon-3 and will use a number of different aldoses as acceptors. The second reaction, catalyzed by *transaldolase*, converts the 7- and 3-carbon sugar phosphates to a 6-carbon and a 4-carbon sugar phosphate. The reaction amounts to the transfer of a dihydroxyacetone group from sedoheptulose-7-P to glyceraldehyde-3-P to form fructose-6-P and to leave erythrose-4-P. This reaction does not involve thiamine pyrophosphate, and the reaction mechanism is rather similar to that of the aldolase of the glycolytic pathway. These reactions constitute a pathway for the conversion of the pentose phosphate formed by the oxidative steps to fructose-6-

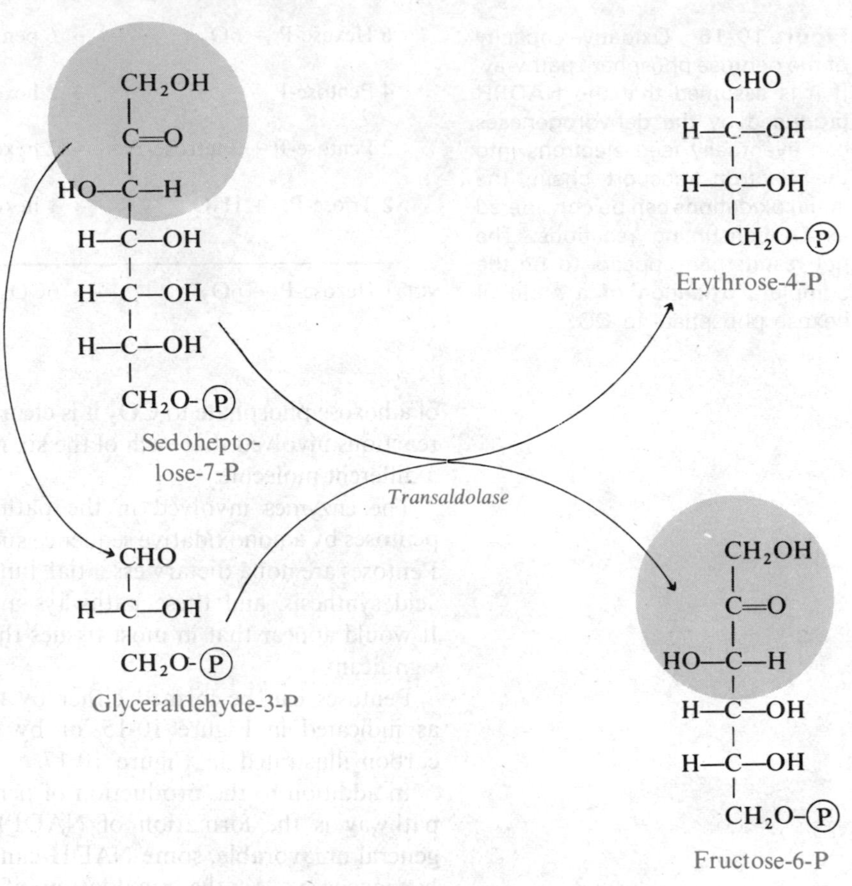

CH_2OH
|
C=O
|
HO—C—H
|
H—C—OH
|
H—C—OH
|
H—C—OH
|
CH_2O-(P)

Sedohepto-
lose-7-P

Transaldolase

CHO
|
H—C—OH
|
CH_2O-(P)

Glyceraldehyde-3-P

CHO
|
H—C—OH
|
H—C—OH
|
CH_2O-(P)

Erythrose-4-P

CH_2OH
|
C=O
|
HO—C—H
|
H—C—OH
|
H—C—OH
|
CH_2O-(P)

Fructose-6-P

P, with the simultaneous formation of a 4-carbon sugar phosphate. This intermediate does not pile up, because as indicated in Figure 10-15, transketolase can catalyze a reaction whereby a second mole of xylulose-5-P can donate a 2-carbon fragment to erythrose-4-P to form more fructose-6-P and to leave the remaining carbons of the pentose phosphate as glyceraldehyde-3-P. The glyceraldehyde-3-P which is formed can equilibrate with dihydroxy acetone phosphate under the influence of triose phosphate isomerase, and aldolase can convert these to fructose-1,6-diphosphate. The action of fructose diphosphatase will hydrolyze the phosphate from the 1-position of this sugar phosphate to form fructose-6-P, which can be acted upon by phosphoglucoisomerase to regenerate glucose-6-P, which was the starting point of the cycle.

Function of Pathway

Using the series of reactions illustrated in Figure 10-15, it is possible to show a stoichiometric conversion of glucose-6-P to CO_2 by this pathway (Figure 10-16). Although the net reaction indicates the complete oxidation

Figure 10-16 Oxidative capacity of the pentose phosphate pathway. If it is assumed that the NADPH produced by the dehydrogenases can eventually feed electrons into the electron-transport chain, the initial oxidations can be considered as O_2-consuming reactions. The net result then appears to be the complete oxidation of a mole of hexose phosphate to CO_2.

$$6 \text{ Hexose-P} + 6O_2 \longrightarrow 6 \text{ pentose-P} + 6CO_2 + 6H_2O \text{ (from the 12 } NADP^+\text{-linked reactions)}$$

$$4 \text{ Pentose-P} \longrightarrow 2 \text{ hexose-P} + 2 \text{ tetrose-P (transketolase and transaldolase action)}$$

$$2 \text{ Pentose-P} + 2 \text{ tetrose-P} \longrightarrow 2 \text{ hexose-P} + 2 \text{ triose-P (transketolase action)}$$

$$2 \text{ Triose-P} + H_2O \longrightarrow 1 \text{ hexose-P} + H_3PO_4 \text{ (aldolase and phosphatase action)}$$

NET $1 \text{ Hexose-P} + 6O_2 \longrightarrow 6CO_2 + 5H_2O + H_3PO_4$

of a hexose phosphate to CO_2 it is clear from an inspection of the individual reactions involved that each of the six moles of CO_2 formed has come from a different molecule.

The enzymes involved in the pathway can also be used to produce pentoses by a nonoxidative sequence such as that illustrated in Figure 10-17. Pentoses are not a dietary essential, but organisms do need them for nucleic acid synthesis, and these pathways are responsible for their production. It would appear that in most tissues the nonoxidative pathway is the most significant.

Pentoses can be formed either by the direct oxidation of glucose-6-P, as indicated in Figure 10-15, or by the nonoxidative rearrangement of carbon illustrated in Figure 10-17.

In addition to the production of pentoses, an important function of the pathway is the formation of NADPH. Although the equilibrium is in general unfavorable, some NADH can be formed from NADPH by transhydrogenases. As the reoxidation of the NADH through the electron transport system can be coupled to the formation of ATP, the entire pentose phosphate pathway could be considered a series of reactions for generating ATP. However, the majority of the NADPH formed is probably used to drive NADPH-requiring reductive reactions, and the main function of the pathway is undoubtedly to furnish reduced pyridine nucleotides for this purpose.

Distribution of Pentose Phosphate Pathway

These series of reactions are widespread in microorganisms and in plant and animal tissues. An idea of the quantitative significance of this pathway compared to glycolysis in a tissue can be obtained by measuring the rate of appearance of radioactive CO_2 when glucose labeled in carbon-1 or carbon-6 is used as a substrate. If the only pathway present is glycolysis, the rate of appearance of radioactive CO_2 from the two substrates will be the same. Both carbon-1 and carbon-6 will become carbon-3 of pyruvate before it is oxidized through the tricarboxylic acid cycle (see Chapter 13). However, if there is a significant contribution of the pentose phosphate pathway, the initial rate of radioactive CO_2 formation from glucose labeled

Figure 10-17 Nonoxidative pathway for the production of pentoses by the enzymes of pentose phosphate pathway. All the pentoses of the pathway can be formed from glyceraldehyde-3-P and fructose-6-P by the normal enzymes of the system.

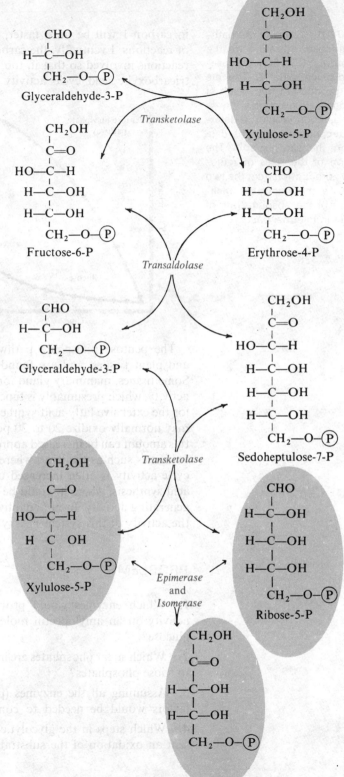

Figure 10-18   Effect of the pathway of glucose utilization on the production of $^{14}CO_2$ from glucose-1-^{14}C and glucose-6-^{14}C. The rate of $^{14}CO_2$ production from each of the radioactive substrates by a tissue preparation would be determined independently, and the data plotted on the same graph. The comparison of the rates of radioactive CO_2 production from the two substrates is the important measurement, not the absolute rate of production from either one.

in carbon-1 will be must faster, because it is released in the first couple of reactions. Eventually, the carbon-6 of glucose will be rearranged by the reactions involved so that it, too, is released as CO_2, even if no glycolytic-tricarboxylic acid cycle activity is present in the tissue (Figure 10-18).

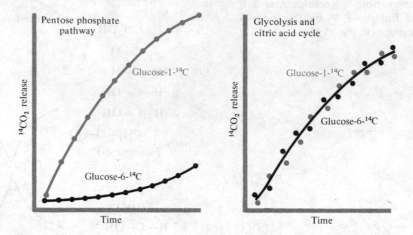

The pentose phosphate pathway enzymes are abundant in microbial and plant tissues, and are found less extensively in mammalian systems. Some tissues, mammary gland for example, have a very high pentose cycle activity, which presumably is functioning as an NADPH-generating system for the extensive fatty-acid synthetic activity needed in this tissue. Rat liver may normally oxidize 20 to 30 percent of its glucose by this pathway, but this amount can be increased appreciably by dietary or hormonal regulation. In tissues such as the liver, where metabolic adaption occurs, the pentose-cycle activity is often increased under conditions such as increased fatty-acid synthesis, where it would be desirable to have an increased NADPH-generating activity. In nonadaptive tissues such as skeletal or heart muscle the activity of this system is very low.

PROBLEMS

1. Which enzymes would probably have carried out some hydrolytic activity on an amylopectin molecule before it is completely degraded to glucose?

2. Which sugar phosphates are intermediates in the conversion of galactose to triose phosphates?

3. Assuming all the enzymes (proteins) were present, what organic cofactors would be needed to convert glucose to ribulose-5-P?

4. Which steps in the glycolytic and pentose phosphate pathways represent an oxidation of the substrates?

5. If glucose labeled in the 1, 3, or 5 position with ^{14}C is added to an anaerobic yeast system producing ethanol and CO_2, where would the label be in the products?

6· The pentose phosphate pathway can be seen as a way to convert glucose to fructose-6-P and erythrose-4-P. If this conversion was occurring, where would these two sugars be labeled if glucose was labeled with ^{14}C in carbon-3?

Suggested Readings

Books

Dickens, F., P. J. Randle, and W. J. Whelan, *Carbohydrate Metabolism and Its Disorders,* 2 vols. Academic Press, New York (1968).

Florkin, M., and E. H. Stotz (eds.). *Carbohydrate Metabolism,* vol. 17 of *Comprehensive Biochemistry.* American Elsevier, New York (1967).

Articles and Reviews

Axelrod, B. Glycolysis, in *Metabolic Pathways,* Greenberg, D. M. (ed.), 3rd ed., vol. 1. Academic Press, New York (1967).

Fischer, E. H., A. Pocker, and J. C. Saari. The Structure, Function, and Control of Glycogen Phosphorylase. *Essays Biochem.* **6**:23–68 (1970).

Lai, C. Y., and B. L. Horecker. Aldolase: A Model for Enzyme Structure-Function Relationships. *Essays Biochem.* **8**:149–178 (1972).

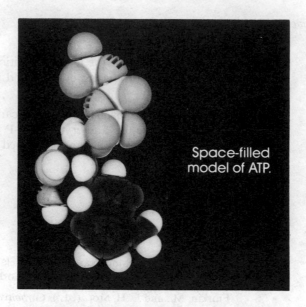

Space-filled
model of ATP.

lipid metabolism

chapter eleven

Ingested lipids are partially hydrolyzed by intestinal lipases and are resynthesized to triglycerides in the intestinal cells. These lipids then pass into the lymphatic system for transport to and storage in other tissues. In the cell, neutral lipids are again hydrolyzed, and the fatty acids are converted to an acyl coenzyme A derivative before they are oxidized. The oxidation of fatty acids occurs in the mitochondria and results in the complete breakdown of the fatty acid to acetyl CoA, which is subsequently oxidized to CO_2 by the tricarboxylic acid cycle. Propionic acid is metabolized by a different pathway which converts it to succinyl CoA before it is completely oxidized. Excessive production of acetyl CoA from fatty acid degradation or an impairment in its utilization can cause an increase in tissue acetoacetate concentration. The urinary excretion of aceto-acetate, or β-hydroxybutyric acid derived from it, causes the metabolic acidosis which is present during starvation or in the disease diabetes.

The metabolism of dietary lipids to CO_2 and water constitutes a major source of energy for higher animals. The adult human consuming a typical American diet obtains roughly 30 percent of his calories from this source. Lipid stored in adipose tissue is also the major long-term energy reserve of the body. Following the depletion of the relatively small amount of glycogen available in muscle and liver, these lipids are utilized to meet caloric demand.

LIPID DIGESTION

Dietary lipids are a complex mixture of the neutral lipids, phospholipids, sterols, and other complex lipids found in plant and animal tissue used as food. There is little, if any, digestion of these dietary lipids in the stomach of higher animals. Ingested lipid passes from the stomach into the small intestine, where it is subjected to the action of bile from the liver and lipase from the pancreas. The bile is an alkaline secretion of the liver which contains the salts of bile acids, mainly cholic and taurocholic acid, cholesterol, and a small amount of other lipids, and various bile pigments. These pigments, which give the bile its yellow-to-green color, are formed when red blood cells are taken up and destroyed by the reticuloendothial system. After rupture of the red blood cell, the hemoglobin is oxidized to methemoglobin (Fe^{3+}), and the porphyrin ring is cleaved by an oxygenase to form bilirubin and biliverdin. These pigments are taken up by the liver and excreted into the bile as the glucuronides. In most animals, bile formed in the liver can be temporarily stored in the gallbladder, but in others, such as the common laboratory rat, the bile duct runs directly from the liver without any appreciable storage capacity. The bile salts lower the surface tension of the digesting food mass (the chyme) and bring about an emulsification of the fatty material which is present. The subsequent action of pancreatic lipase is greatly aided by the formation of an emulsion of small lipid particles which greatly increase the surface area of the insoluble substrate available to the enzyme.

The pancreatic enzyme lipase, which is inactive as it is secreted into the lumen of the small intestine, is readily activated by Ca^{2+} ions. The enzyme is very active with emulsified substrates but inactive toward substrates in true solution. It has its greatest activity toward the esters of the hydroxyl groups at the 1 and 3 position of glycerol and has very little, if any, activity at the 2 position. The final products of the action of pancreatic lipase within the small intestine are therefore 2-monoglycerides, free fatty acids, and a small amount of 1,2-diglycerides (Figure 11-1). These products of lipid digestion also have detergent action and aid in the emulsification of dietary lipid.

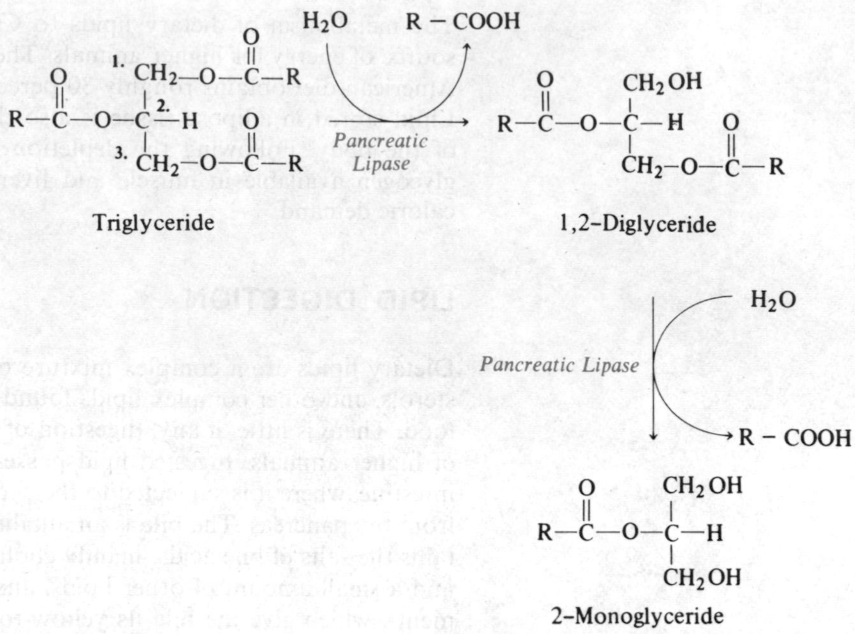

Triglyceride → 1,2–Diglyceride → 2-Monoglyceride (Pancreatic Lipase, H_2O, $R-COOH$)

Figure 11-1 Action of pancreatic lipase on a triglyceride. The major end products of the enzymatic action are free fatty acids and the 2-monoglyceride. Very little free glycerol is formed.

LIPID ABSORPTION

The absorption of the products of lipase digestion from the lumen of the intestine is aided by the bile salts, which form mixed micelles with phospholipids in the lumen of the small intestine. These micelles (see Chapter 4) are colloidal-sized particles (40–100 Å) which, in contrast to the much larger emulsified particles in the chyme, are in equilibrium with the various components which make them up. As pancreatic lipase acts to hydrolyze dietary triglycerides, the resulting monoglycerides and free fatty acids are incorporated into the micellar structures, and are carried to the microvillus region of the mucosal cell. There they diffuse into the cell through the lipid-rich cell membrane. The micelles remaining in the lumen are composed mainly of bile salts, and these are absorbed lower in the digestive tract, and eventually are recycled through the liver and back into the bile.

In the mucosal cell, the free fatty acids are activated by conversion to the corresponding acyl CoA derivatives, and the monoglycerides present are then acylated by these compounds to synthesize first diglycerides and then triglycerides. The small amount of short-chain fatty acids normally present in dietary lipids are not re-esterified in this way but are absorbed unchanged from the mucosal cell and pass directly into the portal blood system.

The resynthesized triglycerides in the mucosal cell are combined with phospholipid, cholesterol and its esters, and a small amount of protein to form a large particle called a chylomicron. These particles, which contain only a few percent protein, are 0.5–1.0 μm in diameter and usually contain

80 to 90 percent triglyceride. The remainder of the lipid is a mixture of phospholipid, cholesterol, and cholesterol esters. The chylomicrons are discharged from the mucosal cell into the extracellular space. Because of their large size, they cannot pass through the membranes of the blood capillaries and instead move into the lymphatic system, where they are eventually dumped into the blood circulation through the thoracic duct. The rapid influx of chylomicrons into the blood is responsible for the turbidity often observed in the plasma following the ingestion of a diet high in lipid. This general process of digestion and absorption of lipids is illustrated in Figure 11-2. In addition to lipase, the pancreatic juice

Figure 11-2 Schematic diagram of lipid digestion and absorption. TG-triglycerides, FFA-free fatty acids, SCFFA=short-chain free fatty acids, MG=monoglycerides, FACoA=fatty acyl derivative of coenzyme A.

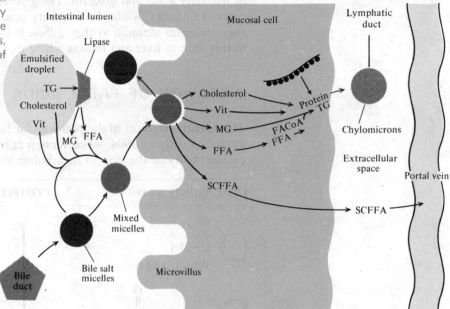

contains phospholipases and cholesterol esterases which function in the digestion and absorption of these compounds. The absorption of the fat-soluble vitamins from the diet is aided by their incorporation into mixed micelles along with products of triglyceride digestion, and in the absence of the bile salts they are very poorly absorbed.

LIPID TRANSPORT AND MOBILIZATION

The removal of the chylomicron particles from the blood stream is a rapid process, and current estimates would indicate that they have a biological half-life of only a few minutes. To a large extent the chylomicrons are taken up by the liver, but some are directly utilized by adipose tissue and other extrahepatic tissues. In all cases it is doubtful that the large particles can penetrate the cell membrane. The triglycerides are hydrolyzed to free fatty acids and glycerol before entering the cells. Once in the cell, the fatty acids can be resynthesized to triglycerides or phospholipids by the pathways to be discussed in Chapter 15. In adipose tissue, heart, and skeletal muscle, this hydrolysis of circulating triglyceride is catalyzed by a specific enzyme called *lipoprotein lipase*, which is present in high concentrations in the capillary bed. In liver and in some extra-hepatic tissues, an enzyme that is specific to the particular organ is probably involved. In the fasting state, the large amount of lipid stored in the adipose tissue can be made available to other tissues. Adipose tissue probably contains a number of lipases which are able to carry out the hydrolysis of the stored triglyceride to free fatty acids and glycerol. The glycerol can diffuse from the cell and enter the plasma, while the free fatty acids are rapidly bound to specific sites on serum albumin as they diffuse from the cell. They are transported in this form to liver and muscle where they can be utilized for energy.

OXIDATION OF FATTY ACIDS

The essential features of the pathway for fatty-acid oxidation were demonstrated as early as 1904, when Knoop carried out a series of experiments on the ability of the dog to metabolize various ω-phenyl aliphatic acids.

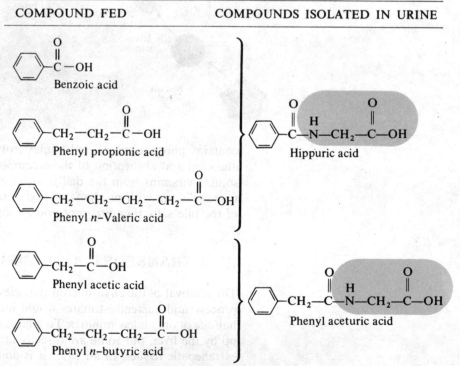

COMPOUND FED	COMPOUNDS ISOLATED IN URINE

Figure 11-3 Observations made by Knoop to establish the β-oxidation scheme.

It was known that when benzoic acid was fed to animals it was detoxified and excreted in the urine as the glycine conjugate, hippuric acid. Knoop fed dogs a number of different aliphatic acids with a phenyl group in the ω position and noted that the feeding of those with even numbers of carbons in the chain led to the excretion of phenylaceturic acid. When phenyl-labeled acids with odd numbers of carbon atoms were fed, hippuric acid was excreted (Figure 11-3). On the basis of these studies, Knoop postulated that fatty acids were metabolized by an oxidation at the β-carbon, followed by a loss of the first two carbons and the formation of a new fatty acid which was two carbons shorter. He was also able to postulate from these data that a reversal of this pathway would explain the observation that the fatty acids found in naturally occurring lipids were predominantly those containing even numbers of carbons. Subsequent work demonstrated that under many conditions acetoacetate was a product of the degradation of fatty acids and that the enzymes involved in the reactions responsible for the complete oxidation of fatty acids to CO_2 were all localized in the mitochondria. The failure of early investigators to find any of the postulated intermediates in the pathway was explained in the early 1950s, when Lynen established that the fatty acids were metabolized, not as the free acids, but as the *S*-acyl derivatives of coenzyme A, which had previously been discovered by Lipmann (Figure 11-4).

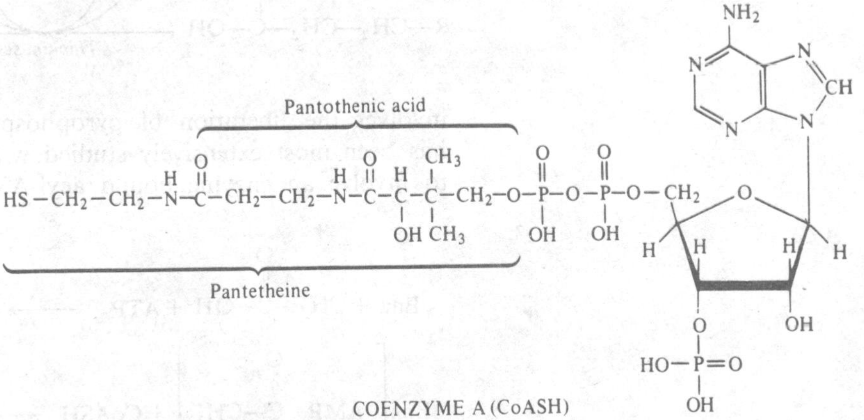

Figure 11-4 Structure of coenzyme A.

Pantothenic Acid This compound was studied for some time as a yeast growth factor. In 1939 Jukes and Woolley demonstrated that a chick antidermatitis factor was not one of the known B vitamins but was pantothenic acid. Its metabolic function is as part of the coenzyme A molecule. A deficiency of the vitamin causes graying of hair in rats and mice and an excretion of porphyrin compounds through the nose in the rat, which results in a condition called "bloody whiskers." An impaired ability to acetylate ingested amines can also be demonstrated in deficient animals. Good sources of the vitamin in the human diet are liver, eggs, and high-protein vegetables. The vitamin is rather widespread in food products, and there is a considerable amount of intestinal synthesis. Because of this a human requirement has not been established, but a deficiency is unlikely, and the average human diet contains about 10 mg per day.

$$HO-CH_2-\underset{\underset{CH_3}{|}}{\overset{\overset{CH_3}{|}}{C}}-\underset{\underset{H}{|}}{\overset{\overset{OH}{|}}{C}}-\overset{\overset{O}{||}}{C}-\underset{H}{N}-CH_2-CH_2-\overset{\overset{O}{||}}{C}-OH$$

Pantothenic acid

Fatty-Acid Activation

The first step in the oxidation of fatty acids is an activation of the fatty acid by coenzyme A. The enzyme involved in the formation of most of the acyl CoA esters is called a *fatty acid thiokinase*, and the energy needed for the formation of the high-energy thio ester is furnished by the degradation of one mole of ATP for each mole of acyl CoA synthesized. The reaction

$$R-CH_2-CH_2-\overset{\overset{O}{||}}{C}-OH \xrightarrow[\textit{Thiokinase}]{} R-CH_2-CH_2-\overset{\overset{O}{||}}{C}-SCoA$$

CoASH   PP
ATP   AMP

involves the liberation of pyrophosphate, and the mechanism, which has been most extensively studied with acetate as a substrate, appears to involve an enzyme bound acyl-AMP intermediate. There are three

$$Enz + CH_3-\overset{\overset{O}{||}}{C}-OH + ATP \longleftrightarrow Enz\left[AMP-\overset{\overset{O}{||}}{C}-CH_3\right] + PP$$

$$Enz\left[AMP-\overset{\overset{O}{||}}{C}-CH_3\right] + CoASH \longleftrightarrow CH_3-\overset{\overset{O}{||}}{C}-SCoA + AMP + Enz$$

different enzymes with specificity for fatty acids of short-, intermediate-, or long-chain length. These function to activate the various fatty acids presented to the tissues. Most tissues contain *pyrophosphatases*, which hydrolyze the pyrophosphate produced, and the action of this enzyme will aid in forcing the reaction in the direction of acyl CoA synthesis even if the concentration of free fatty acids is low.

Although there is some thiokinase activity in the mitochondria, it is essentially a cytoplasmic enzyme, and the majority of the fatty acids are converted to CoA derivatives in the cytoplasm. The mitochondrial membrane is not very permeable to the CoA derivatives of the fatty acids, and it has been demonstrated that the addition of carnitine will stimulate the oxidation of fatty acids by isolated mitochondria. The mitochondrial membrane is permeable to acyl carnitine molecules, and it appears that this compound serves as the physiological carrier for the movement of acyl CoA derivatives of fatty acids generated outside the mitochondria into the mitochondria for oxidation (Figure 11-5).

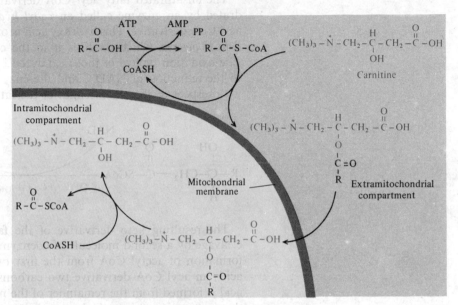

Figure 11-5 Action of carnitine in transporting fatty acids into the mitochondria. After deacylation, the free carnatine can diffuse back across the mitochondrial membrane.

Oxidation of the Acyl CoA Derivatives

The acyl CoA derivatives formed by the thiokinase reaction are acted upon by a series of *acyl CoA dehydrogenases*, which oxidize the saturated

$$R-CH_2-CH_2-\overset{\overset{\displaystyle O}{\|}}{C}-SCoA \underset{\textit{Acyl CoA dehydrogenase}}{\overset{\text{Enz—FAD} \qquad \text{Enz—FADH}_2}{\rightleftharpoons}} R-\overset{H}{\underset{\underset{H}{|}}{C}}=C-\overset{\overset{\displaystyle O}{\|}}{C}-SCoA$$

fatty-acid derivative to the *trans-α-β* unsaturated acyl CoA derivative. This oxidative enzyme utilizes an enzyme-bound FAD as a coenzyme in the reaction. The participation of a flavin coenzyme rather than a pyridine nucleotide is typical of enzymes that form carbon-carbon double bonds. As was the case with the activating enzymes, different enzymes seem to oxidize the CoA derivatives of fatty acids with different chain lengths.

271

$$R-\overset{\overset{\displaystyle H}{|}}{C}=\overset{\overset{\displaystyle H}{|}}{\underset{\displaystyle H}{C}}-\overset{\overset{\displaystyle O}{||}}{C}-SCoA \xleftarrow[\text{\textit{Enoyl hydratase}}]{H_2O} R-\overset{\overset{\displaystyle OH}{|}}{\underset{\displaystyle H}{C}}-CH_2-\overset{\overset{\displaystyle O}{||}}{C}-SCoA$$

The unsaturated fatty acyl CoA derivative which is formed is next hydrated by an enzyme called an *enoyl hydratase* to form an L-β-hydroxy acyl CoA derivative. This hydroxy acid is oxidized by a *β-hydroxy acyl CoA dehydrogenase*, which oxidizes it to the corresponding β-keto acyl CoA. The oxidation typical of those catalyzed by pyridine nucleotides, is linked to the reduction of NAD$^+$, and the enzyme is thought to be nonspecific with respect to the chain length of the fatty-acid derivative involved.

$$R-\overset{\overset{\displaystyle OH}{|}}{\underset{\displaystyle H}{C}}-CH_2-\overset{\overset{\displaystyle O}{||}}{C}-SCoA \underset{\text{\textit{β-OH acyl CoA dehydrogenase}}}{\overset{NAD^+ \qquad NADH+H^+}{\rightleftharpoons}} R-\overset{\overset{\displaystyle O}{||}}{C}-CH_2-\overset{\overset{\displaystyle O}{||}}{C}-SCoA$$

The resulting keto derivative of the fatty acid undergoes a thiolytic cleavage by a second molecule of coenzyme A. This cleavage results in the formation of acetyl CoA from the first two carbons of the original fatty acid; an acyl CoA derivative two carbons shorter than the original fatty acid is formed from the remainder of the molecule. The enzyme catalyzing the reaction is a *β-keto thiolase*, and there are probably a number of these

$$R-\overset{\overset{\displaystyle O}{||}}{C}-CH_2-\overset{\overset{\displaystyle O}{||}}{C}-SCoA \xrightarrow[\text{\textit{β-Ketothiolase}}]{CoASH \qquad CH_3-\overset{\overset{\displaystyle O}{||}}{C}-SCoA} R-\overset{\overset{\displaystyle O}{||}}{C}-SCoA$$

enzymes with specificity for fatty acids of varying chain lengths. The equilibrium of this reaction is such that the thiolytic cleavage is favored, and the reaction is not one which is useful for the stepwise synthesis of fatty acids.

These series of reactions amount to a cyclic pathway which causes the stepwise degradation of fatty acids, two carbons at a time (Figure 11-6), until the entire fatty acid has been converted to acetyl CoA. With the exception of the thiokinase, the rest of the reactions in the pathway for fatty-acid oxidation are located in the mitochondria, and the reduced flavin and pyridine nucleotide generated are reoxidized by the mitochondrial electron transport chain.

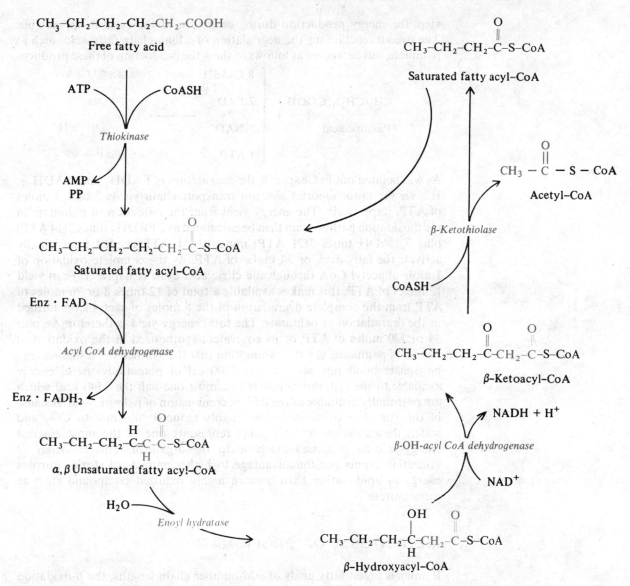

Figure 11-6 Reactions of the β-oxidation pathway. The cycle is shown as a pathway for the conversion of hexanoic acid to butyryl CoA which could then serve as a substrate for the flavin-linked dehydrogenase.

ENERGETICS OF THE REACTION

Each turn of this degradation cycle yields a mole of acetyl CoA and produces 1 mole of $FADH_2$ and one of $NADH + H^+$. As the system requires an input of energy in the form of ATP only in the initial activation step, the energy production during each turn of the cycle is appreciable. The overall reaction for the degradation of a long-chain fatty acid, such as palmitate, can be written as follows to show the production of these products.

$$CH_3(CH_2)_{14}COOH + \begin{cases} 8 \text{ CoASH} & 8 \text{ acetyl CoA} \\ 7 \text{ FAD} & 7 \text{ FADH}_2 \\ 7 \text{ NAD}^+ & 7 \text{ NADH} + H^+ \\ 1 \text{ ATP} & 1 \text{ AMP} + PP \end{cases}$$

Palmitic acid

As was pointed out in Chapter 8, the reoxidation of $FADH_2$ and $NADH + H^+$ via the mitochondrial electron transport chain yields 2 and 3 moles of ATP, respectively. The energy yield from the oxidation of palmitate in the β-oxidation pathway can then be calculated as 7 $FADH_2$ times 2(14 ATP) plus 7 NADH times 3(21 ATP) minus the 1 ATP needed to originally activate the fatty acid, or 34 moles of ATP. As the complete oxidation of 1 mole of acetyl CoA through the citric acid cycle (Chapter 13) can yield 12 moles of ATP, this makes available a total of 12 times 8 or 96 moles of ATP from the complete degradation of the 8 moles of acetyl CoA formed in the degradation of palmitate. The total energy yield is therefore 96 plus 34 or 130 moles of ATP or its equivalent synthesized in the oxidation of 1 mole of palmitate. On the assumption that the synthesis of a high-energy phosphate bond represents about 7300 cal of potentially useful energy available to the cell, this amounts to almost one-half the 2340 kcal which are potentially available as the ΔG^0 of combustion of palmitic acid. Because of this complete oxidation of the highly reduced substrate to CO_2 and water, the oxidation of fatty acids represents one of the most efficient energy-yielding processes available to the organism. This efficiency of utilization points out the advantage to higher animals of storing surplus energy as lipid, rather than a more highly oxidized compound such as carbohydrate.

METABOLISM OF PROPIONATE

If animals ingest fatty acids of odd-number chain lengths, the β-oxidation pathway can proceed in a normal manner, with the exception that the final thiolase cleavage results in the formation of propionyl CoA and acetyl CoA. The propionyl CoA formed is, of course, incapable of undergoing another oxidation and cleavage. Although the need to oxidize odd-number chain-length fatty acids does not arise often, all animals are faced with a need to metabolize a small amount of dietary propionic acid and have a pathway to handle it. This is particularly important in ruminants, which are able to use large amounts of cellulose and other polysaccharides not normally utilized by monogastric animals. In the rumen of these animals (cattle, sheep, goats) an extensive bacterial population metabolizes the ingested carbohydrate and excretes a mixture of acetic, propionic, butyric, and a small amount of other short-chain fatty acids, which are rapidly absorbed by the animal. About 20 percent of this mixture is usually propionic acid, and the metabolism of propionic acid is, therefore, quantitatively an important pathway in these animals. The predominant pathway involved (Figure 11-7) is one which converts propionic acid to succinyl CoA, an

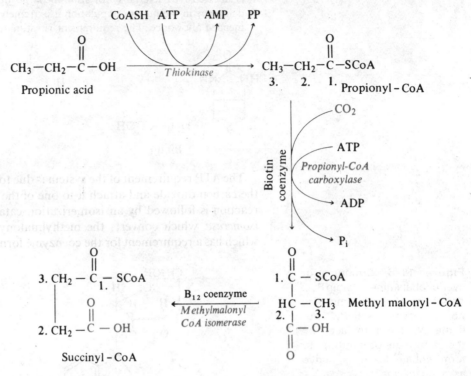

Figure 11-7 Metabolism of propionic acid. The requirement for ATP in the propionyl CoA carboxylase reaction is to form the carboxyl-biotin intermediate. It can be seen from the numbered carbons, that the isomerase causes a migration of the carbonyl CoA group rather than the carboxyl group of methyl malonyl CoA.

intermediate of the citric acid cycle. Propionate is activated by a *thiokinase* in the same manner as indicated for long-chain fatty-acid activation, and the propionyl CoA formed can be carboxylated to methyl malonyl CoA by an enzyme, *propionyl CoA carboxylase*, which uses as a cofactor the coenzyme form of the water-soluble vitamin biotin.

Biotin Observations that diets containing raw egg white as a protein source were toxic to experimental animals were made around 1920. By 1939 it was shown that a factor which would alleviate this toxicity when added to the diet was identical to an unidentified yeast growth factor. The chemical structure of the vitamin was established by 1942. The vitamin functions as a coenzyme in a number of carboxylation reaction, and in the coenzyme form the vitamin is usually bound to the enzyme through the ε-amino group of lysine. This biotin-lysine peptide can be isolated by hydrolysis from biotin-containing enzymes, and is called biocytin. The substance in raw egg white responsible for the toxicity is a basic protein, *avidin*, which has a binding constant on the order of 10^{21} for biotin. A deficiency of the vitamin results in various types of skin lesions in most species, including man. The vitamin is widely distributed in common foods, and there is an extensive bacterial synthesis in the large intestine. For these reasons a biotin deficiency in the human population is extremely unlikely, and there is no recommended allowance. The requirement is estimated to be around 0.2 mg per day.

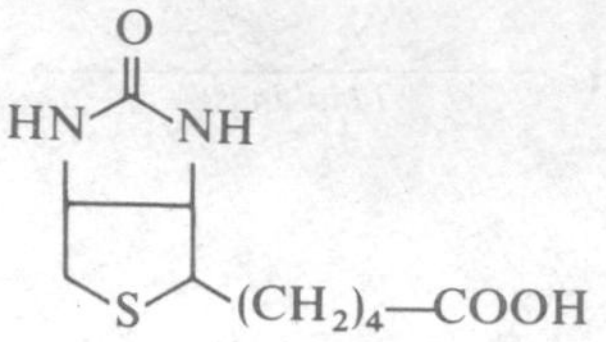

Biotin

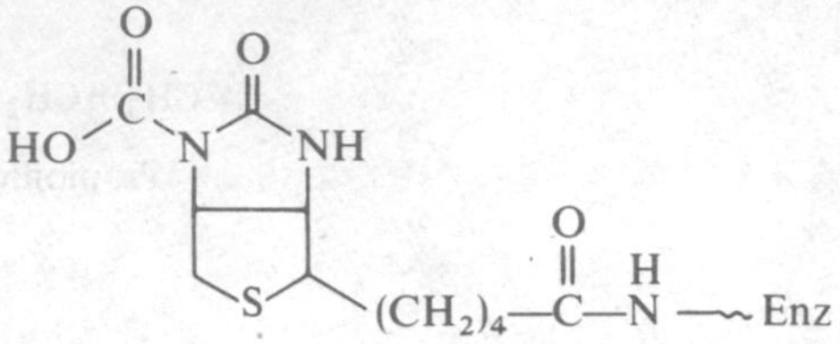

Enzyme bound
carboxyl-biotin

The ATP requirement of the system is due to the energy needed to activate the carbon dioxide and attach it to one of the ring nitrogens of biotin. This reaction is followed by an isomerization catalyzed by *methylmalonyl CoA isomerase*, which converts the methylmalonyl CoA to succinyl CoA and which has a requirement for the coenzyme form (Figure 11-8) of vitamin B_{12}.

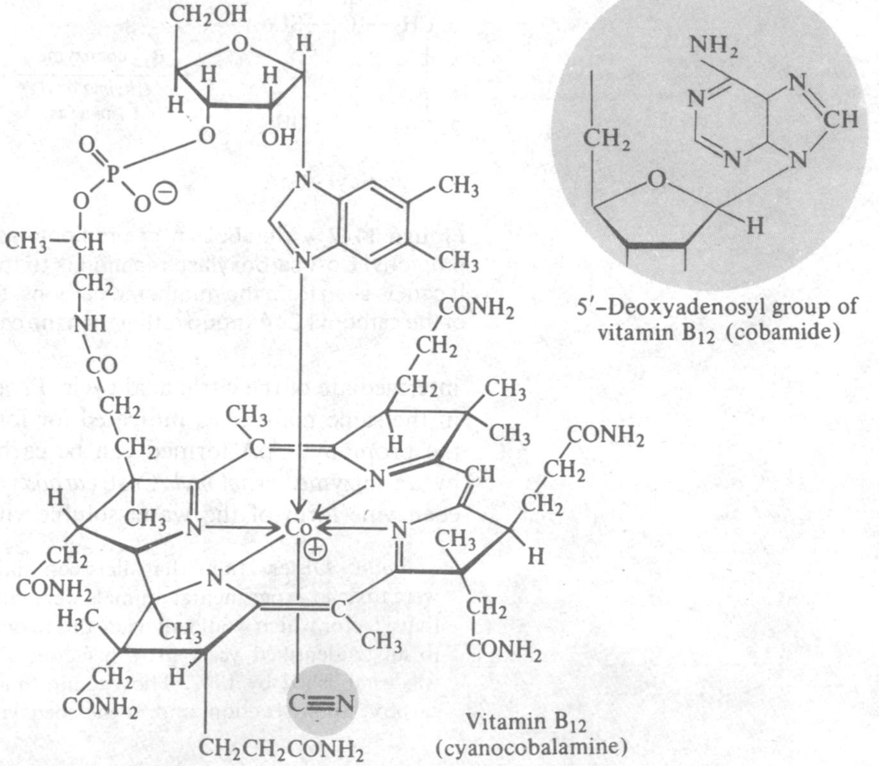

Figure 11-8 Structure of cyanocobalamine, vitamin B_{12}. This form of the vitamin has no coenzyme activity, but if the cyanide is replaced by the deoxyadenosyl group to form cobamide, it is active as a coenzyme.

5'–Deoxyadenosyl group of vitamin B_{12} (cobamide)

Vitamin B_{12}
(cyanocobalamine)

Vitamin B_{12} This factor was the last of the commonly accepted water-soluble vitamins to be characterized. It was isolated from a liver extract in 1948 as a crystalline material that would cure pernicious anemia in humans. The form of the vitamin usually isolated from natural products is cyanocobalamine. The coenzyme active forms of the vitamin are those where some group other than cyanide furnishes one of the coordination bonds to the corrin-bound cobalt. Vitamin B_{12} activity is widely distributed in animal products, is produced by many bacteria, but is not found in plant material. The human disease, pernicious anemia, is caused by the lack of a low molecular weight mucoprotein (intrinsic factor) which is produced in the stomach and is required for the intestinal absorption of the vitamin. Very high oral doses of the vitamin are effective in the absence of intrinsic factor, and the vitamin is effective in small doses if injected. There is little possibility of a vitamin B_{12} deficiency developing in the human, since the recommended allowance is only 5 μg per day.

The reaction is a complex one, and amounts to the migration of the thioester group of methylmalonyl CoA to the methyl carbon, and not the migration of the free carbonyl group, as appears from simply looking at the equation. The succinyl CoA formed in this reaction can then enter into the citric acid cycle. Although there are other pathways of propionate metabolism, they are important chiefly in microbial and plant tissue, and the pathway described here is the predominant one in mammalian tissue.

OXIDATION OF UNSATURATED FATTY ACIDS

A rather large percentage of the fatty acids present in natural sources consists of unsaturated fatty acids. These acids are metabolized by essentially the same β-oxidation pathway with the aid of a couple of additional enzymes. The stepwise removal of two carbons from oleic acid will lead to the formation of a β-γ-*cis*-unsaturated acyl CoA. There is an enzyme which can convert this to the α,β-*trans* unsaturated acyl CoA (Figure 11-9).

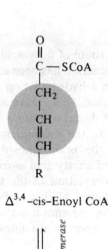

$\Delta^{3,4}$–cis-Enoyl CoA

Isomerase

$\Delta^{2,3}$–trans-Enoyl CoA

$\Delta^{2,3}$–cis-Enoyl CoA   D-3-Hydroxyacyl CoA   L-3-Hydroxyacyl CoA

Figure 11-9 Additional enzymes needed to metabolize unsaturated fatty acids by the β-oxidation pathway. The isomerase interconverts a cis 3,4-double bond and a trans 2,3-double bond; the epimerase interconverts a D and L hydroxyl group.

277

This compound is the normal substrate for the enoyl hydratase, which is the next enzyme in the degradative pathway.

Fatty acids with more than one point of unsaturation present an additional problem. The second double bond turns out to be in the correct position—α,β to the carboxyl group—but is a *cis* rather than a *trans* isomer. This acyl CoA derivative can be hydrated by the normal enoyl hydratase, but the resulting hydroxy acid has the D rather than the L configuration. There is, however, an enzyme which catalyzes the epimerization of this hydroxy acid to form the correct substrate for the β-hydroxy fatty acyl CoA dehydrogenase (Figure 11-9). The combination of these two additional enzymes allows the oxidation of unsaturated fatty acids to proceed in much the same manner as for saturated fatty-acid oxidation.

ALTERNATE PATHWAYS OF FATTY-ACID METABOLISM

Although the β-oxidation pathway is the major route of fatty-acid utilizations, other oxidative reactions are important. A reaction has been identified in both plant and animal tissues which results in a hydroxylation of long-chain fatty acids to form the α-hydroxyl fatty acids. These serve as a substrate for an oxidase that catalyses an oxidative decarboxylation to form an odd-chain fatty acid one carbon shorter than the original molecule. These are important constituents of some of the brain phospholipids. Plant tissues also have a relatively high activity of a fatty acid ω-oxidation enzyme system. This system begins with a hydroxylation of the terminal methyl group of a fatty acid and then the oxidation of this carbon to a carboxylic acid to form a dicarboxylic acid. These enzymes are also found in liver microsomes, but their metabolic significance is not known.

ACETOACETATE METABOLISM

A disease state called *ketosis* is associated with an increased blood and urine concentration of acetoacetic acid, its reduction product, β-hydroxy butyric acid, and its decarboxylation product, acetone. The disease develops under those conditions where there is an impairment of carbohydrate metabolism—as, for example, in diabetes, or where large amounts of fatty acids are being metabolized during the ingestion of high-fat diets or during starvation. In severe cases the ketotic condition is also associated with a metabolic acidosis. This is the result of a loss of Na^+ ion from the body during the urinary excretion of acetoacetic and β-hydroxy butyric acid.

The basic metabolic defect in ketosis is an inability of the liver to utilize acetyl CoA at a sufficient rate to prevent its accumulation. An increase in acetyl CoA concentration will encourage the formation of acetoacetyl

CoA by the same condensation reactions needed to initiate fatty acid synthesis. The increase in acetoacetyl CoA concentration results in a subsequent increase in free acetoacetate. There is a deacylase in liver which will form free acetoacetate from its CoA ester, but this represents a minor pathway for formation of the free keto acid, and most of its formation is through the intermediate formation of β-hydroxy-β-methyl-glutaryl CoA (Figure 11-10). This compound is a normal intermediate in sterol biosyn-

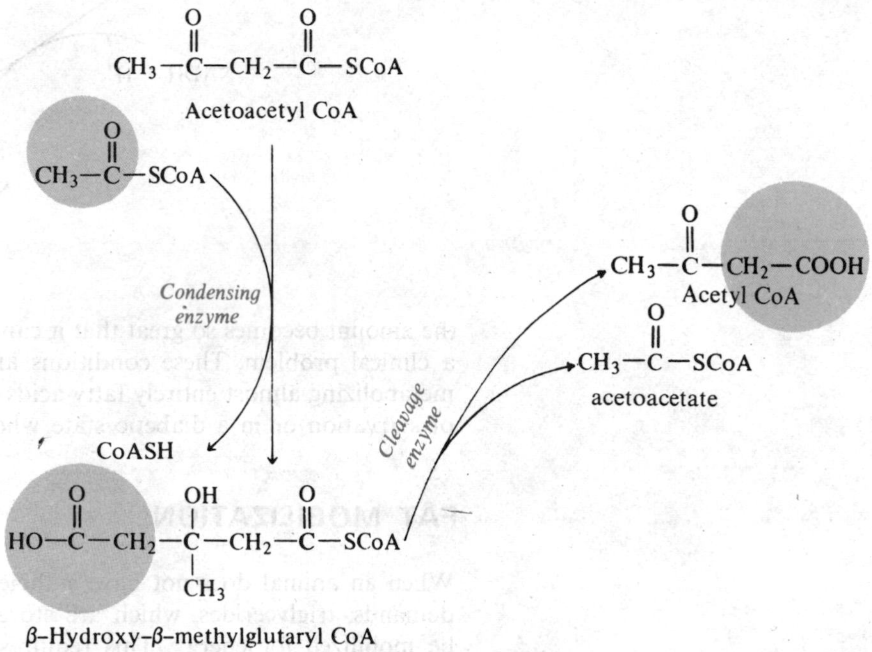

Figure 11-10 Major pathway for the formation of free acetoacetic acid in the liver.

thesis, but when its concentration is increased, it breaks down to acetyl CoA and free acetoacetate.

The free keto acid can then be reduced or decarboxylated by the enzymatic reactions indicated in Figure 11-11. The ketone bodies which are formed can readily diffuse into the extracellular fluid and are eventually carried by the blood to the other tissues of the body. The extrahepatic tissues can reoxidize β-hydroxy butyrate to acetoacetate and, unlike the liver, have an enzyme system which can reform acetoacetyl CoA so that it can reenter the main pathway of lipid oxidation.

It should be clear that the formation of acetoacetate by the liver and its reutilization by the extrahepatic tissues is a normal process. Only when

279

Figure 11-11 Formation of acetone and β-hydroxybutyric acid from acetoacetic acid.

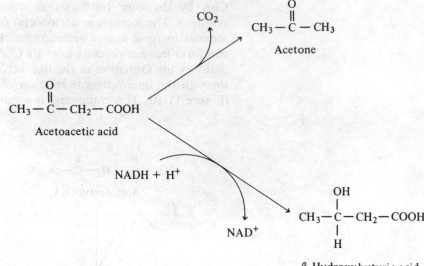

Acetone

Acetoacetic acid

β-Hydroxybutyric acid

the amount becomes so great that it cannot be metabolized does it become a clinical problem. These conditions are usually found when the liver is metabolizing almost entirely fatty acids for energy, for example, in periods of starvation or in a diabetic state when glucose cannot be utilized.

FAT MOBILIZATION

When an animal does not have sufficient energy intake to satisfy energy demands, triglycerides, which are stored in the adipose tissue cells, will be mobilized for energy. This requires the hydrolysis of the stored triglycerides, and the rate-limiting step in the mobilization process is the hydrolysis of the triglyceride by a specific lipase that increases in activity during starvation. Both the free fatty acids and the glycerol resulting from the hydrolysis diffuse from the adipose tissue to the blood. Adipose tissue does not have an active glycerol kinase, and α-glycerol phosphate for triglyceride synthesis in this tissue can be formed only from glucose metabolism.

The free fatty acids leaving the adipose tissue are complexed to serum albumin and carried to other tissues. The fatty acids can be taken up by muscle tissue and metabolized to CO_2, or taken up by the liver. Some of the liver fatty acids are completely oxidized; some are converted to ketone bodies which can be metabolized for energy in the extrahepatic tissues. The liver can also convert the free fatty acids into triglycerides and then excrete them back into the plasma as lipoproteins. The lipid in these lipoproteins can be utilized for energy in other tissue or it can be redeposited in the adipose tissue as triglyceride. Some of these relationships between lipid metabolism in various tissues are illustrated in Figure 11-12.

Figure 11-12 Major pathways involved in the movement of lipid between the adipose tissue, liver, and extrahepatic tissue, and the metabolic fate of lipids in these tissues.

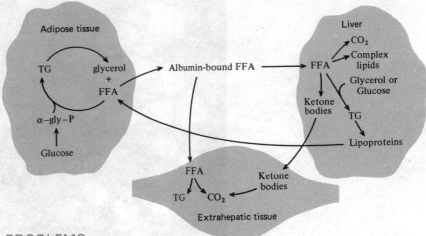

PROBLEMS

1. Contrast the route by which a molecule of butyric acid and a molecule of oleic acid, both of which were injected in the triglyceride form, would get from the intestine to a liver cell.

2. If *n*-heptanoic acid labeled with ^{14}C in carbons 2, 4, and 6 were metabolized to acetyl CoA and succinyl CoA, where would the labeled carbons be?

3. Write the overall reaction for conversion of octanoic acid to acetyl CoA by ATP, FAD, NAD^+, and CoASH.

4. The coenzyme form of five vitamins would be required in the conversion of pentanoic acid to acetyl CoA and succinate. What are the vitamins?

5. The metabolism of palmitic acid labeled in the 3 position in a ketotic animal would result in the formation of radioactive acetone. Where would it be labeled?

Suggested Readings

Books

Dawson, R. M. C., and D. N. Rhodes (eds.). *Metabolism and Physiological Significance of Lipids.* Wiley, New York (1964).

Florkin, M., and E. H. Stotz (eds.). *Lipid Metabolism,* vol. 18 of *Comprehensive Biochemistry.* American Elsevier, New York (1967).

Articles and Reviews

Green, D. E., and D. W. Allmann, Fatty Acid Oxidation, in *Metabolic Pathways,* Greenberg, D. E. (ed.), 3rd ed., vol. 2, pp. 000–000. Academic Press, New York (1968).

Greville, G. D., and P. K. Tubbs. The Catabolism of Long-Chain Fatty Acids in Mammalian Tissues. *Essays Biochem.* **4**:155–212 (1968).

Krebs, H. A., D. H. Williamson, M. W. Bates, M. A. Page, and R. A. Hawkins. The Role of Ketone Bodies in Caloric Homeostasis, in *Advances in Enzyme Regulation,* Weber, G. (ed.), vol. 12, pp. 000–000. Pergamon Press, Oxford (1974).

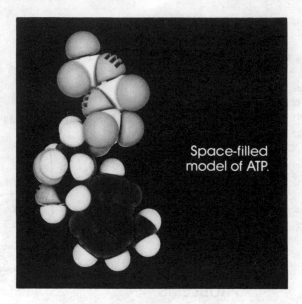

Space-filled
model of ATP.

protein and nucleotide metabolism

chapter twelve

Ingested proteins are hydrolyzed to free amino acids by the combined action of gastric, pancreatic, and intestinal proteases, and the free amino acids liberated are absorbed from the small intestine. Some amino acids are used for synthesis of tissue proteins, and the remainder are metabolized by the release of the amino group as ammonia. Aquatic animals excrete this toxic compound; in most terrestrial vertebrates ammonia is detoxified by conversion to urea, while in birds and reptiles it is converted to uric acid. After the removal of the amino group, the carbon skeletons of the amino acids are metabolized by conversion to intermediates of carbohydrate or lipid metabolism. Higher animals require about one-half the common amino acids in their diet and can form the rest from other dietary sources. The nutritional value of a protein depends on the relative abundance and balance of these "essential amino acids" in the particular protein.

The ingestion of proteins to furnish amino acids for tissue protein synthesis is essential for higher animals, and the provision of an adequate supply of amino acids is a nutritional problem of critical importance. Amino acids not needed for tissue protein synthesis are subjected to deamination and the carbon skeleton is metabolized for energy. Small amounts of nucleic acids are also ingested as food components, and they are subjected to hydrolysis, absorption, and degradation.

PROTEIN DIGESTION

The amount of protein ingested each day by a human consuming a typical American diet is in the range of from 50 to 100 g. The digestion of this protein begins in the stomach, where the acidic nature of the gastric contents (pH 2–3) activates the proenzyme or zymogen, pepsinogen. At a low pH, pepsinogen, which is secreted from the gastric mucosa, is converted to the proteolytic enzyme, *pepsin*, by the hydrolytic removal of a small peptide. The cleavage of the peptide can also be catalyzed by pepsin itself. Once the activation reaction is started, it proceeds rapidly. Pepsin is an endopeptidase which cleaves peptide bonds in the interior of a protein substrate; it shows some specificity for peptide bonds where at least one of the amino acid residues is aromatic. A second enzyme, *rennin*, is found in the stomach of some young animals, particularly calves; it is probably not present in humans. It causes a rapid coagulation of ingested casein and also possesses a limited amount of proteolytic activity.

Pancreatic Proteases

Very little formation of free amino acids occurs by the action of pepsin, and most of the digestion of protein is carried out in the small intestine by pancreatic enzymes. Three pancreatic proenzymes, *trypsinogen*, *chymotrypsinogen*, and *procarboxypeptidases*, are secreted into the intestine, where they are activated by proteolytic cleavage. Trypsinogen can be activated to trypsin by trypsin itself or by a proteolytic enzyme, *enterokinase*, which is secreted from the mucosal cells of the small intestine. The activation of trypsinogen consists simply of the hydrolytic removal of a hexapeptide, Val-Asp$_4$-Lys, from the amino terminal end, whereas the conversion of chymotrypsinogen to chymotrypsin occurs in three distinct steps, with specific peptides split each time. These activations are summarized in Figure 12-1, and the specificity of the pancreatic proteases is indicated in Table 12-1. The combined action of these digestive enzymes is to degrade the ingested protein molecules to free amino acids and small peptides. As in the case of carbohydrate digestion, the final hydrolysis takes place by the action of hydrolases of the intestinal mucosa. Some of this hydrolytic action occurs in the lumen of the small intestine, but the major portion probably occurs as the peptide comes in contact with the epithelial border or within the mucosal cell. One enzyme responsible for

Figure 12-1 Activation of protein-digesting enzymes.

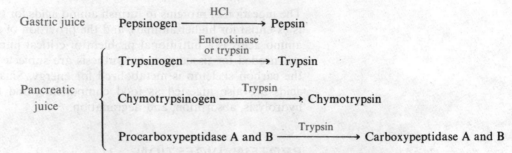

much of the peptidase activity is a rather nonspecific exopeptidase, *leucine aminopeptidase*. However, a number of other poorly characterized dipeptidases are involved in the final steps of the digestive process.

Table 12-1 Specificity of Protein Digesting Enzymes

Enzyme	Type	Specificity for Amino Acid Residues
Pepsin	Endopeptidase	Aromatic or leucine
Trypsin	Endopeptidase	Lysine or arginine[a]
Chymotrypsin	Endopeptidase	Aromatic
Carboxypeptidase A	Exopeptidase	C-terminal aromatic or neutral
Carboxypeptidase B	Exopeptidase	C-terminal arginine or lysine
Leucine aminopeptidase	Exopeptidase	N-terminal leucine or neutral

[a] Has an almost absolute requirement for a basic group on the amino acid residue furnishing the carboxyl group. The rest of the enzymes are not nearly this specific.

Amino Acid Absorption

There is little evidence to indicate that intact proteins, or even relatively small peptides, can be absorbed by the intestine of the adult animal. Absorption of free amino acids occurs in the small intestine by an energy-requiring active transport system. The rate of absorption differs for various amino acids. Although each individual amino acid does not have a specific transport system, there are a number of distinctly different transport systems, each of which handles a group of structurally related amino acids. The amino acids are absorbed from the intestine into the portal vein and must therefore pass through the liver before being made available to the other tissues of the body.

GROSS PROTEIN METABOLISM

Early ideas regarding protein metabolism were based on observations of the urinary excretion products which were observed following the ingestion of proteins. The major end products of nitrogen metabolism in higher animals were found to be urea, creatinine, and uric acid. Early workers in the field believed that most of the amino acids absorbed from the diet did not enter into tissue protein metabolism, but were catabolized for energy and the nitrogen excreted as urea. This constituted the *exogenous* metabolism of protein. A smaller amount, the *endogenous* metabolism, which

Urea   Creatinine   Uric acid

was thought to represent that amount of amino acid nitrogen which was somehow involved in the wear and tear of tissue proteins was reflected by creatinine excretions. These ideas were based on the observations that urea excretion, but not creatinine excretion, was influenced by varying the amount of protein in the diet. It is now known that the uniformity of creatinine excretion is the result of a constant formation of creatinine by the breakdown of some of the phosphocreatinine present in muscle. The rate of this reaction is roughly proportional to muscle mass, and has nothing to do with amino acid metabolism.

A real understanding of the details of amino acid metabolism was not possible until the availability of isotopically labeled compounds. When ^{15}N as a stable isotope of nitrogen became available, it was used to demonstrate that even in nongrowing animals dietary amino acids labeled with ^{15}N were incorporated into tissue proteins and that tissue proteins were being constantly broken down and resynthesized. These observations led to what is now known as the *dynamic equilibrium* theory of body proteins. That is, tissue proteins are constantly being broken down and resynthesized, and the amino acids released from these proteins are continually equilibrating with those which are absorbed from dietary protein. This relationship is illustrated in Figure 12-2.

Figure 12-2 Relationship between ingested proteins and tissue proteins. Amino acids from both sources form a common metabolic pool which can provide amino acids for resynthesis of proteins or which can be metabolized to other end products.

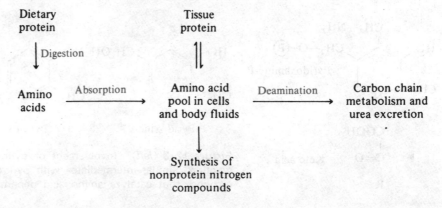

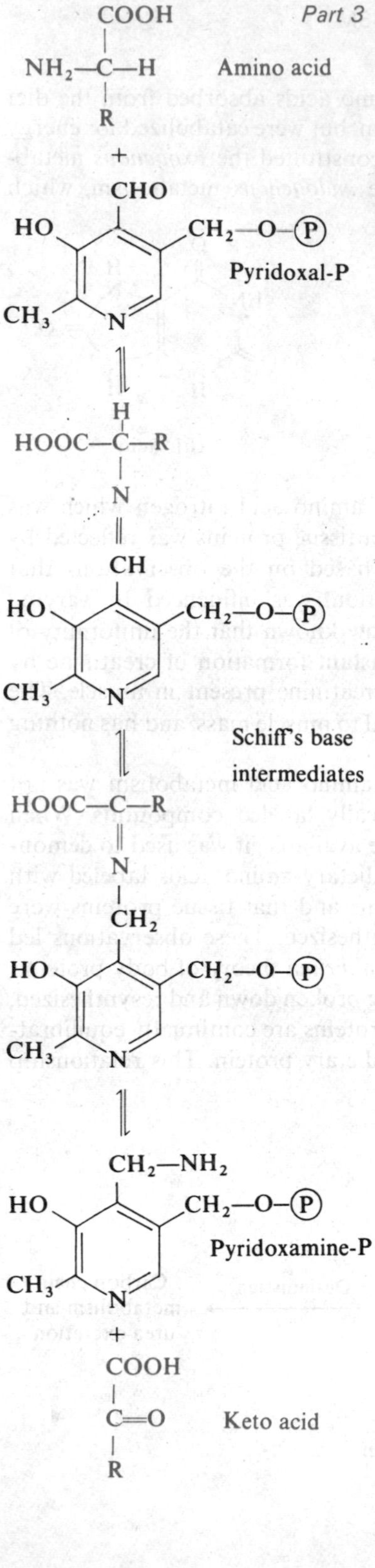

COOH
|
NH$_2$—C—H   Amino acid
|
R

+

CHO

HO   CH$_2$—O—Ⓟ

CH$_3$ N   Pyridoxal-P

‖

H
|
HOOC—C—R
|
N
‖
CH

HO   CH$_2$—O—Ⓟ

CH$_3$ N

Schiff's base
intermediates

‖

HOOC—C—R
‖
N
|
CH$_2$

HO   CH$_2$—O—Ⓟ

CH$_3$ N

‖

CH$_2$—NH$_2$
HO   CH$_2$—O—Ⓟ

CH$_3$ N   Pyridoxamine-P

+

COOH
|
C=O   Keto acid
|
R

AMINO ACID METABOLISM

Although the majority of ingested amino acids in growing animals is used for protein synthesis, there is a continual metabolism of the excess. In the adult animal, a large quantity of amino acids is constantly being metabolized for energy.

The general metabolic pathways involved are those which remove the amino group and shunt the carbon skeleton of the amino acid to some other metabolic pathway.

Transamination The most widespread reaction of amino acid catabolism is transamination. The generalized reaction is the transfer of an amino group from one amino acid to a keto acid, converting it to an amino acid with the formation of a new keto acid from the original amino acid. The mechanism of the reaction involves the participation of an enzyme-bound

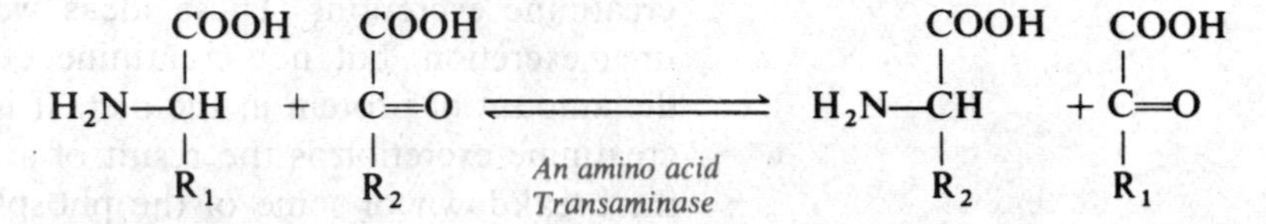

COOH   COOH   COOH   COOH
| | | |
H$_2$N—CH + C=O ⇌ H$_2$N—CH + C=O
| | *An amino acid* | |
R$_1$ R$_2$ *Transaminase*   R$_2$ R$_1$

pyridoxal phosphate to form an intermediate Schiff's base or aldamine with the amino acid (Figure 12-3). The other amino acid-metabolizing enzymes which have a pyridoxal phosphate requirement form a similar complex during the course of the reaction.

Pyridoxine (Vitamin B$_6$) This compound was identified as a required factor for the nutrition of the rat by Gyorgy in 1934 and was isolated by a number of other groups by 1938. The aldehyde (pyridoxal), amino (pyridoxamine), and alcohol (pyridoxine) forms of the vitamin all have biological activity for the human. Pyridoxal phosphate is the metabolically active coenzyme form of the vitamin. A deficiency of the vitamin usually results in some form of dermatitis and a specific anemia. There is no established human requirement, but a dietary allowance of 2 mg per day is recommended. Good sources of the vitamin are liver, meat, milk, and cereal grains.

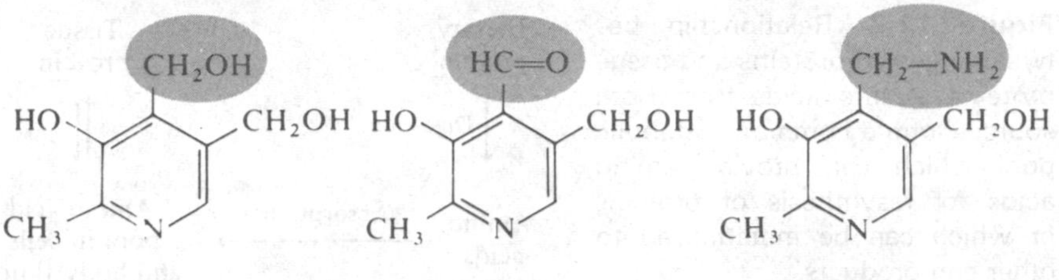

CH$_2$OH   HC=O   CH$_2$—NH$_2$
HO   CH$_2$OH   HO   CH$_2$OH   HO   CH$_2$OH

CH$_3$ N   CH$_3$ N   CH$_3$ N

Pyridoxine   Pyridoxal   Pyridoxamine

Figure 12-3 (*left*) Involvement of pyridoxal phosphate in a transaminase reaction. Similar intermediates with pyridoxal phosphate are formed by the enzymes that catalyze amino acid decarboxylations and racemizations.

286

The most active transaminase in mammalian tissues is the glutamic-oxaloacetate transaminase, which produces α-ketoglutarate and aspartate, and the glutamate-pyruvate transaminase, which yields α-ketoglutarate and alanine. Even though these are specific transaminases, it is now apparent that there is some transaminase activity for almost all the amino acids and that this enzyme system offers a way to convert all the amino acids to their corresponding keto acids.

L - *Glutamate Dehydrogenase* Probably the most active amino acid-metabolizing enzyme in mammalian tissue is the one catalyzing the oxidative deamination of glutamate. The reaction is readily reversible and can utilize either NAD^+ or $NADP^+$. Which coenzyme is the most active depends on the source of the enzyme and, in some cases, on whether or not

L-Glutamic acid   *Glutamic dehydrogenase*   α-Ketoglutaric acid

the reaction is studied in the presence of the allosteric modifiers of the enzyme—ADP or ATP. The reaction is particularly important because of the ability of the various transaminase systems to use α-ketoglutaric acid to form glutamic acid, which can then be metabolized by this enzyme. Although there are specific dehydrogenases for other amino acids in microbial systems, this is the only one with high activity in mammalian tissue.

Serine Dehydratase This enzyme carries out the nonoxidative deamination of L-serine to yield pyruvate. Along with a threonine dehydratase, which carries out the analogous reaction to yield α-ketobutyrate, it is found in rather high concentrations in most mammalian liver. In some tissues these activities are apparently catalyzed by the same enzyme; in others two different proteins are involved. Both of them require pyridoxal phosphate as a coenzyme.

Serine   *Serine dehydratase*   Pyruvic acid

Decarboxylations Another series of reactions which are of physiological importance are the pyridoxal phosphate-dependent amino acid decarboxylases. Enzymes of this type are widespread; of particular importance are

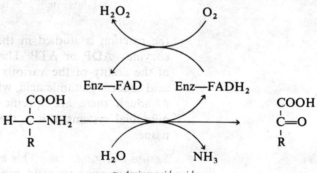

$$
\begin{array}{ccc}
\text{COOH} & & \text{NH}_2 \\
| & & | \\
\text{H}_2\text{N}-\text{C}-\text{H} & \xrightarrow[\text{decarboxylase}]{\text{Glutamic acid}} & \text{CH}_2 \\
| & (\text{CO}_2) & | \\
\text{CH}_2 & & \text{CH}_2 \\
| & & | \\
\text{CH}_2 & & \text{CH}_2 \\
| & & | \\
\text{COOH} & & \text{COOH}
\end{array}
$$

Glutamic acid   γ-Amino butyric acid

the decarboxylases, which act specifically on histidine and tryptophan to produce the very physiologically active compounds, histamine and tryptamine.

Amino Acid Oxidases The amino acid oxidase with the highest activity in mammalian kidney and liver is a D-*amino acid oxidase*. It is a FAD-containing enzyme which catalyzes the reaction shown below. The H_2O_2 which is produced by the reaction is, in most cases, rapidly broken down

$$
\begin{array}{c}
\text{H}_2\text{O}_2 \qquad\qquad \text{O}_2 \\
\\
\text{Enz}-\text{FAD} \qquad \text{Enz}-\text{FADH}_2 \\
\\
\text{COOH} \qquad\qquad\qquad\qquad \text{COOH} \\
| \qquad\qquad\qquad\qquad\qquad | \\
\text{H}-\text{C}-\text{NH}_2 \longrightarrow \qquad\qquad \text{C}=\text{O} \\
| \qquad\qquad\qquad\qquad\qquad | \\
\text{R} \qquad\qquad\qquad\qquad\qquad \text{R} \\
\\
\text{H}_2\text{O} \qquad\qquad \text{NH}_3
\end{array}
$$

D-*Amino acid oxidase*

by intracellular catalase to H_2O. The enzyme will oxidize most of the D-amino acids but has the greatest activity toward D-proline and the neutral amino acids as substrates. Acidic and basic amino acids are oxidized by this enzyme at a relatively slow rate. Very little is known about the metabolic function of this enzyme, and although D-amino acids are not rare in natural products, from a quantitative dietary standpoint they are not very important. It has been suggested that the function of the enzyme is to remove potentially toxic D-amino acids from cells, but this has not been shown to be necessary; large quantities of some of the D-amino acids can be tolerated by most species.

A generalized L-amino acid oxidase has been purified from avian liver, but it is not as active as the D-amino acid oxidase, and has been very difficult to find in other species. It carries out essentially the same reaction as the D-amino acid oxidase.

Glutaminase and Asparaginases   Much of the glutamic acid and aspartic acid in tissues is found not as the free acid, but as the corresponding amide, glutamine or asparagine. There are widespread tissue enzymes which are able to hydrolyze these amino acids to the corresponding free acid. This reaction, which amounts to the hydrolysis of an amide bond, proceeds completely to the right as written, and there are separate enzymes required for the synthesis of these compounds.

$$
\begin{array}{ccc}
\text{COOH} & & \text{COOH} \\
| & & | \\
\text{H}_2\text{N}-\text{C}-\text{H} & & \text{H}_2\text{N}-\text{CH} \\
| & \xrightarrow[\text{Glutaminase}]{\text{H}_2\text{O} \qquad \text{NH}_3} & | \\
\text{CH}_2 & & \text{CH}_2 \\
| & & | \\
\text{CH}_2 & & \text{CH}_2 \\
| & & | \\
\text{C}-\text{NH}_2 & & \text{C}-\text{OH} \\
\| & & \| \\
\text{O} & & \text{O} \\
\text{Glutamine} & & \text{Glutamic acid}
\end{array}
$$

NITROGEN EXCRETION

The method of excretion of excess nitrogen varies in different species. Most aquatic animals, presumably because they are readily able to dilute this rather toxic compound, excrete ammonia and are called *ammonotelic* animals. Mammals and most other terrestrial vertebrates that excrete urea are called *ureotelic* animals, and those which excrete uric acid as the major nitrogen excretion product are described as *uricotelic*. This latter group, birds and reptiles, often have a limited water intake, and uric acid, because of its low solubility, is excreted as a semisolid suspension of crystals.

Urea Biosynthesis

The deamination reactions which have been discussed convert excess amino acids to the corresponding keto acids and liberate ammonia. In most mammals, this toxic compound is excreted as urea, which is formed by a series of metabolic reactions located in the liver. The pathway is indicated in Figure 12-4. These reactions constitute a cyclic system which forms urea from arginine and then resynthesizes arginine from ornithine by the addition of two moles of ammonia and one of CO_2.

The first mole of ammonia added to ornithine is initially converted to the very reactive compound, carbamyl phosphate, by the addition of CO_2. This is a reaction which, at least in mammalian tissues, requires two moles of ATP. The action of this enzyme, *carbamyl phosphate synthetase*, is complex, and it requires the presence of *N*-acetyl-glutamate, presumably as an allosteric activator. The carbamyl phosphate synthesized by this reaction reacts with ornithine to form citrulline, with the loss of a mole of phosphate. The enzyme catalyzing the reaction is *ornithine transcarb-*

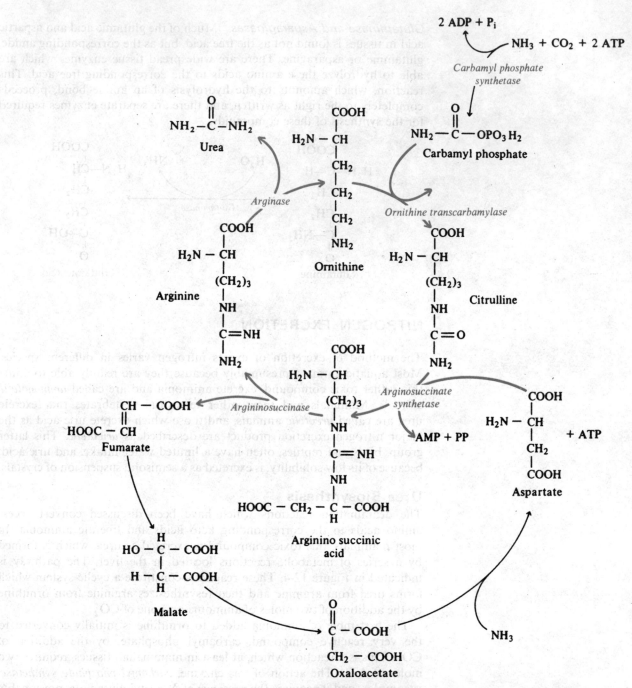

Figure 12-4 Enzymes and intermediates of the urea cycle. The fumarate, which is produced by argininosuccinate breakdown, can be converted by enzymes of the citric acid cycle (Chapter 13) to oxaloacetate, and this can be transaminated to reform aspartic acid.

amylase, and the reaction has an equilibrium which lies far in the direction of citrulline formation.

The second mole of nitrogen is introduced into the urea cycle by a reaction involving the condensation of citrulline and aspartic acid to form argininosuccinic acid. The reaction is catalyzed by *argininosuccinate synthetase,* and the energy for the formation of this complex acid is furnished by the breakdown of a mole of ATP to AMP and pyrophosphate. The equilibrium for the reaction is strongly in the direction of argininosuccinate formation.

The enzyme *argininosuccinase* catalyzes a readily reversible cleavage of argininosuccinate to arginine and fumarate. Figure 12-4 also indicates that the fumarate which is formed can readily enter the series of reactions of the citric acid cycle to form oxaloacetate. This compound acts as the keto acid acceptor for a number of amino acids in a transaminase reaction and thus generates more aspartic acid.

The key enzyme in the urea cycle, *arginase,* catalyzes a hydrolytic cleavage of urea from arginine and regenerates ornithine. Although many tissues are able to synthesize arginine and have the other enzymes important to the pathway, they do not form urea. Only in the liver of the animals that form urea as a nitrogen excretion product is arginase found in high concentrations.

ENERGY UTILIZATION FROM AMINO ACIDS

Most of the amino acids can undergo a transamination reaction, and the pathway for their oxidation may be very direct. For example, alanine, aspartic acid, and glutamic acid will be transaminated to the corresponding keto acids, pyruvate, oxaloacetate, and α-ketoglutarate. Each of these can be directly oxidized by the reactions of the citric acid cycle (Chapter 13). For many of the other amino acids, the pathway to the formation of oxidizable intermediates is not so direct. Before ^{14}C-labeled amino acids were available, it was difficult to determine the pathway of metabolism of these compounds, and various indirect means were used. One technique was to feed large amounts of an amino acid to starved animals to see if this resulted in an increase in blood glucose or in acetoacetate (ketone bodies). On this basis, amino acids were divided into two groups, those which were glucogenic and those which were ketogenic. It is now clear that the glucogenic group includes those which are able to convert the deaminated carbon chain to pyruvate or Krebs cycle intermediates, and the ketogenic group includes those which can be degraded to acetyl CoA. Only one amino acid, leucine, is completely ketogenic. Four other amino acids, isoleucine, lysine, phenylalanine, and tyrosine, are both ketogenic and glucogenic. The rest of the amino acids do not furnish carbons to either acetyl CoA or acetoacetate, and are only glucogenic. As such, they provide a very important source of carbon for glucose synthesis during periods of starvation.

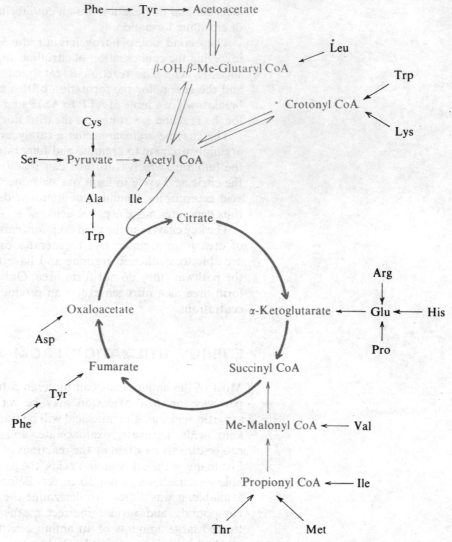

Figure 12-5 Metabolic fate of the common amino acids. Following deamination, the carbon skeletons of the amino acids become intermediates in other metabolic pathways. Note that some amino acids contribute carbon to more than one metabolic pool.

A description of the metabolic fate of each of the amino acids is beyond the scope of this book. Figure 12-5 provides an indication of the major route of metabolism of the various amino acids, and some of the interconversions that occur. The pathway in Figure 12-6 for the metabolism of isoleucine, which is both ketogenic and glucogenic, indicates the types of reactions involved.

A number of hereditary diseases are associated with abnormalities of amino acid metabolism. One of the most common is phenylketonuria, which is associated with a defect in phenylalanine metabolism. The first step in phenylalanine degradation is its conversion to tyrosine, and if the enzyme catalyzing this step is missing, the amino acid and its corresponding

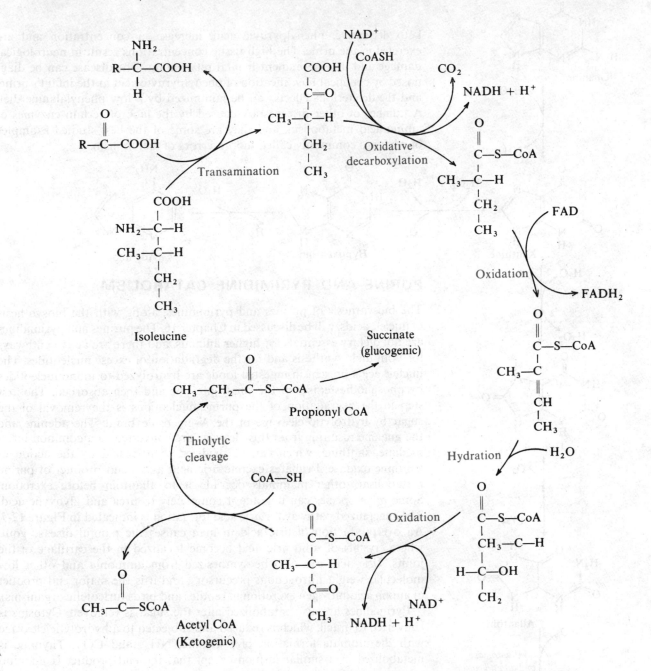

Figure 12-6 Simplified pathway for the degradation of isoleucine. The types of reactions involved are typical of those involved in the metabolism of the other amino acids. Isoleucine is partially converted to propionyl CoA, and then to succinate, a good precursor for gluconeogenesis, and partially converted to acetyl CoA.

keto derivative, phenylpyruvic acid, increase in concentration and are excreted in the urine. The high tissue concentrations result in neurological damage and in a permanent mental retardation. The disease can be diagnosed by chemical identification of phenylpyruvic acid in the infant's urine, and the deleterious effects can be minimized by a low phenylalanine diet. A number of other diseases are caused by the lack of certain enzymes of amino acid metabolism, and they are some of the best-studied examples of what are commonly called inborn errors of metabolism.

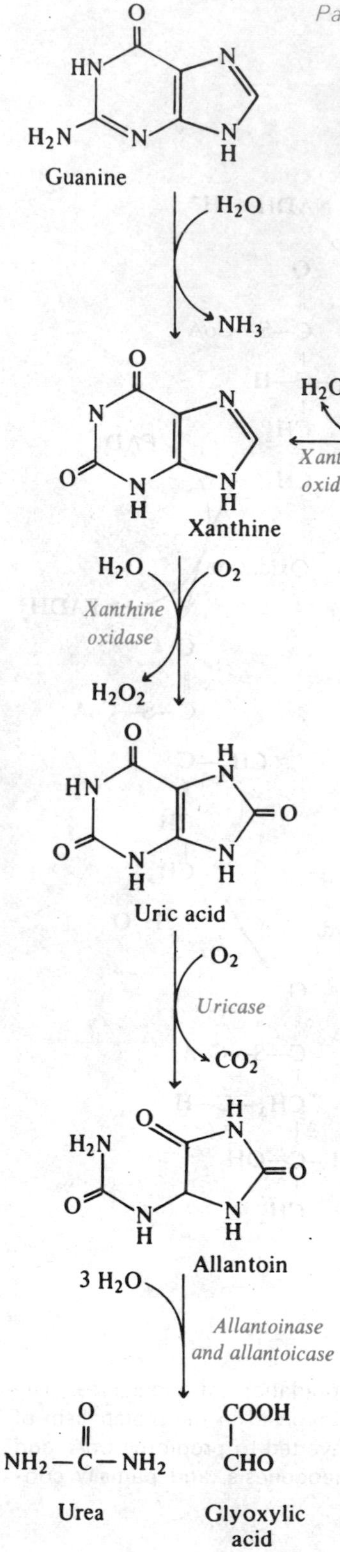

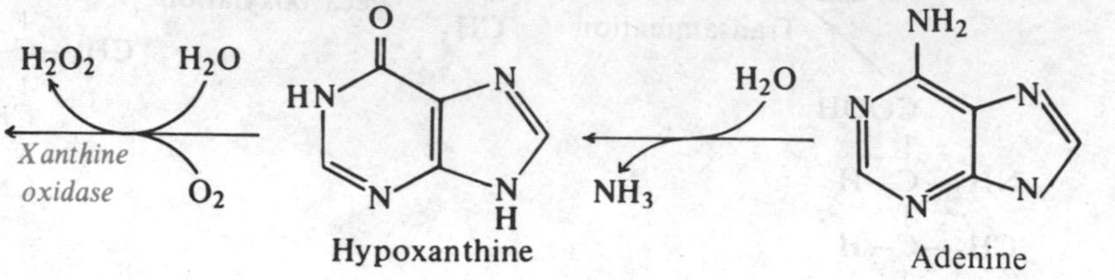

PURINE AND PYRIMIDINE CATABOLISM

The biosynthesis of purines and pyrimidines, along with the biosynthesis of nucleic acids, will be discussed in Chapter 16. The purines and pyrimidines are not dietary essentials for higher animals, and there are active pathways for both their synthesis and for the degradation of excess nucleotides. The nucleic acids present in ingested foods are hydrolyzed to mononucleotides by phosphodiesterases in the small intestine and then absorbed. The first step in the degradation of the purine nucleotides is the removal of the sugar by hydrolytic cleavage of the N-glycoside bond. The adenine and the guanine resulting from this cleavage are converted to a common intermediate, xanthine, which can be oxidized to uric acid by the action of xanthine oxidase. Primates excrete uric acid as an end product of purine metabolism; other mammals degrade it to allantoin before excretion. Some other species can degrade it completely to urea and glyoxylic acid. The generalized pathway for uric acid formation is indicated in Figure 12-7. An overproduction of uric acid in man causes the painful disease gout, where crystals of solid uric acid become localized in the cartilage of the joints. Uric acid can also be synthesized from ammonia and other low molecular weight nitrogenous precursors, and it is the major end product of amino acid nitrogen excretion in reptiles and birds (uricotelic organisms).

Pyrimidines are also catabolized after the sugar is removed. Cytosine is converted to uracil, which is reduced and subjected to a hydrolytic cleavage with the ultimate formation of β-alanine, NH_3, and CO_2. Thymine is metabolized in a similar fashion except that the end product is β-amino isobutyric acid rather than β-alanine (Figure 12-8).

Figure 12-7 Simplified diagram of the metabolism of purines. The conversion of adenine to hypoxanthine can occur at the level of the free base, the nucleoside or the nucleotide. The enzyme xanthine oxidase catalyzes the oxidation of both xanthine and hypoxanthine as well as a large number of other aldehydes.

Figure 12-8 Simplified metabolic pathway for the degradation of pyrimidines.

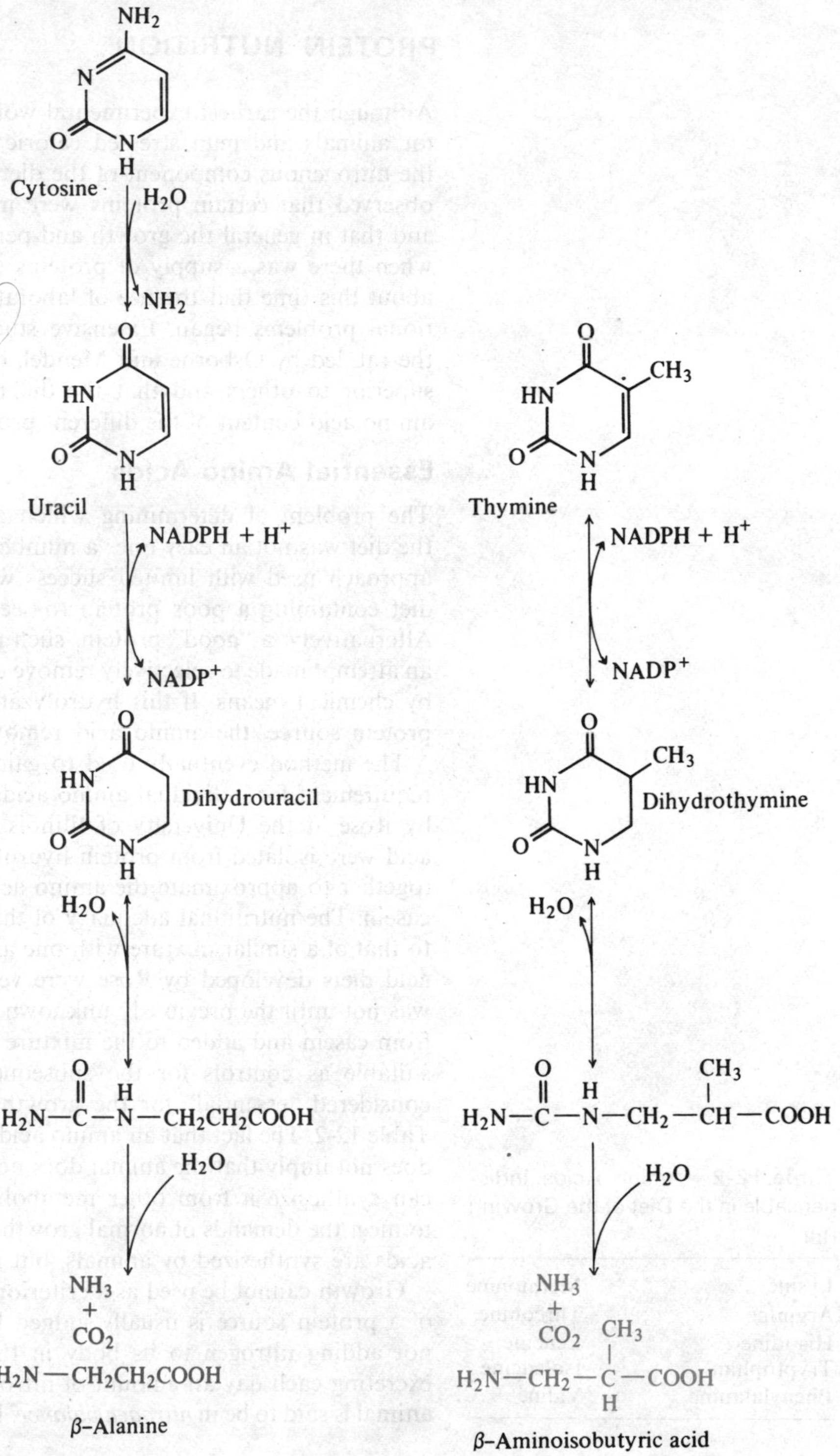

Cytosine

H_2O

NH_3

Uracil

$NADPH + H^+$

$NADP^+$

Dihydrouracil

H_2O

$H_2N - C - N - CH_2CH_2COOH$

H_2O

NH_3
$+$
CO_2

$H_2N - CH_2CH_2COOH$

β–Alanine

Thymine

$NADPH + H^+$

$NADP^+$

Dihydrothymine

H_2O

$H_2N - C - N - CH_2 - CH - COOH$

H_2O

NH_3
$+$
CO_2

$H_2N - CH_2 - C - COOH$

β–Aminoisobutyric acid

PROTEIN NUTRITION

Although the earliest experimental work on the adequacy of various diets for animals and man stressed caloric content, it was soon realized that the nitrogenous component of the diet was of extreme importance. It was observed that certain proteins were in some manner better than others, and that in general the growth and performance of animals was enhanced when there was a supply of proteins of animal origin in the diet. It was about this time that the use of laboratory animals for the study of nutritional problems began. Extensive studies of the protein requirement of the rat, led by Osborne and Mendel, established that some proteins were superior to others and that the difference must be in the variations of amino acid content of the different proteins.

Essential Amino Acids

The problem of determining which amino acids were indispensable in the diet was not an easy one; a number of procedures were used. An early approach used with limited success was to add single amino acids to a diet containing a poor protein to see if the growth could be improved. Alternatively, a "good" protein, such as casein, could be hydrolyzed and an attempt made to selectively remove one amino acid from the hydrolyzate by chemical means. If this hydrolyzate could no longer serve as a good protein source, the amino acid removed would be considered essential.

The method eventually used to gain most of the knowledge about the requirement for individual amino acids was the laborious process applied by Rose at the University of Illinois. Sufficient amounts of each amino acid were isolated from protein hydrolyzates so that they could be mixed together to approximate the amino acid composition of a protein such as casein. The nutritional adequacy of this mixture would then be compared to that of a similar mixture with one amino acid removed. The first amino acid diets developed by Rose were very inferior to hydrolyzed casein. It was not until the previously unknown amino acid, threonine, was isolated from casein and added to the mixture that these amino acid diets became suitable as controls for the subsequent experiments. The amino acids considered "essential" for the growth of the young rat are indicated in Table 12-2. The fact that an amino acid is not considered a dietary essential does not imply that the animal does not need it, but rather that the animal can synthesize it from other metabolic intermediates at a sufficient rate to meet the demands of normal growth. In fact, some of the essential amino acids are synthesized by animals, but not in adequate amounts.

Growth cannot be used as a criterion of need in the adult, and adequacy of a protein source is usually judged by *nitrogen balance*. If an animal is not adding nitrogen to its body in the form of new protein, it must be excreting each day an amount of nitrogen equal to that ingested. Such an animal is said to be in *nitrogen balance*. If one essential amino acid is lacking,

Table 12-2 Amino Acids Indispensable in the Diet of the Growing Rat

Lysine	Methionine
Arginine	Threonine
Histidine	Leucine
Tryptophan	Isoleucine
Phenylalanine	Valine

the tissues cannot use the other amino acids for protein synthesis either, and they will be catabolized and the nitrogen excreted. This excess excretion then results in a negative nitrogen balance; that is, more nitrogen is excreted than ingested. The amino acids essential in the diet of the adult human have been determined in this fashion and found to be nearly the same as those needed by the rat. The adult human does not require arginine and histidine. In the sense that all amino acids found in body proteins could be considered essential to the animal, the term "indispensable" is preferred by many to designate those amino acids which must be included in the diet.

Protein Quality

The relative nutritional worth of various proteins is usually measured by their *biological value*. This is an experimentally determined indication of the percentage of protein nitrogen absorbed by the animal which is actually retained in the body. The biological value of a number of proteins is indicated in Table 12-3. It can be seen that, in general, animal proteins are good proteins from a nutritional standpoint, whereas many common plant proteins are poor. The low biological value of plant seed proteins is mainly the consequence of their low tryptophan, methionine, and lysine content, and their biological value can be greatly improved by the addition of these amino acids. The amount of protein needed by an animal is that amount which will supply sufficient amounts of all the indispensable amino acids and sufficient excess nitrogen to synthesize the nonessential amino acids. For a protein of good biological value the adult human requirement has often been considered to be on the order of one gram of protein per kg body weight, but recent evidence indicates that it might be considerably lower than this.

At the present time protein malnutrition is a serious problem in many parts of the world. Although it can take a number of different forms, protein malnutrition is often characterized by a complex nutritional disease called *Kwashiorkor*. Most of these areas have a severe lack of animal protein, and attempts are now being made to supplement or develop adequate mixtures of available plant proteins to cure this disease. The cure is complicated by the fact that vitamin and mineral deficiencies are often associated with the protein deficiency, and in many forms of infant protein deficiencies, there is also a total caloric deficiency.

Table 12-3 Biological Value of Some Common Proteins

Protein	Biological Value[a]	
	Man	Growing Rat
Whole egg	0.94	0.87
Egg white	0.91	0.97
Beef	0.67	0.76
Casein	0.68	0.69
Penut flour	0.56	0.54
Wheat gluten	0.42	0.40

[a] This value is a measurement of the fraction of the absorbed food nitrogen retained in the body. Those proteins with a higher biological value are "better" proteins. (Reprinted with permission from *Mammalian Protein Metabolism*, Vol. 2, H. N. Munro and J. B. Allison, eds., Academic Press, 1964, p. 57.)

PROBLEMS

1. If glutamic dehydrogenase is the only L-amino acid oxidative deaminating enzyme found in high concentration in most mammalian tissues, how can ammonia be liberated from almost all free amino acids at a rapid rate?

2. What are the direct metabolic precursors of the two nitrogen atoms of urea?

3. Why might a deficiency of biotin result in an impairment of isoleucine metabolism, but not of leucine metabolism?

4. What would be the products of the partial transaminase reaction obtained by reacting pyridoxal phosphate and alanine with the enzyme?

5. Pathways are known for the degradation of purines to urea. Would the ingestion of diets high in nucleic acids by a human result in an increased urea excretion?

Suggested Readings

Books

Davidson, J. N. *Biochemistry of Nucleic Acids,* 7th ed. Methuen, Wiley, London (1972).

Meister, A. *Biochemistry of the Amino Acids,* 2nd ed., vols. 1 and 2. Academic Press, New York (1965).

Articles and Reviews

Cohen, P. P., and G. W. Brown, Jr. Ammonia Metabolism and Urea Biosynthesis, in *Comparative Biochemistry,* Florkin, M., and H. S. Mason (eds.), vol. 11, pp. 161–294. Academic Press, New York (1961).

Greenberg, D. M., and W. W. Rodwell. Carbon Catabolism of Amino Acids, in *Metabolic Pathways,* Greenberg, D. (ed.), 3rd ed., vol. 3, pp. 000–000. Academic Press, New York (1969).

Knox, W. E. Phenylketonuria, in *The Metabolic Basis of Inherited Disease,* Stanbury, J. B., J. B. Wyngaarden, and D. S. Fredrickson (eds.), 3rd ed., pp. 266–295. McGraw-Hill, New York (1972).

Space-filled
model of ATP.

tricarboxylic acid cycle and energy production in metabolism

chapter thirteen

Pyruvate from carbohydrate metabolism can be
converted to acetyl coenzyme A, and large
amounts of acetyl coenzyme A are formed from
fatty-acid degradation. This key metabolic
intermediate is then oxidized to CO_2 by the
mitochondrial enzymes of the tricarboxylic acid
cycle. In this process acetyl coenzyme A is
condensed with oxaloacetic acid to form citric
acid and this 6-carbon acid is subsequently
subjected to a series of oxidations which
regenerate oxaloacetate and yield CO_2, ATP, and
reduced coenzymes. As the reduced coenzymes can
be reoxidized by the mitochondrial electron-
transport chain to form ATP, this series of
reactions constitute the major pathway of both
CO_2 generation and energy production in aerobic
organisms. The amount of energy animals obtain
from their diet can be measured by determining
the composition of the diet and can be related to
energy expenditure. About one-half the energy a
person needs is utilized for basal metabolism,
while the rest is required for physical activity.

In most organisms the major portion of cellular energy is generated by the oxidation of carbohydrates and fatty acids. It has already been indicated that fatty acids can be readily degraded to acetyl coenzyme A, and that the Embden-Meyerhof pathway can convert glucose or the other common sugars to pyruvic acid. Although a small amount of ATP is generated by the glycolytic pathway and an appreciable amount of NADH is formed by the β-oxidation of fatty acids, the efficient energy conversion, characteristic of aerobic organisms, is due to the complete oxidation of pyruvate or acetyl CoA to CO_2 and water.

PYRUVIC DEHYDROGENASE SYSTEM

The major energy-yielding pathway of aerobic metabolism, the citric acid cycle, is a mechanism for the complete oxidation of acetate in the form of acetyl CoA. Pyruvate, generated by carbohydrate breakdown, is converted to this intermediate by the action of the pyruvate dehydrogenase complex. The overall reaction catalyzed by this mitochondrial enzyme is shown below. The reaction does not occur in a single step; rather, three enzymes

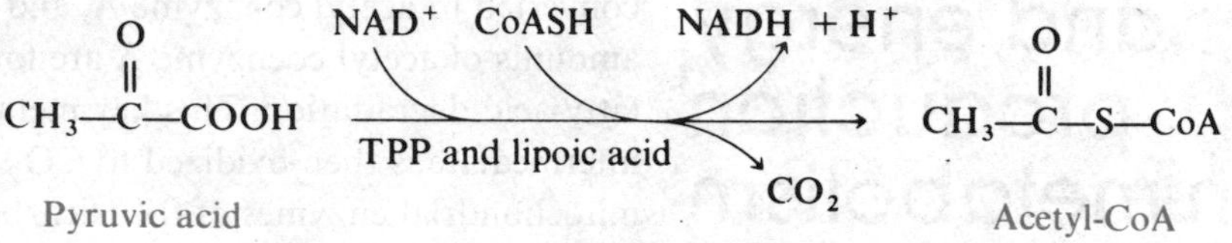

$$
\underset{\text{Pyruvic acid}}{CH_3-\overset{\overset{\displaystyle O}{\|}}{C}-COOH} \quad \xrightarrow[\text{TPP and lipoic acid}]{\text{NAD}^+ \quad \text{CoASH} \quad \text{NADH} + \text{H}^+ \quad \searrow \quad CO_2} \quad \underset{\text{Acetyl-CoA}}{CH_3-\overset{\overset{\displaystyle O}{\|}}{C}-S-CoA}
$$

Pyruvic dehydrogenase
complex

and five separate coenzymes are involved. The mechanism of the reaction is indicated in Figure 13-1; it can be seen from these reactions that the NADH produced in the reaction has not been formed by the direct oxidation of pyruvate. Rather, it has been generated during the reoxidation of the reduced flavin that was formed when reduced lipoic acid was oxidized. The coenzyme A and the NAD$^+$ involved are freely dissociable from the enzyme complex, but the thiamine pyrophosphate, FAD, and lipoic acid are tightly bound—the latter by a covalent bond to a lysine residue.

The complex consists of three different enzymes: a *pyruvate dehydrogenase*, a *dihydrolipoyl transacetylase*, and a *dihydrolipoyl dehydrogenase*. The complex has been purified from many sources, but the preparation most extensively studied has been obtained from *E. coli*. This complex has a molecular weight of about 4 million, and the distinct morphology of the 72 protein molecules involved has been elucidated (see Chapter 5). It is thought that the long side chain of lipoic acid (attached to the dihydrolipoyl transacetylase) actually moves between the other two enzymes, where it can be alternately acetylated, deacetylated, oxidized, and recycled (Figure

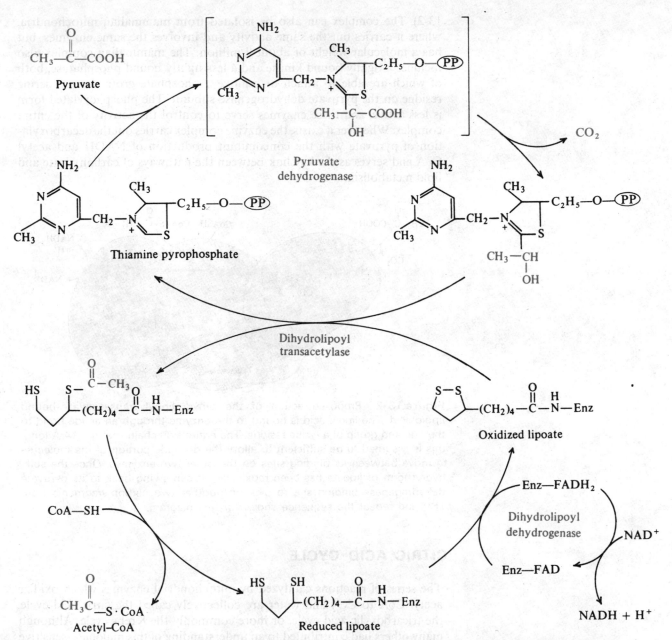

Figure 13-1   Mechanism of action of the pyruvic dehydrogenase complex. Note that the three enzymes have distinct functions. The *pyruvate dehydrogenase,* through the use of thiamine pyrophosphate, catalyzes a decarboxylation of pyruvate, and the transacetylase then transfers the two carbon fragment formed from thiamine pyrophosphate to coenzyme A with lipoic acid as an intermediate carrier and oxidizing agent. The *dihydrolipoyl dehydrogenase* then functions to reoxidize the reduced lipoic acid which is formed.

13-2). The complex can also be isolated from mammalian mitochondria, where it carries out the same activity and involves the same enzymes, but has a molecular weight of about 7 million. The mammalian complex also includes a tightly bound kinase and a less tightly bound phosphatase, both of which are able to attach or remove a phosphate group from a serine residue on the pyruvate dehydrogenase subunit. The phosphorylated form is less active, and these enzymes serve to control the activity of the entire complex. Wherever it exists, the enzyme complex carries out the decarboxylation of pyruvate with the concomitant production of NADH and acetyl CoA and serves as the key link between the pathways of carbohydrate and lipid metabolism.

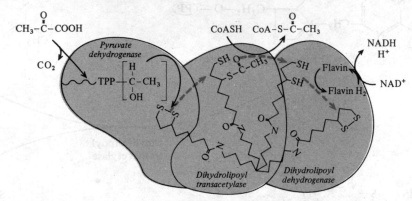

Figure 13-2 Proposed action of the dihydrolipoyl *trans*acetylase-bound lipoic acid. The lipoic acid is bound to the enzyme through an amide bond to the ϵ-amino group of a lysine residue. The entire side chain is about 14 Å long; this is assumed to be sufficient to allow the disulfide portion of the molecule to move between its binding sites on the other two enzymes. Once the sulfhydryl form of lipoate has been reoxidized, it can swing back to its pyruvate dehydrogenase binding site to pick up another two carbon fragments from TPP and repeat the sequence shown in this diagram.

CITRIC ACID CYCLE

The series of reactions catalyzed by mitochondrial enzymes which oxidize acetyl CoA to CO_2 and water are collectively called the citric acid cycle, the tricarboxylic acid cycle, or more commonly the Krebs cycle. Although many others had contributed to an understanding of this malonate-sensitive system which could oxidize several different dicarboxylic acids, Krebs contributed a number of key ideas. He was the first to show the catalytic nature of citrate and to realize that the conversion of the 6-carbon acids to 4-carbon acids could be made a cyclic process if they were subsequently condensed with a 2-carbon intermediate to reform a 6-carbon acid. Although it was some time before experimental evidence supported all Krebs' postulations, they were eventually shown to be essentially correct.

The reactions involved are shown in Figure 13-3. They constitute a mechanism for sequentially condensing acetyl CoA with oxaloacetic acid to form citrate, oxidizing the citrate to succinate with the liberation of 2 moles of CO_2, and converting succinate to oxaloacetate, thus regenerating the original acetyl CoA acceptor. Only catalytic amounts of the intermediate acids are needed to carry out the oxidation of large amounts of acetyl CoA and the formation of the reduced coenzymes needed for ATP production in aerobic cells.

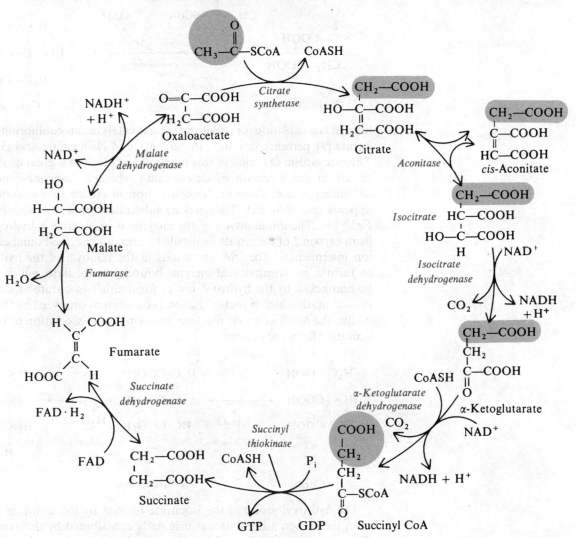

Figure 13-3 Intermediates and enzymes of the citric acid cycle (Krebs cycle). The carbons of citrate that come from acetyl CoA are shaded. It can be seen that these are not lost as CO_2 during the first turn of the cycle. After the formation of succinate, it is no longer possible to distinguish the carbons that originated from acetyl CoA, as succinate is a symmetrical compound.

303

Enzymes of Citric Acid Cycle

The first enzyme, *citrate synthetase* or the *citrate condensing enzyme*, catalyzes a reaction which is essentially irreversible with a $\Delta G^{0'}$ of -7.7 kcal/mole in the direction of citric acid formation. The reaction catalyzed is an aldol condensation, and the concerted cleavage of the thiol ester bond to coenzyme A drives the reaction in the direction of citrate formation. The citrate

$$
\begin{array}{ccc}
\text{O} & \text{O} & \text{H}_2\text{C—COOH} \\
\| & \| & | \\
\text{C—COOH} & \text{CH}_3\text{—C—SCoA} \quad \text{CoASH} & \text{HO—C—COOH} \\
| & \xrightarrow[\text{enzyme}]{\text{Citrate-condensing}} & | \\
\text{CH}_2\text{—COOH} & & \text{H}_2\text{C—COOH}
\end{array}
$$

Oxaloacetic acid   Citric acid

formed is a substrate for *aconitase*, which catalyzes an equilibrium between citrate (91 percent), isocitrate (6 percent), and *cis*-aconitic acid (3 percent). The mechanism of action of this reaction has received a great deal of study, chiefly in the direction of determining whether *cis*-aconitic acid is an obligatory intermediate in the conversion of citrate to isocitrate. It now appears that it is not. The enzyme-substrate complex is stabilized by a Fe^{2+} ion. The initial action of the enzyme is to abstract a hydrogen atom from carbon 2 of the citrate molecule to form an enzyme-bound carbonium ion intermediate. The iron atom aids in the removal of the hydroxyl ion to form a *cis*-aconitic acid enzyme-bound intermediate, which can then be reattacked by the hydroxyl ion to form either isocitrate or citrate. The *cis*-aconitate, which is part of the total equilibrium catalyzed by the enzyme, is thus the result of a side reaction involving the dissociation of bound *cis*-aconitate from the enzyme.

$$
\begin{array}{ccccc}
\text{H}_2\text{C—COOH} & & \text{H}_2\text{C—COOH} & & \text{H}_2\text{C—COOH} \\
| & & | & & | \\
\text{HOC—COOH} & \longleftrightarrow & \text{C—COOH} & \longleftrightarrow & \text{HC—COOH} \\
| & \text{H}_2\text{O} & \| & \text{H}_2\text{O} & | \\
\text{H}_2\text{C—COOH} & & \text{HC—COOH} & & \text{HOC—COOH} \\
& \text{Aconitase} & & \text{Aconitase} & | \\
& & & & \text{H}
\end{array}
$$

Citrate   *cis*-Aconitate   Isocitrate

The hydroxyl group of the isocitrate formed by the action of aconitase is on the carbon atom that was originally contributed by the oxaloacetate. This was not realized until investigators began studying the metabolism of radioactive citrate. It was then shown that the apparently symmetrical molecule, citrate, was being metabolized in an asymmetrical fashion. The confusing result was ultimately explained by Ogston in what is generally known as the three-point attachment theory. This theory explains that although a molecule may be symmetrical, the enzyme-substrate complex formed by the molecule may be asymmetrical if certain rules are followed. The central feature is that if a molecule with a meso carbon, such as citrate, makes an attachment with at least three different sites on an enzyme, it is possible to distinguish between what appear to be similar functional groups. This is illustrated in Figure 7-14. A number of such examples are now known, and compounds with meso carbons, which are metabolized in an asymmetric fashion, are fairly common in metabolic reactions.

Aconitase is the enzyme in the tricarboxylic acid cycle that is inhibited by fluoroacetate (Figure 13-4). Fluoroacetate can be activated to fluoro-

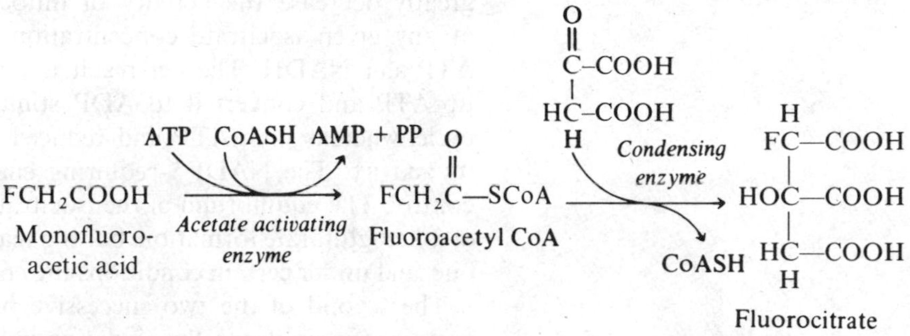

Figure 13-4 Sequential action of the acetate-activating enzyme and the condensing enzyme on monofluoroacetic acid to produce fluorocitric acid. The fluoroacetic acid itself does not inhibit the reactions of the citric acid cycle, and because of this, these reactions have been called a *lethal synthesis.*

acetyl CoA, and fluorocitrate can, therefore, be formed from fluoroacetate by the citrate synthetase system. Fluorocitrate is a classical competitive inhibitor of aconitase, and fluoroacetate poisoning therefore results in a large accumulation of unoxidized citrate in tissues.

The first oxidative enzyme in the pathway, *isocitrate dehydrogenase*, oxidizes the hydroxyl group of citrate to a keto group and causes the simultaneous decarboxylation of the compound to α-ketoglutaric acid. There is no evidence that the intermediate oxidation product, which would be oxalosuccinic acid, is present as a true intermediate in the conversion of isocitrate to α-ketoglutarate. However, in some forms of the enzyme which have been isolated, there is evidence that oxalosuccinic acid may exist in an enzyme-bound form.

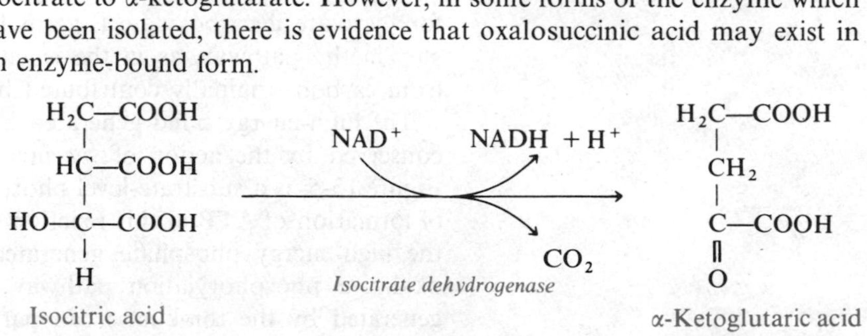

The enzyme requires a metal ion, either Mg^{2+} or Mn^{2+}, and there is both an NAD^+- and an $NADP^+$-linked enzyme in most cells. Although the situation was confusing for some time, it is now clear that the NAD^+-linked enzyme is located in the mitochondria, and that this is the enzyme responsible for most of the isocitrate oxidation of the tricarboxylic acid cycle. The $NADP^+$-linked enzyme is located in both mitochondria and cytoplasm, and its function is more likely that of producing NADPH to drive biosynthetic reactions. The reaction catalyzed by the mitochondrial enzyme is one of the rate-limiting reactions in the Krebs cycle and is subject to strong allosteric control by ADP. High concentrations of adenosine diphosphate stimulate the action of the enzyme, and low concentrations greatly decrease the activity of mitochondrial isocitrate dehydrogenase at any given isocitrate concentration. The enzyme is also inhibited by ATP and NADH. The net result is that conditions in the cell which use up ATP and convert it to ADP stimulate the activity of the citric acid cycle, whereas high ATP and reduced pyridine nucleotide levels decrease its activity. The $NADP^+$-requiring enzyme is not subject to this type of control. The equilibrium of the isocitrate dehydrogenase is in the direction of α-ketoglutarate formation, but the reaction can be considered a reversible one, and under certain conditions the enzyme catalyzes the reverse reaction.

The second of the two successive oxidations in the tricarboxylic acid cycle is very similar to the oxidation involved in the conversion of pyruvate to acetyl CoA. It is an oxidative decarboxylation of an α-keto acid to yield CO_2 and the acyl CoA derivative—in this case succinyl CoA—of the new acid which is formed. The enzyme, the *α-ketoglutarate decarboxylase complex*, requires thiamine pyrophosphate, lipoic acid, CoA, NAD^+, and enzyme-bound FAD. The mechanism is of the same type as that shown

$$\underset{\text{α-Ketoglutaric acid}}{\begin{array}{c} H_2C-COOH \\ | \\ CH_2 \\ | \\ C-COOH \\ \| \\ O \end{array}} \quad \xrightarrow[\text{TPP and lipoic acid}]{\overset{\displaystyle NAD^+ \quad CoASH \qquad NADH + H^+}{\qquad \qquad \qquad \qquad}} \quad \underset{\text{Succinyl-CoA}}{\begin{array}{c} CH_2-COOH \\ | \\ CH_2-C-S-CoA \\ \| \\ O \end{array}}$$

α-Ketoglutarate decarboxylase complex

CO_2

for pyruvate decarboxylase (Figure 13-1). The CO_2 which is lost at this step in the pathway, as in the isocitrate dehydrogenase reaction, comes from carbon originally contributed by oxaloacetate and not acetyl CoA.

The high-energy bond generated by the oxidation of α-ketoglutarate is conserved by the action of *succinyl thiokinase*. This reaction, shown in Figure 13-5, is a substrate-level phosphorylation reaction. The mechanism of formation of ATP in this reaction is very different from that of most of the high-energy phosphate generated in the mitochondria through the oxidative phosphorylation pathway discussed in Chapter 8. The GTP generated by the thiokinase can participate in a reaction with ADP to

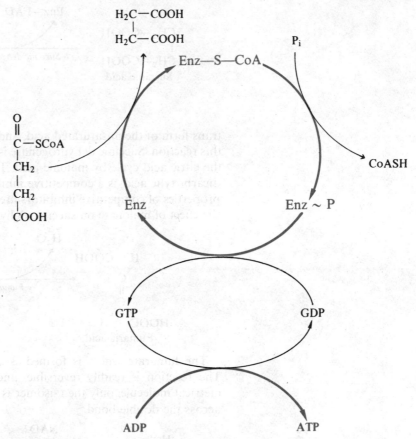

Figure 13-5 Reactions involved in the action of succinyl thiokinase, a substrate-level phosphorylation. It has been shown that the enzyme-phosphate intermediate in this system is a 3-phosphohistidine residue of the protein. A second enzyme can then catalyze a *trans*phosphokinase reaction to transfer the phosphate from GTP to ADP.

generate ATP and reform GDP, which can once again act as an acceptor of inorganic phosphate.

The reactions of the citric acid cycle up to this point have converted the 6-carbon tricarboxylic acid, citrate, to the 4-carbon dicarboxylic acid, succinate. The subsequent steps are oxidations, which regenerate oxalo-acetate. The next enzyme, *succinate dehydrogenase*, oxidizes succinate to fumarate by the use of an enzyme-bound flavin adenine dinucleotide. The enzyme appears to form an integral part of the mitochondrial membrane, and in addition to the covalently bound FAD it contains 4 moles of labile sulfur and 4 moles of nonheme iron.

As indicated in Chapter 8, this enzyme system transfers electrons to co-enzyme Q, bypassing one of the potential phosphorylation sites of the electron transport chain. The product of the reaction, fumaric acid, is the

Enz—FAD   Enz—FADH$_2$

CH$_2$—COOH   H   COOH

 | *Succinic dehydrogenase*   C

CH$_2$—COOH ||

Succinic acid   C

HOOC   H

Fumaric acid

trans form of the unsaturated acid, and none of the cis isomer results from this reaction. Succinic dehydrogenase is the site of the classical inhibition of the citric acid cycle by malonic acid. This compound, which is a 3-carbon dicarboxylic acid, is a competitive inhibitor of the enzyme, and the general properties of competitive inhibitors are often demonstrated by determining the effect of malonate on succinic dehydrogenase.

H$_2$O

H   COOH   COOH

  C   *Fumarase*   HO—CH

  || |

  C   CH$_2$

HOOC   H   COOH

Fumaric acid   Malic acid

The fumarate which is formed is hydrated to L-malate by *fumarase*. The reaction is readily reversible, and although fumaric acid is a symmetrical molecule, only the L-isomer is formed as a mole of water is added across the double bond.

NAD$^+$ NADH + H$^+$ O

H ||

HO—C—COOH   C—COOH

   | *Malate dehydrogenase* |

H$_2$C—COOH   H$_2$C—COOH

Malic acid   Oxaloacetic acid

The final reaction in the cycle is the oxidation of L-malate by the enzyme, *malate dehydrogenase*. This is another NAD$^+$-requiring enzyme, and it accomplishes the regeneration of oxaloacetate, which can again condense with acetyl CoA to start a subsequent turn of the cycle.

Distribution of Pathway

This series of reactions is the major aerobic degradative pathway of metabolism in a large number of tissues and organisms. The system has been so extensively studied that demonstrating its existence in a cellular preparation involves a number of standard approaches. If a tissue is oxidizing a substrate such as pyruvate through this series of reactions, the addition of small amounts of the di- and tricarboxylic acids of the pathway should stimulate this oxidation, and the oxidation should be inhibited by the addition of malonate. In the presence of malonate, however, citrate should be converted to succinate. The occurrence of the citric acid cycle in a tissue can also be verified by assaying for the presence of the various enzymes of the pathway and by detecting the metabolic intermediates in the pathway.

Cellular Localization   The enzymes of the cycle are located exclusively in the mitochondria. The two flavoprotein-containing complexes, the succinate and the α-keto acid dehydrogenases, are found as an integral part of the inner membrane. The other enzymes are located in the mitochondrial matrix enclosed by the inner membrane. The reduced FAD which is generated can thus feed electrons directly into the electron-transport chain, and the soluble NAD^+-linked dehydrogenases in the matrix produce NADH, which can readily be oxidized by the flavoprotein complex of the electron-transport chain.

Energy Yield from Pathway

A simplified diagram of the citric acid cycle showing the key reactions is illustrated in Figure 13-6. Looked at in this way, the cycle appears to be a

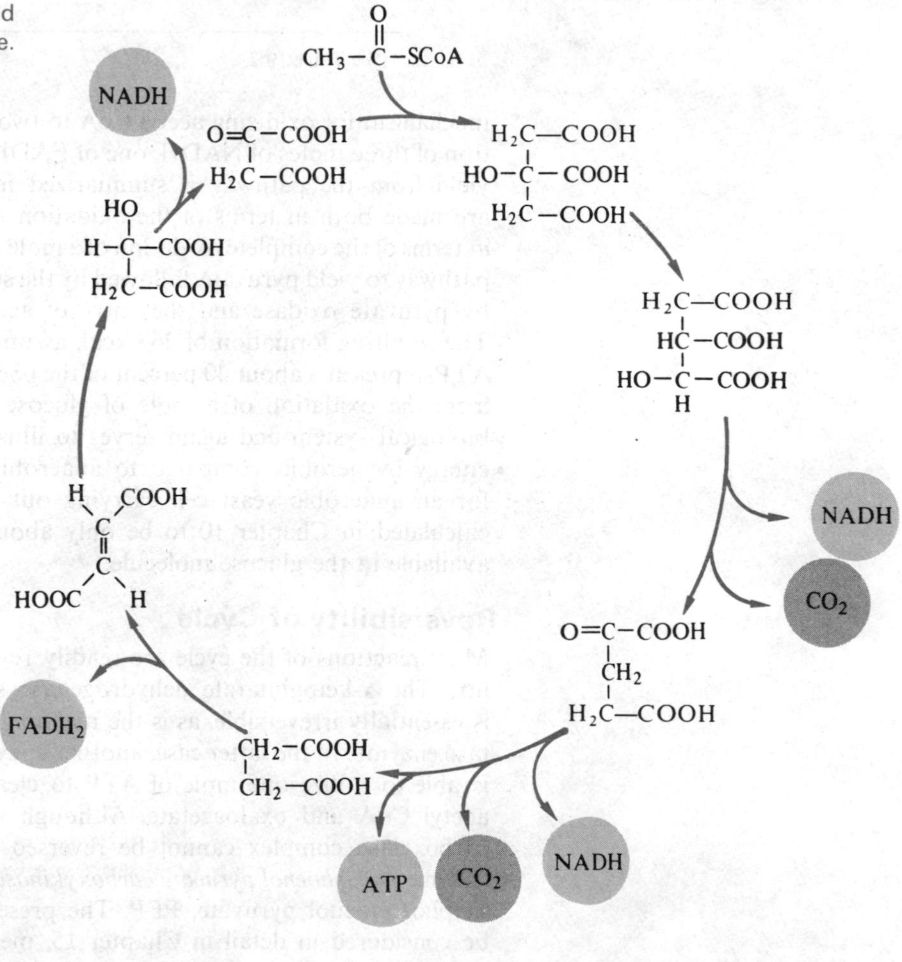

Figure 13-6 Key reactions and products of the citric acid cycle.

309

Table 13-1 Energy Yield from Aerobic Glucose Oxidation

Reactions	Moles of ATP Produced[a]	
	Per Pyruvate	Per Hexose
Citric Acid Cycle		
3 NADH produced	9	
1 FADH produced	2	
1 ATP produced	1	
	—	
	12	24
Pyruvate to Acetyl CoA		
1 NADH produced	3	6
Glycolysis		
Net ATP from glucose to pyruvate	1	2
1 NADH produced	3	6
		—
		38

[a] See Table 10-2.

mechanism for oxidizing acetyl CoA to two moles of CO_2, with the formation of three moles of NADH, one of $FADH_2$, and one of ATP. The energy yield from the pathway is summarized in Table 13-1. The calculations are made both in terms of the oxidation of one mole of acetyl CoA and in terms of the complete oxidation of a mole of glucose through the glycolytic pathway to yield pyruvate, followed by the subsequent oxidation of pyruvate by pyruvate oxidase and the entry of acetyl CoA into the Krebs cycle. The resulting formation of 267 kcal, assuming a $\Delta G^{0'}$ of 7.3 kcal/mole for ATP, represents about 40 percent of the 686 kcal/mole potentially available from the oxidation of a mole of glucose. This is an extremely efficient biological system and again serves to illustrate the efficient utilization of energy by aerobic, compared to anaerobic, organisms. The energy yield for an anaerobic yeast cell carrying out an alcoholic fermentation was calculated in Chapter 10 to be only about 2 percent of that potentially available in the glucose molecule.

Reversibility of Cycle

Most reactions of the cycle are readily reversible, but some key steps are not. The α-ketoglutarate dehydrogenase system, which yields succinate, is essentially irreversible, as is the reaction catalyzed by the citrate synthetase enzyme. In the latter case, another enzyme, the *citrate cleavage enzyme*, is able to utilize one mole of ATP to cleave citrate and to convert it to acetyl CoA and oxaloacetate. Although the action of the pyruvate decarboxylase complex cannot be reversed to yield pyruvate directly, the enzyme *phosphoenol pyruvate carboxykinase* is able to convert oxaloacetate to phosphoenol pyruvate, PEP. The presence of this enzyme, which will be considered in detail in Chapter 15, means that the cycle can serve as an effective means of integrating and interconverting metabolites from a number of sources. Fatty acids feed carbon into the cycle as acetyl CoA or succinate; amino acids contribute oxaloacetate, α-ketoglutarate, pyruvate, and acetyl CoA, and these compounds can be converted by the enzymes of the cycle to various other intermediates. Carbon can be shunted out of the cycle as phosphoenol pyruvate, or as amino acids through the transamination of oxaloacetate or α-ketoglutarate.

MALATE OR GLYOXYLATE CYCLE

By a series of reactions which will be discussed in detail in Chapter 15, carbon from oxaloacetate shunted out of the Krebs cycle as phosphoenol pyruvate can be converted to glucose. By this pathway, carbon atoms which came from acetyl CoA can get into glucose. There can be, however, no net synthesis of glucose from fatty acids in animal tissues by this pathway.

Figure 13-7 Glyoxylate cycle. This modification, utilizing some of the citric acid cycle enzymes and two additional enzymes, allows the continual formation of succinate and the regeneration of oxaloacetic acid. Through reactions discussed in Chapter 15, succinate can serve as a precursor for carbohydrate synthesis.

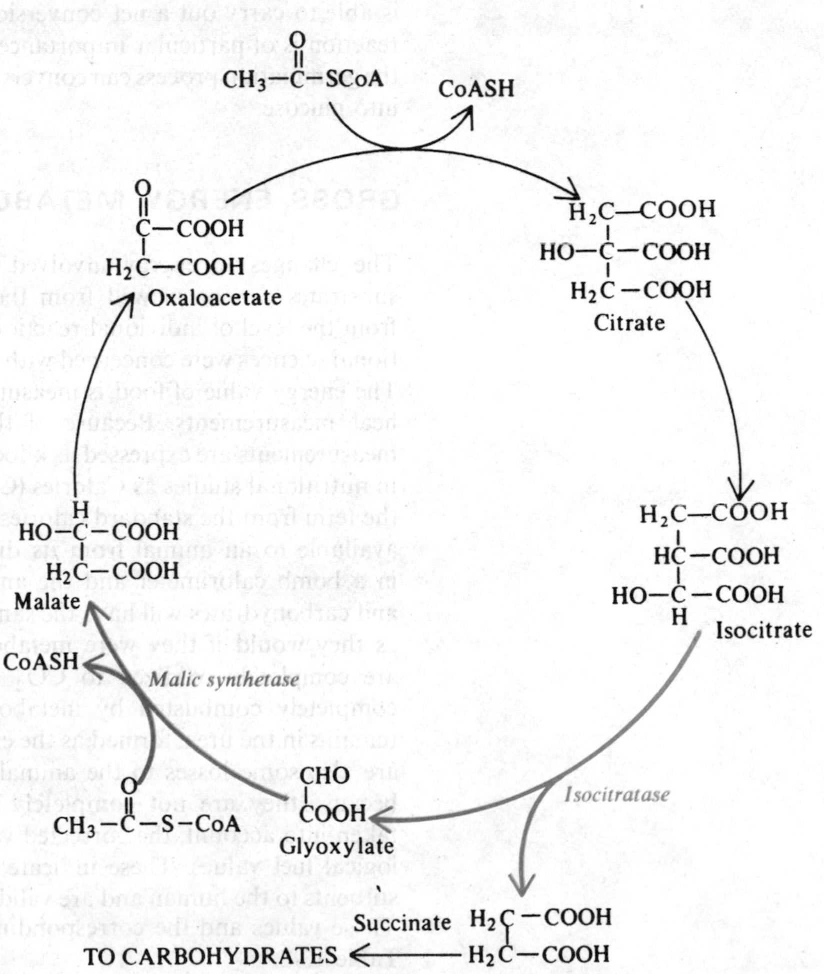

Two molecules of CO_2 are lost by the tricarboxylic acid cycle reactions between the introduction of acetyl CoA (two atoms of carbon) and the formation of oxaloacetate. This being the case, if no mechanism for replenishment is provided, a constant removal of oxaloacetate would represent a net loss of carbon from the cycle, and the concentration of all the intermediates would be depleted. In most plant and many microbial systems, this problem is circumvented by the action of two additional enzymes, *isocitratase* and *malate synthetase*. Isocitratase converts isocitrate to succinate and glyoxylate; malate synthetase catalyzes a reaction similar to the citrate synthetase enzyme in that it condenses acetyl CoA and glyoxylate to form malate. The presence of these two enzymes, plus the enzymes of the citric acid cycle, means that a cycle can be started with oxaloacetate, a total of four carbons can be added (as two acetyl CoA molecules), succinate can be converted to oxaloacetate and shunted off toward glucose formation, and oxaloacetate can be regenerated from glyoxylate to renew the cycle. These reactions are illustrated in Figure 13-7. By this mechanism, a cell is able to carry out a net conversion of fatty acid carbon to glucose. This reaction is of particular importance in seeds of higher plants, which during the germination process can convert the fatty acids liberated from triglyceride into glucose.

GROSS ENERGY METABOLISM

The changes in energy involved in the metabolism of energy-yielding substrates can be viewed from the level of the intact animal as well as from the level of individual reactions. Some of the earliest studies in nutritional sciences were concerned with the energy value of foods, or *calorimetry*. The energy value of food is measured in units of calories, just as are other heat measurements. Because of the mass of material usually involved, measurements are expressed as kilocalories (kcal) or sometimes, particularly in nutritional studies as Calories (C), written with a capital C to distinguish the term from the standard calories (c). To determine the amount of energy available to an animal from its diet, dietary components can be burned in a bomb calorimeter and the amount of liberated heat measured. Fats and carbohydrates will have the same caloric value in the bomb calorimeter as they would if they were metabolized in the body. In both cases they are completely oxidized to CO_2 and water. Proteins, however, are incompletely combusted by metabolic reactions; some energy value still remains in the urea, formed as the end product of protein catabolism. There are also some losses to the animal of the caloric value of ingested foods, because they are not completely absorbed. When these two losses are taken into account, the corrected values are what nutritionists call physiological fuel values. These indicate the caloric value of various food constituents to the human and are valid for foods relatively low in fiber content. These values and the corresponding heats of combustion are indicated in Table 13-2.

Table 13-2 Energy Value of Classes of Foods

Food	Heat of Combustion[a]	Psychological Fuel Values[a]
Carbohydrates	4.1	4
Fat	9.4	9
Protein	5.6	4

[a] kcal/g

Measurement of Heat Production

The amount of energy available to an animal from the diet can, therefore, be calculated from the composition of the diet. A great deal of this type of work has been done, and tables of the caloric value of different foods are readily available. The science of calorimetry also deals with the amount of heat produced by an animal, and a measurement of this value is not so readily obtained, although it can be determined by what is called direct calorimetry. In this method, the heat actually given off by an animal confined to a closed environment is trapped and measured. This can readily be done for small animals, but for man or domestic animals, the size of the apparatus needed makes the entire process very complex and expensive. Because of this, only a limited amount of the equipment needed to make these measurements is available.

The heat production of an animal can also be determined by a technique called indirect calorimetry. This technique requires the measurement of O_2 utilized by the animal, and the calculations are based on the concept of the respiratory quotient, or RQ, where

$$RQ = \frac{\text{vol } CO_2 \text{ produced}}{\text{vol } O_2 \text{ consumed}}$$

The RQ for the oxidation of any compound can be calculated from the equation for its oxidation. The respiratory quotients for glucose and lauric acid are calculated in Figure 13-8. For carbohydrates in general,

OXIDATION OF LAURIC ACID

$$CH_3(CH_2)_{10}COOH + 17O_2 \longrightarrow 12CO_2 + 12H_2O$$

17 moles × 22.4 liters/mole = 381 liters (O_2)
12 moles × 22.4 liters/mole = 268 liters (CO_2)

$$RQ = \frac{268 \text{ liters } CO_2}{381 \text{ liters } O_2} = 0.70$$

OXIDATION OF GLUCOSE

$$C_2H_{12}O_6 + 6O_2 \longrightarrow 6CO_2 + 6H_2O$$

6 moles × 22.4 liters/mole = 134 liters (CO_2 & O_2)

$$RQ = \frac{134 \text{ liters } CO_2}{134 \text{ liters } O_2} = 1.0$$

the RQ is about 1.0; for fats, around 0.7; and for proteins, about 0.8. With a knowledge of the caloric value per gram and the equation for the oxidation of the compound, it is possible to calculate the calories produced per liter of O_2 consumed. Using glucose for example, which has a caloric value of 3.76 kcal/g, the calculation would be

$$C_6H_{12}O_6 + 6O_2 \longrightarrow 6CO_2 + 6H_2O$$

$$\frac{180 \text{ g/mole glucose} \times 3.76 \text{ kcal/g}}{6 \text{ moles } O_2/\text{mole glucose}} = 113 \text{ kcal/mole } O_2$$

$$\frac{113 \text{ kcal/mole}}{22.4 \text{ liters/mole}} = 5.05 \text{ kcal/liter of } O_2$$

As this value will depend on the amount of oxygen needed to oxidize a compound, it is directly related to the RQ. It is therefore possible to construct tables of kcal/liter of O_2 for substances with differing RQs, and to calculate heat production by a measurement of the RQ. In most cases, 90 to 95 percent of an animal's heat production is from the oxidation of fat or carbohydrate, and the heat production from protein metabolism is ignored. If more accurate values are needed, a correction for the contribution of protein to the RQ can be made after a measurement of urinary nitrogen excretion.

Utilization of Energy

Energy utilization, and therefore heat production, in an animal is due both to what is called the basal metabolism as well as to various energy-requiring processes beyond this. The basal metabolism of an animal is that amount of energy expended in a postabsorptive resting state and is the energy required for the maintenance of cellular and tissue processes, the work of the kidneys, and the muscular work associated with breathing. Basal metabolism is quite commonly measured in clinical medicine by indirect calorimetry. The resting subject breathes into a closed system. The O_2 consumption is measured, and the energy expenditure is calculated by assuming an RQ of 0.82. This has been found to be the RQ of an average subject in a postabsorptive state, and is equivalent to 4.83 kcal/liter of O_2 consumed.

The data in Table 13-3 indicate that the heat production per unit of body weight is inversely related to body weight over a wide range of body sizes and is rather closely related to body surface area. Body surface area is difficult to measure, but it can be shown that heat production is directly proportional to the body weight to the three-fourths power. For many species the basal metabolic rate, BMR, is about 70 kcal/kg$^{3/4}$ day. In general, younger animals have a higher BMR than adults, and males higher than females. The energy requirement for the basal metabolism of a 180-lb 6-foot college-age male is about 1900 kcal/day, and for a 5'6″,

120-lb female about 1300 kcal/day. The basal metabolic rate is very much influenced by thyroid activity, and determination of thyroid status is one of the main clinical uses of such information.

Table 13-3 Relationship of Heat Production to Body Size

	Body Weight, kg	Metabolism/kg of Body Weight per Day, kcal	Metabolism/M^2 of Body Surface per Day, kcal
Horse	441.0	11.3	948
Pig	128.0	19.1	1078
Man	64.3	32.1	1042
Dog	15.2	51.5	1039
Goose	3.5	66.7	969
Fowl	2.0	71.0	943
Mouse	0.018	212.0	1188

Source: Dr. G. Lusk, *The Elements of the Science of Nutrition*, 4th ed., Saunders, Philadelphia, 1928.

The energy requirement beyond that needed to satisfy basal metabolism is the amount needed to perform muscular work, and that associated with the increase in heat production which follows the ingestion of a meal. This increase is called the *heat increment* or, sometimes, the *specific dynamic action* of food. It can be as much as 30 percent of the calories ingested in the case of a high-protein diet, but is much lower for diets containing mainly fats and carbohydrates. The biochemical basis of the production is not really well understood. It is the heat produced by the work of digesting and metabolizing the ingested food, and seems to be mainly involved with those processes needed to maintain tissue homeostasis in the face of the large amount of metabolites being absorbed.

The energy required by an animal beyond that needed to satisfy its basal metabolism and heat increment is extremely variable. It depends largely on the physical activity of the individual. The values for energy expenditure of various types of exercise are presented in Table 13-4. The total energy expenditure of the average-size college student might vary considerably, but for most it will be on the order of 2800–3000 kcal/day for men and 2000–2300 kcal/day for women. The energy requirement of an individual performing manual labor can be much in excess of these figures.

If a person does not consume as many calories each day as he expends, the deficit must be made up by metabolizing body fat; conversely, an oversupply of calories will be retained in the body as newly synthesized adipose tissue. Although some people do have a problem in maintaining their desired weight, the majority are able to match dietary intake of calories to energy expenditure with no conscious effort, and the ability of the appetite to regulate food intake is, for the most part, very good. A few simple calculations will reveal that a consistent overconsumption of calories by

Table 13-4 Energy Expenditure for Various Activities

Activity	Energy Expenditure, kcal/hr
Sitting	96
Playing cards	144
Driving a car	168
Walking 2 miles/hr	180
Walking 4 miles/hr	336
Cross-country running	636
Playing tennis	426
Vigorous swimming	660
Playing squash	612
Shoveling dirt	480
Sawing wood	540
Hard manual labor	654

[a] Value averages are for a 65–75 kg male taken from different sources.

as little as 10 percent per day will lead to an energy excess sufficient to result in the synthesis of between 1 and 2 pounds of excess fat per month. An inspection of the data in Table 13-4 will also reveal that, with fat providing 9 kcal/g or about 4000 kcal/lb, the amount of exercise needed to metabolize a pound of excess fat is appreciable, and the easiest way to lose weight is to decrease intake rather than to increase energy expenditure.

PROBLEMS

1. Explain how the oxidation of pyruvate to acetyl CoA and CO_2 is coupled to the reduction of NAD^+ even though the pyridine nucleotide does not interact with the same subunit of the pyruvate dehydrogenase complex.

2. Because of an active α-ketoglutarate transaminase system, carbon from glucose is rapidly incorporated into glutamic acid through the glycolytic and citric acid cycle enzymes. If glucose-2-^{14}C is used, which carbon of glutamic acid is the first to become labeled?

3. Through the action of transaminases and enzymes of the citric acid cycle, carbon from glutamic acid can be found in aspartic acid. Where would the label be found in aspartic acid if glutamic acid labeled in the α carbon is used?

4. Calculate the respiratory quotient for the oxidation of palmitic acid.

5. Although it is often stated that animals cannot synthesize carbohydrate from fatty acids, it is an experimental observation that the administration

316

of radioactive palmitic acid to a starving animal would result in the appearance of radioactive glucose in the plasma. How can this be rationalized?

6. How many grams of a diet containing 30 percent fat, 20 percent protein, and 50 percent carbohydrate would be required to furnish 2200 Cal?

Suggested Readings

Books

Goodwin, T. W. (ed.) *The Metabolic Roles of Citrate*. Academic Press, London (1968).

Lowenstein, J. M. (ed.) *Citric Acid Cycle: Control and Compartmentation*. Dekker, New York (1969).

Articles and Reviews

Kornberg, H. L. Anaplerotic Sequences and Their Role in Metabolism, in *Essays in Biochemistry*, Campbell, P. N., and G. D. Greville (eds.), vol. 2, pp. 1–32. Academic Press, New York (1966).

Krebs, H. A. The History of the Tricarboxylic Acid Cycle. *Perspect. Biol. Med.* **14**:154–170 (1970).

Linn, T. C., J. W. Pelley, F. H. Pettit, F. Hucho, D. D. Randall, and L. J. Reed. Purification and Properties of the Component Enzymes of the Pyruvate Dehydrogenase Complexes from Bovine Kidney and Heart. *Arch. Biochem. Biophys.* **148**:327–342 (1972).

Lowenstein, J. M. The Tricarboxylic Acid Cycle, in *Metabolic Pathways*, Greenberg, D. (ed.), vol. 1. Academic Press, New York (1967).

PART
FOUR

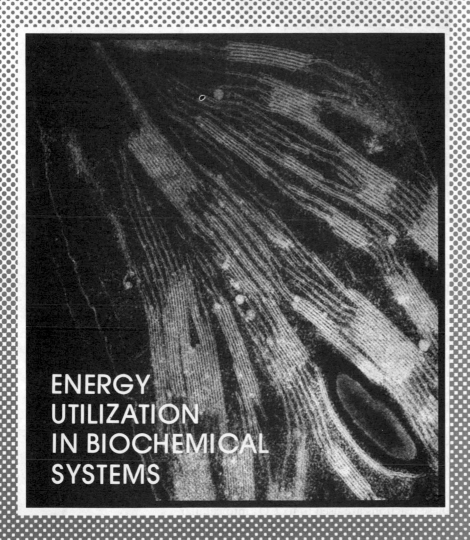

ENERGY
UTILIZATION
IN BIOCHEMICAL
SYSTEMS

photosynthesis   chapter fourteen

The photosynthetic process which occurs in green plants utilizes light energy to convert H_2O and CO_2 to O_2 and carbohydrates. The light energy is used to drive an electron from the light-absorbing chlorophyll molecule to an electron acceptor with a very negative redox potential. This acceptor can reduce $NADP^+$ to form NADPH, and the electron deficit in the chlorophyll can be made up by the oxidation of water to liberate O_2. Alternatively the electron can be returned to the chlorophyll molecule by passing through a series of electron carriers during which ATP is generated through the process of photosynthetic phosphorylation. The ATP and NADH formed by these reactions are used to fix CO_2 into cellular carbohydrates. Many of the enzymes involved in this conversion are the same ones that are utilized in the glycolytic and pentose phosphate pathway of carbohydrate metabolism.

PHOTOSYNTHETIC PROCESS

Directly or indirectly, the biological conversion of electromagnetic energy into chemical energy through the photosynthetic process provides all living organisms with their sources of energy. The expenditure of energy through mechanical work and the expenditure invariably associated with the coupling of anabolic reactions to catabolic reactions require that organisms have available a constant supply of energy. Without the photosynthetic process, there would be a gradual conversion of all the metabolizable carbon sources in the environment to CO_2.

The overall photosynthetic process, as it occurs in green plants, is illustrated by the following equation:

$$2H_2O + CO_2 \xrightarrow[\text{chlorophyll}]{\text{light energy}} (CH_2O) + H_2O + O_2$$

Water is the electron donor used to reduce CO_2, and the process is associated with the evolution of molecular oxygen. The overall reaction is associated with a $\Delta G^{0'}$ of $+118,000$ cal/mole of CO_2 reduced. Although the reaction in this equation is the one commonly known as the photosynthetic process, the photosynthetic bacteria are able to use a number of electron donors other than H_2O. Some purple bacteria, for example, use H_2S, and the corresponding reaction in that system would be:

$$2H_2S + CO_2 \xrightarrow[\text{chlorophyll}]{\text{light energy}} (CH_2O) + H_2O + 2S$$

Chlorophyll

Irrespective of variations in the details of the system, the photosynthetic process in all organisms is associated with excitation of the Mg^{2+}-containing porphyrin compound, chlorophyll, by light energy. A number of different chlorophyll pigments of slightly differing structure are found in various plants. Chlorophyll *a* (Figure 14-1) is found in all oxygen-producing organisms; chlorophyll *b* is also found in green plants and algae; in other forms of algae, chlorophyll *a* is associated with chlorophyll *c* or *d*. There are also some forms specific to bacteria, and small amounts of a specific chlorophyll molecule called P700 appear to be associated with the major forms of chlorophyll in various organisms. In the cell, the green chlorophyll pigments are usually found associated with other plant pigments such as carotenoids.

In eucaryotic cells, the chlorophyll is found localized in an organelle called the chloroplast. Chloroplasts are subcellular structures about 4 μm long and similar in morphology to mitochondria. They have an outer membrane similar to that of the mitochondrion and an inner membrane which has been formed into a series of vesicles called thylakoid disks. A

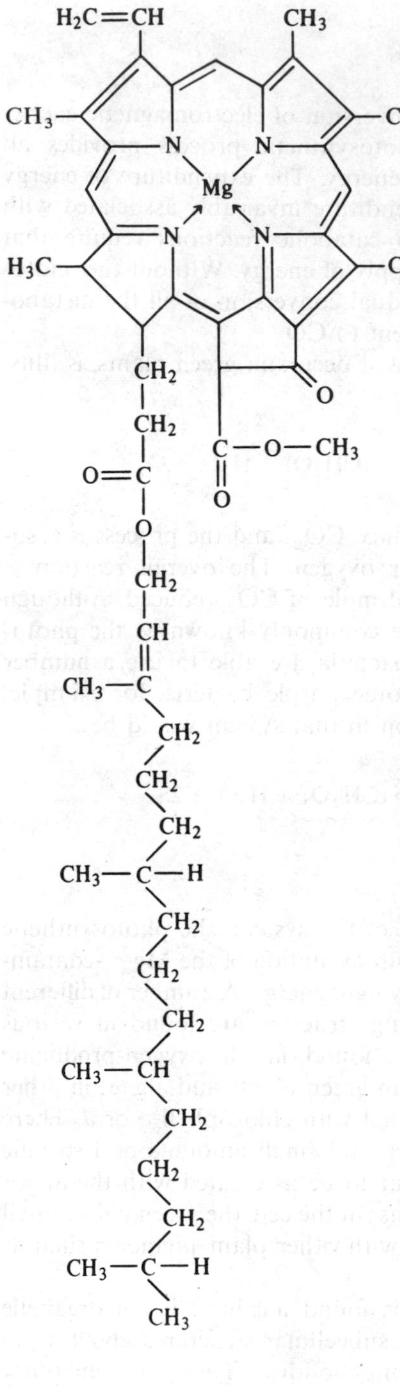

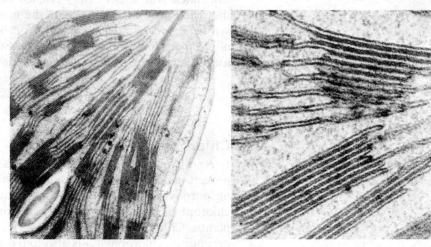

Figure 14-1 Structure of chlorophyll *a*. In chlorophyll *b*, the ethyl group on the one pyrrole ring is -CHO group. The long side chain esterified to the propionic acid residue is a phytol group.

number of these disks are stacked on top of one another to form a structure called a *granum* (Figure 14-2). It is in these grana that the chlorophyll pigments are found and the primary photochemical events occur.

Early in the study of photosynthesis it was realized that the process could be conveniently divided into a *light reaction*, which produced the energy, and the actual CO_2 fixation or *dark reaction*, which required energy. Investigation of the two processes proceeded quite independently, and they will be discussed separately.

Figure 14-2 Chloroplast from bean plant leaf. The lamellar membranes and the membrane stacks are clearly visible. (left) Part of a chloroplast; mitochondrion is also visible at top (×75,000); (right) Higher magnification of the membrane stacks (×300,000). (Courtesy of T. E. Weier)

DARK REACTION

Any real understanding of the nature of the reactions involved in the fixation of CO_2 and its reduction to the level of a carbohydrate developed very slowly until $^{14}CO_2$ was available to follow the reaction. The pathway

$H_2C - OH$
|
$C = O$
|
$HC - OH$
|
$HC - OH$
|
$H_2C - O - \text{P}$

Ribulose — 5 —P

   ATP

Phosphoribulose kinase

   ADP

$H_2C - O - \text{P}$
|
$C = O$
|
$HC - OH$
|
$HC - OH$
|
$H_2C - O - \text{P}$

Ribulose—1, 5—diP

CO_2

$CH_2 - O - \text{P}$
|
$C - OH$
‖
$C - OH$
|
$HC - OH$
|
$CH_2 - O - \text{P}$

        O
        ‖ $CH_2 - O - \text{P}$
$HO - C — C - OH$
         $C = O$
       $H - C - OH$
         $CH_2 - O - \text{P}$

H_2O

Ribulose diphosphate carboxydismutase

COOH
|
2 $HC - OH$
|
$H_2C - O - \text{P}$

3-Phosphoglyceric acid

was elucidated largely by Calvin and his coworkers. In their system, $^{14}CO_2$ was added to an algal suspension, and after a very short time the reaction was stopped and the radioactivity in different cellular metabolites was determined. The early experiments indicated that the carboxyl atom of 3-phosphoglyceric acid was one of the first compounds to become labeled. After numerous attempts to find a 2-carbon acceptor which would react with CO_2 to form 3-phosphoglyceric acid had failed, it was realized that the acceptor is a 5-carbon compound, ribulose-1,5-diphosphate. This sugar diphosphate can be formed from ribulose-5-P by a specific enzyme, *phosphoribulokinase* (Figure 14-3). The enzyme *ribulose diphosphate carboxydismutase*, which catalyzes the actual fixation of CO_2, is a very high molecular weight enzyme (MW = 550,000), which makes up about 15 percent of the total protein of the chloroplast. The compounds shown in brackets in the figure were postulated by Calvin to be the intermediates in the dismutase reaction. The enol form of the pentose diphosphate is presumably enzyme-stabilized, and has been difficult to identify. Recently it has been possible to obtain direct experimental evidence that this intermediate does exist.

Carbohydrate Synthesis

The product of the dismutase reaction, 3-phosphoglyceric acid, can be converted to glucose-6-P by a series of reactions which amount to essentially a reversal of the glycolytic pathway. The first two reactions in the pathway are energy-requiring, the conversion of 3-phosphoglyceric acid to 1,3-diphosphoglyceric acid by *phosphoglycerate kinase*, and the reduction of this compound by *glyceraldehyde-3-phosphate dehydrogenase*. The latter enzyme is found as an NADPH-requiring system in green plants rather than being an NAD^+-linked system, as it is in animal tissues.

The glyceraldehyde-3-P formed by these two energy-requiring reactions can serve as the source of carbon for the synthesis of both glucose, the end product of the reaction series, and ribulose-5-P, needed to regenerate the CO_2 acceptor. The enzymes involved are the *transketolase*, *pentose isomerase*, and *pentose epimerase* of the hexose monophosphate pathway, the enzymes of the glycolytic pathway, and some specific phosphatases. The reaction pathways involved are indicated in Figure 14-4. Note that in this pathway *aldolase* is functioning with both glyceraldehyde-3-P and erythrose-4-P as the aldo sugar, and that none of the reactions are energy-

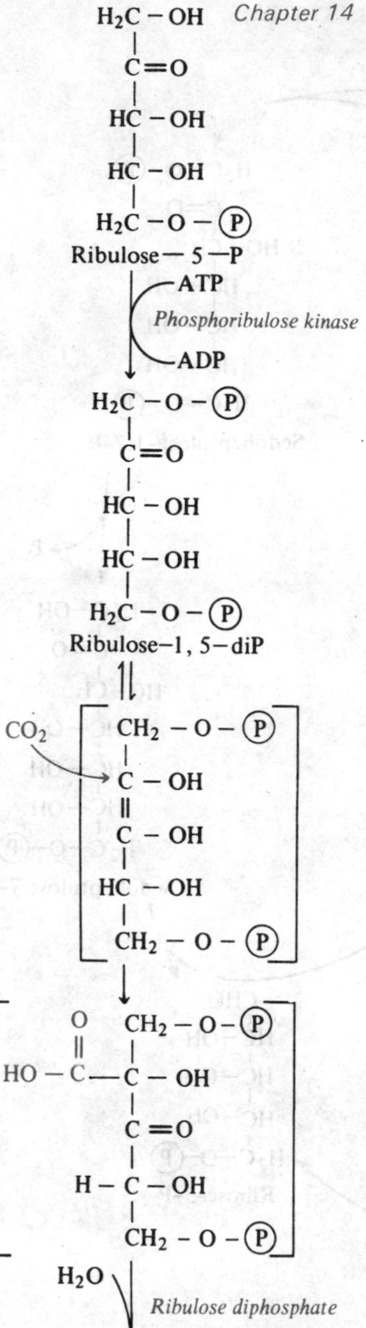

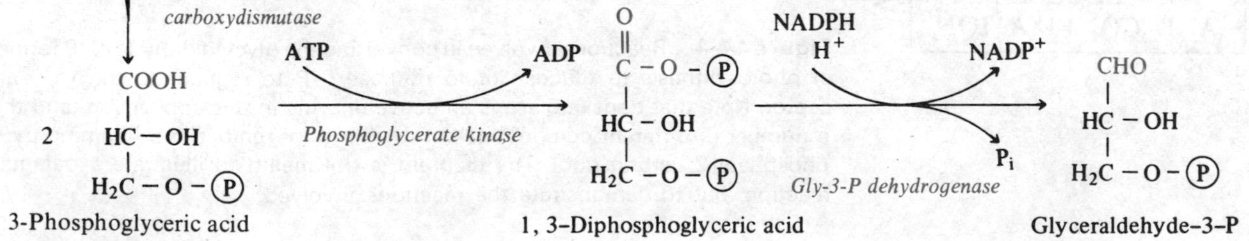

Figure 14-3 Fixation of CO_2 in photosynthesis. Two moles of glyceraldehyde-3-P are formed from each mole of ribulose-1,5-P. The compounds shown in brackets are hypothetical intermediates that cannot be isolated.

ATP   ADP   NADPH H$^+$ NADP$^+$

               O
               ‖
            $C - O - \text{P}$ CHO
            | |
            $HC - OH$ $HC - OH$
            | P$_i$ |
            $H_2C - O - \text{P}$ $H_2C - O - \text{P}$

Phosphoglycerate kinase   *Gly-3-P dehydrogenase*

         1, 3-Diphosphoglyceric acid   Glyceraldehyde–3–P

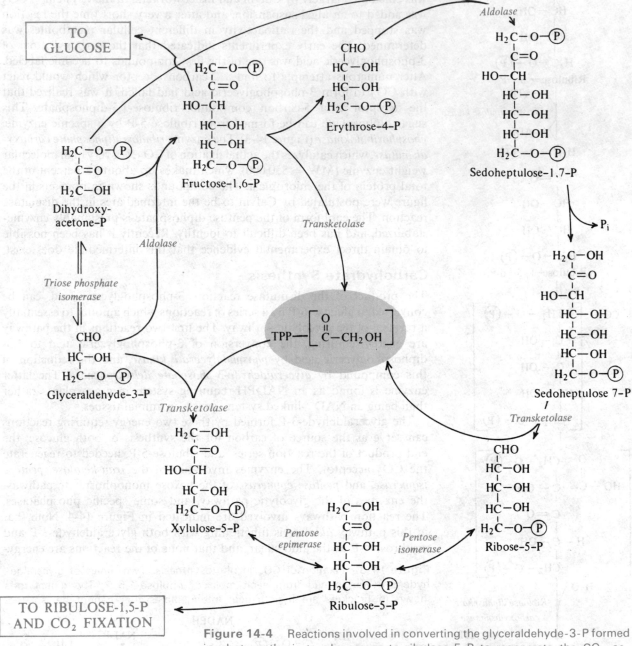

Figure 14-4 Reactions involved in converting the glyceraldehyde-3-P formed in photosynthesis to glucose or to ribulose-5-P to regenerate the CO_2 acceptor. Note that transketolase is an active enzyme in this conversion, and that a number of different compounds can feed into or remove the thiamine pyrophosphate-2 carbon pool. The diagram is not meant to illustrate a balanced reaction, but to demonstrate the reactions involved.

requiring. The energy requirement in the photosynthetic process is for ATP to regenerate ribulose-1,5-diphosphate and to form 1,3-phosphoglyceric acid, and for NADPH to reduce the latter compound.

CO_2 Fixation in Tropical Grasses

The path of carbon in the photosynthetic process was originally formulated by Calvin on the basis of experiments in algae. It was later shown that tropical grasses such as sugar cane and corn fix CO_2 by a different mechanism, now often called the C-4 acid pathway. In leaves from these plants, the initial incorporation of CO_2 is into oxaloacetate, malate, and aspartate. The predominant reaction is a carboxylation of phosphoenolpyruvate to form oxaloacetate, which rapidly equilibrates with malate and aspartate. The 4-carbon acids which are formed are not converted to carbohydrate, but diffuse to a different part of the leaf, where they are decarboxylated, probably by the *malic enzyme*, to yield CO_2 and pyruvate. The CO_2 is then reutilized by the *carboxydismutase* system, and the pyruvate is converted to phosphoenolpyruvate by an enzyme which uses both ATP and inorganic phosphate (Figure 14-5). The chief advantage of this system

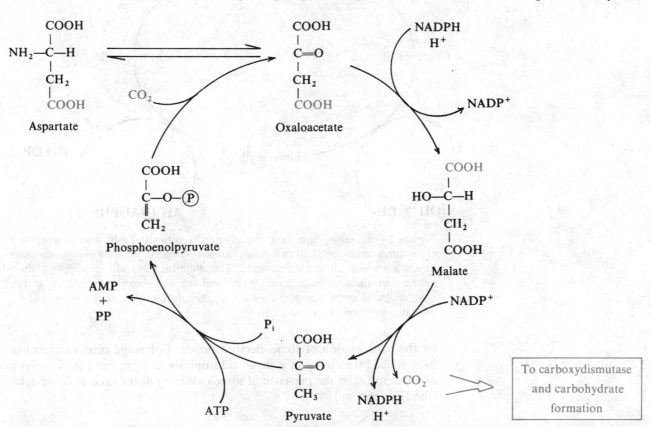

Phosphoenolpyruvate

Oxaloacetate

Aspartate

Oxaloacetate

Malate

Phosphoenolpyruvate

AMP
+
PP

P_i

Pyruvate

ATP

To carboxydismutase and carbohydrate formation

Figure 14-5 Path of carbon in the C-4 pathway of CO_2 fixation.

325

appears to be its ability to operate at a rapid rate at low partial pressures of CO_2. The initial carboxylase enzyme has a much higher affinity for CO_2 than the Calvin cycle dismutase, and it therefore traps CO_2 for the second enzyme to use.

LIGHT REACTION

The basic reaction in the light phase of the photosynthetic process is illustrated in Figure 14-6. A chlorophyll molecule is excited, the excited molecule gives up an electron, and the electron deficiency in the system is overcome

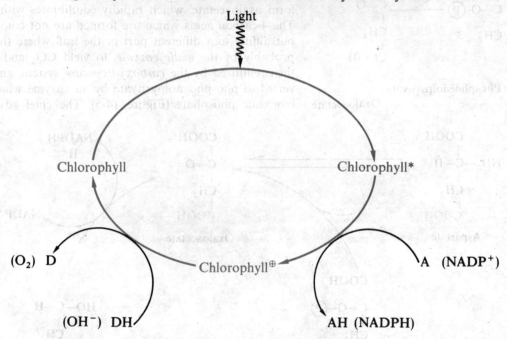

Figure 14-6 Basic light reaction of photosynthesis. Light energy excites a chlorophyll molecule such that it can lose an electron, which is eventually used to reduce some acceptor molecule. The electron-deficient chlorophyll then extracts an electron from some donor, and the cycle can be repeated. In the case of green plant photosynthesis, the ultimate electron acceptor is $NADP^+$ and the electron donor is H_2O.

by the participation of some electron donor. For some time biochemists have studied the ability of isolated chloroplasts to carry out the photolysis of water to O_2 in the presence of some oxidizing agent such as ferric salts (the Hill reaction):

$$2H_2O + 2A \xrightarrow[\text{chlorophyll}]{\text{light energy}} 2AH_2 + O_2$$

In the 1950s it was shown that $NADP^+$ would function in this capacity as a *Hill reagent*. It was known at this time that both NADPH and ATP were required for photosynthesis, and the ability of the chloroplast to reduce $NADP^+$ offered an explanation for the source of the ATP. It was assumed that some of the NADPH formed in the chloroplast was reoxidized by the mitochondria and that the ATP required was formed by oxidative phosphorylation. It was later demonstrated by Frenkel and by Arnon that chloroplasts alone could form ATP from ADP and phosphate and that NADPH was not a required intermediate. The generation of ATP in the chloroplast came to be known as *photosynthetic phosphorylation*.

Flow of Electrons

Assuming that the appropriate enzymes are present, electrons will flow in a spontaneous fashion from a redox pair with a more negative potential to a pair with a more positive potential (Chapter 8). In the photosynthetic process, however, water is oxidized to O_2 ($E'_0 = +0.82$ volts) and NADP is reduced to NADPH ($E'_0 = -0.32$ volts). The light energy absorbed by the chlorophyll molecule must therefore be used to drive this unfavorable reaction. It is now known that the production of ATP by the process of photosynthetic phosphorylation can take two different forms. In *cyclic photophosphorylation*, which produces only ATP, the electrons removed from the chlorophyll molecule by light energy are returned to a chlorophyll molecule. In *noncyclic photophosphorylation*, which produces both ATP and NADPH, the electrons excited from chlorophyll by light energy are eventually transferred to $NADP^+$ to reduce it to NADPH.

It is now clear that plant chloroplasts contain two different types of light-gathering systems. One of these, photopigment system I (PS I), contains only chlorophyll *a*, and the other, photopigment system II (PS II), contains chlorophyll *a* and some second type of chlorophyll which varies somewhat depending on the species of green plant. These systems contain other light-absorbing pigments such as the carotenoids, and the PS I system contains about one molecule of P700 pigment for every 200–400 chlorophyll molecules. P700 pigment is the chlorophyll molecule which is actually able to absorb the light energy protons and in the process lose an electron. The chlorophyll and other pigments of these systems, as well as the lipids and electron carriers associated with them, are grouped together in the chloroplast to form a distinct structural and functional unit called a *quantosome*.

The chloroplasts also contain a series of cytochromes, heme proteins similar to those found in mitochondria; a reducible copper protein, plastocyanin; and two quinones, phylloquinone, or vitamin K_1, and plastoquinone, which has a structure similar to coenzyme Q. If the process of photophosphorylation is to be capable of reducing $NADP^+$, an electron carrier must be present which has a more negative potential than NADPH. At least one such protein, called *ferredoxin*, has been identified. Ferredoxin is a protein with a low molecular weight (6–12,000 in different species) which contains oxidizable and reducible iron atoms bound to labile sulfur atoms in the protein. Ferredoxins have been isolated from various species, and although they differ slightly, they all have standard reduction potentials about 0.1 volts more negative than the $NADP^+/NADPH$ pair.

Photophosphorylation

The relationship of the various electron carriers and of photopigment systems is illustrated in Figure 14-7. The figure indicates that for the noncyclic pathway, the following events occur: photopigment system II absorbs light energy and gives up a pair of electrons, which are used to reduce an electron acceptor with an absorption maximum at 550 nM. The electron

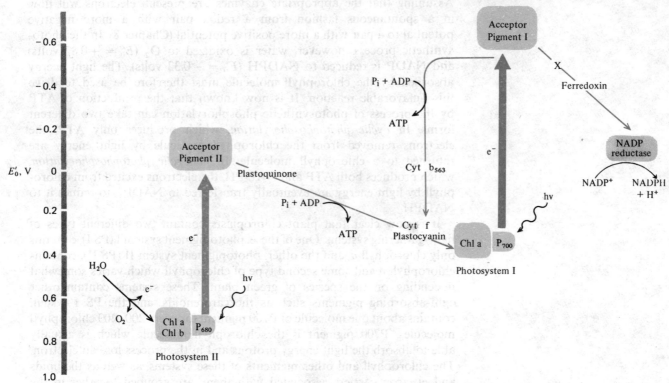

Figure 14-7 Photophosphorylation system. The pathway of the electrons is detailed in the text. The entire scheme represents the process of noncyclic photophosphorylation and it accounts for the liberation of O_2 and the production of ATP and NADPH. In cyclic photophosphorylation, electrons move from photopigment system I to its acceptor, to the electron transport system and back to photopigment system I. The only product is ATP, and no O_2 is liberated. The thermodynamically favorable flow of electrons would be toward acceptors with a more positive redox potential, and light energy is utilized in the photosynthetic system to move electrons toward a more negative potential.

deficit in the chlorophyll system is filled by an electron which comes from the oxidation of water. The liberation of O_2 in the photosynthetic process occurs at this step. The electrons are transferred from the primary acceptor through plastoquinone and a b-type cytochrome to photopigment system I. This transport of electrons through this short electron transport system is coupled to the phosphorylation of ADP, which probably occurs between plastoquinone and the cytochrome. The electrons from this cytochrome are used to reduce the Cu containing protein plastocyanin and to satisfy the electron deficit in photopigment system I, which results when it is excited by light to pass electrons to the photopigment system I acceptor. This acceptor, which is probably a bound form of ferredoxin, has a redox potential which is probably about 0.1 volts more negative than ferredoxin and which can readily reduce ferredoxin by transferring the electrons on to it. Ferredoxin then reacts with a flavoprotein, which can couple its reoxidation to the reduction of $NADP^+$. The net result of this *noncyclic photophosphorylation* has been the absorption of light by the two separate pigment systems and the production of NADPH, ATP, and O_2.

The process of *cyclic photophosphorylation* utilizes only the photopigment system I part of the pathway. The bound ferredoxin is reduced by accepting electrons released by photoexcitation of photopigment system I. This reduced protein then reduces the major ferredoxin species of the system, which reduce a series of cytochromes which eventually fill the electron deficit in photopigment system I. No O_2 is liberated, and no NADPH is formed. During this process, the flow of electrons through the cytochrome system is coupled to the formation of ATP. The process therefore involves the continual movement of electrons around this cycle and the production of ATP from ADP and phosphate.

When only the reactions detailed in Figures 14-3 and 14-4 are considered (that is, the dark reactions), the overall equation for the reduction of one mole of CO_2 in photosynthetic process is

$$CO_2 + 2NADPH + 2H^+ + 3ATP + 2H_2O \longrightarrow (CH_2O) + 2NADP^+ + 3ADP + 3P_i$$

It can be seen from this that to drive the dark reactions requires ATP and NADPH in the ratio of 3:2. If the coupled phosphorylation in the noncyclic pathway is assumed to produce only 1 ATP per pair of electrons, it is clear that the system does not produce sufficient ATP in relation to the NADPH to operate the carbon-reduction cycle. The discrepancy in the ATP required can then be made up by the cyclic operation of the system for a portion of the time.

The process which has been described (Figure 14-7) occurs in green plants and algae. Similar cyclic and noncyclic photophosphorylation occurs in photosynthetic bacteria, but the process differs in that only one pigment system is involved and the electron donor is not water. The electron donor is an oxidizable organic molecule such as a dicarboxylic acid, or a compound such as H_2S which can be oxidized to elemental sulfur.

329

PROBLEMS

1. If radioactive $^{14}CO_2$ is made available to a photosynthesizing system, what carbons in glucose will be the first to become labeled? Why will there be some label in other carbons?

2. Which compounds are products of noncyclic photophosphorylation, but not of cyclic photophosphorylation?

3. Before ferredoxins were discovered, the existence of such a compound in the photosynthetic process was postulated. What was the evidence for this?

4. In the C-4 pathway of CO_2 fixation (Figure 14-5) the formation of phosphoenol pyruvate is coupled to the breakdown of ATP. Why would plants use this enzyme instead of pyruvate kinase, which carries out the same reaction with no loss of high-energy phosphate?

Suggested Readings

Books

Gregory, R. P. F. *Biochemistry of Photosynthesis.* Wiley-Interscience, London (1971).

Rabinowitch, E., and X. Govindjee. *Photosynthesis.* Wiley, New York (1969).

Articles and Reviews

Arnon, D. I. The Light Reaction of Photosynthesis. *Proc. Nat. Acad. Sci. USA* **68**:2883–2892 (1971).

Bassham, J. A. The Path of Carbon in Photosynthesis. *Sci. Amer.* **206**:88–100 (1962).

Hill, R. The Biochemists' Green Mansions: The Photosynthetic Electron Transport Chain in Plants, in *Essays in Biochemistry,* Campbell, P. N., and G. D. Greville (eds.), vol. 1, pp. 121–152. Academic Press, New York (1965).

Levine, R. P. The Mechanism of Photosynthesis. *Sci. Amer.* **221**:58–70 (1969).

biosynthesis of carbohydrates and lipids

chapter fifteen

Animals need glucose for nervous tissue metabolism and when dietary intake and glycogen breakdown cannot maintain normal blood glucose concentrations, it can be synthesized from noncarbohydrate precursors by the liver and kidney cortex. When dietary carbohydrates are being absorbed faster than they can be oxidized, the excess will be stored as glycogen. The enzymes involved utilize UDP-glucose as an activated form of glucose to make an α-1,4-linked polyglucan, and a separate enzyme makes the α-1,6-branch points. Fatty acids are synthesized in the cytoplasm from acetyl CoA by a multienzyme system that requires NADPH as the reductant. One of the first steps in this reaction is a carboxylase which forms malonyl CoA as an intermediate in the reaction. Phosphatidic acid, which can be formed by the acylation of α-glycerol phosphate by long-chain acyl CoA's, serves as an intermediate in both triglyceride and complex lipid synthesis. The common phospholipid, phosphatidyl choline, can be formed from phosphatidic acid and CDP-choline, and many other complex lipids are formed from nucleosidediphospho activated compounds.

Of all the metabolic pathways in carbohydrate synthesis, the one which is physiologically most important in higher animals is undoubtedly the pathway involved in the formation of glucose from noncarbohydrate precursors. The metabolic fate of glucose in animal tissues is varied; it can be oxidized for energy, deposited as glycogen in liver or muscle, metabolized to other carbohydrate compounds, or metabolized to furnish carbon for lipid synthesis. The drain on the circulating concentration of blood glucose is balanced by a number of mechanisms. Glucose or other sugars that can be converted to glucose are absorbed from the small intestine, and glucose can be released to the circulation by the breakdown of glycogen (glycogenolysis). When these sources of glucose are exhausted, blood glucose can be replenished by the formation of glucose from noncarbohydrate precursors such as amino acids, glycerol, and lactic acid. This process, gluconeogenesis, is of extreme importance to the animal because of the need for glucose in brain metabolism. Under normal conditions, almost all the energy requirement of the brain is met by glucose utilization. When the blood glucose concentration falls to a low level, the brain cells cannot obtain sufficient glucose to meet their energy demands, and a coma results.

GLUCONEOGENESIS

The overall pathway of carbon in gluconeogenesis, and the metabolic precursors of glucose which are involved, are indicated in Figure 15-1. These are reactions that occur almost exclusively in the liver and kidney

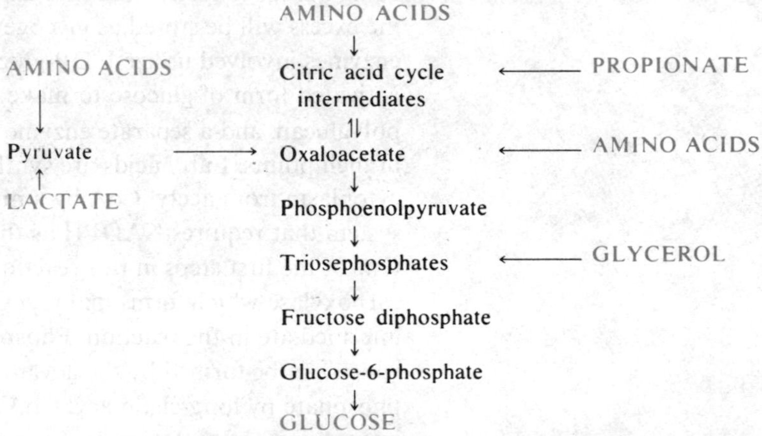

Figure 15-1 Major precursors and intermediates in the pathway of gluconeogenesis. As carbon is removed from the citric acid cycle as oxaloacetate, any compound that can contribute carbon to an intermediate of the cycle can serve as a precursor for carbohydrate synthesis. Most of the glucogenic amino acids enter into the gluconeogenic pathway at this site.

cortex. Because of the mass of tissue involved, however, about 90 percent of the total glucose-forming capacity of the body is localized in the liver. The other tissues of the body lack some of the enzymes needed, particularly those which form phosphoenol pyruvate from pyruvate, and the specific glucose-6-phosphatase that can form free glucose from glucose-6-phosphate.

One major source of carbon for glucose synthesis is lactic acid. When animals are undergoing periods of heavy exercise, muscle glycogen is depleted with the concomitant formation of lactic acid. Much of the lactic acid formed during these periods can be resynthesized to muscle glycogen when the tissues again become aerobic. The necessary enzymes for the conversion of lactate to glucose are not present in sufficient concentration in muscle, and it was realized fairly early in the study of muscle metabolism that lactic acid readily diffused out of the muscle into the plasma. The resynthesis of glucose from lactate is carried out in the liver, and the glucose formed is released to the blood and taken up by the muscle and resynthesized into glycogen. This transfer of carbon from muscle to liver and back is called the *Cori cycle*, and is illustrated in Figure 15-2.

Figure 15-2   Cori cycle.

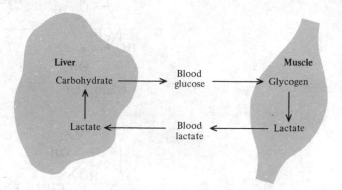

The second major source of carbon for gluconeogenesis is that from the degradation of the glucogenic amino acids. As indicated in Chapter 12, all the amino acids except leucine furnish at least part of their carbon skeleton to either Krebs cycle intermediates or pyruvate. Amino acids therefore become the major source of carbon for glucose synthesis under fasting conditions.

As indicated in Figure 15-1, fatty acids do not contribute to a net synthesis of glucose, but glycerol from the degradation of neutral fats is a third source of carbon for gluconeogenesis. There is a specific *glycerol kinase* which forms glycerol-3-phosphate, and this can be oxidized by a NAD^+-linked *glycerol phosphate dehydrogenase* to dihydroxyacetone phosphate.

$$
\begin{array}{ccc}
\text{CH}_2\text{OH} & \text{CH}_2\text{OH} & \text{CH}_2\text{OH} \\
| & | & | \\
\text{HC—OH} & \text{HO—C—H} & \text{C}=\text{O} \\
| & | & | \\
\text{CH}_2\text{OH} & \text{CH}_2\text{O-}\textcircled{P} & \text{CH}_2\text{O-}\textcircled{P}
\end{array}
$$

Glycerol → (ATP → ADP) *Glycerol kinase* → Glycerol-3-P → (NAD^+ → $NADH + H^+$) *Glycerol phosphate dehydrogenase* → Dihydroxyacetone-P

The dihydroxyacetone phosphate formed by this reaction is in equilibrium with glyceraldehyde-3-P, which is an intermediate on the pathway to glucose formation.

Enzymes in Gluconeogenesis

The reactions involved in the synthesis of glucose from pyruvate are for the most part catalyzed by the same enzymes that are involved in the glycolytic degradation of glucose. Three of the reactions catalyzed by glycolytic enzymes are, however, essentially irreversible: the enzymes involved are *hexokinase, phosphofructokinase,* and *pyruvate kinase.* Specific phosphatases, *glucose-6-phosphatase* and *fructose diphosphatase,* accomplish the reversal of the first two reactions.

Glucose-6-P — *Glucose-6-phosphatase* (H_2O → P_i) → Glucose

Fructose-1,6-P — *Fructose diphosphatase* (H_2O → P_i) → Fructose-6-P

These enzymes hydrolytically remove the phosphate from these sugars and thus accomplish the reversal of the two reactions involved in their synthesis. They do not, however, result in the resynthesis of the high-energy phosphate required for their formation. The concentration of both of these enzymes is very low in muscle compared to liver.

The equilibrium of the pyruvic kinase reaction lies far in the direction of pyruvate formation, and there is no mammalian enzyme which can catalyze the conversion of pyruvate to phosphoenol pyruvate at a significant rate. To bypass this apparent block in the conversion of pyruvate to glucose, a more elaborate metabolic transformation is used. A two-step reaction series can convert pyruvate to phosphoenol pyruvate. *Pyruvate carboxylase* carboxylates pyruvate to oxaloacetate in a reaction that is a typical biotin-dependent (see Figure 11-7), ATP requiring carboxylation. The second enzyme, *phosphoenol pyruvate carboxykinase,* which requires GTP (or ITP) *but not biotin,* decarboxylates the oxaloacetic acid to form phosphoenol pyruvate. The carboxyl group which is removed from oxaloacetate by the second enzyme is the one that was added to pyruvate by the original carboxylation. This interjection of the carbon from pyruvate into the oxaloacetic acid

$$\text{Pyruvate} \xrightarrow[\text{carboxylase}]{\text{ATP } CO_2 \quad\quad\quad ADP \text{ } P_i} \text{Oxaloacetate} \xrightarrow[\text{carboxykinase}]{\text{GTP} \quad\quad\quad GDP}_{CO_2} \text{Phosphoenolpyruvate}$$

Pyruvate   Oxaloacetate   Phosphoenolpyruvate

pool before it is converted to phosphoenol pyruvate is of extreme physiological importance. It means that any amino acid which can be converted to a citric acid cycle intermediate, *and therefore to oxaloacetate*, can serve as a precursor for gluconeogenesis.

In most species, pyruvic carboxylase is a mitochondrial enzyme, whereas PEP carboxykinase is a cytoplasmic enzyme. This would appear to pose a problem in the gluconeogenic pathway because oxaloacetic acid does not readily permeate the mitochondrial membrane. A number of possible mechanisms have been suggested to overcome this problem. The oxaloacetate could be transaminated to aspartate, which moves out of the mitochondria at a reasonable rate. The aspartate could then be reconverted to oxaloacetate in the cytoplasm by a second transamination. Another possibility is that the citrate-condensing enzyme could react oxaloacetate with acetyl CoA to form citrate. After diffusion across the mitochondrial membrane, citrate could be reconverted to oxaloacetate and acetyl CoA by the action of the citrate cleavage enzyme.

Although these two mechanisms may contribute, it appears that the major route for the movement of oxaloacetate out of the mitochondria is that illustrated in Figure 15-3. Oxaloacetate is reduced to malate in the

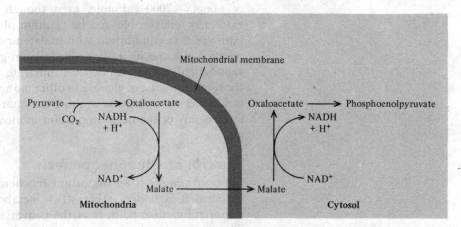

Figure 15-3 Transfer of oxaloacetate out of the mitochondria. The mitochondrial membrane is much more permeable to malate than oxaloacetate, and the high-mitochondrial and low-cytosol NADH levels favor this reaction series.

mitochondria, where the NADH concentration is high, and malate, after diffusion to the cytoplasm, can then be reoxidized to oxaloacetate. The latter reaction will be favored by the high NAD^+ concentration in the cytoplasm. The net result of all these reactions is a movement of pyruvate carbon out of the mitochondria into the cytoplasm, where the rest of the enzymes needed for glucose synthesis are located.

Energy Requirements

The overall synthesis of glucose from pyruvate requires an input of energy in the form of six moles of high-energy phosphate compounds (ATP or GTP) and two moles of NADH.

2 [Pyruvate →(ATP) Oxaloacetate →(GTP) PEP → P-glycerate →(NADH, ATP) Triose-P]

Malate ⇌ Fumarate

GLUCOSE ←(P_i) Glucose-6-P ←(P_i) Fructose-di-P

Because of the input of energy into the system to drive some of the reactions, the overall synthesis of glucose by these reactions is associated with a ΔG^0 of about -7000 cal/mole, even though some individual reactions are extremely unfavorable. As the diagram illustrates, the oxaloacetate in the pathway is in equilibrium with malate and fumarate. This means that the position of a specific radioactive label in gluconeogenic precursors will be randomized in the symmetrical fumarate pool. The reactions involved in the transformation of glucose to other monosaccharides have been discussed in Chapter 10. By the use of those enzymatic reactions, the phosphorylated form of any of the other important monosaccharides can be formed from glucose.

Control of Gluconeogenesis

Gluconeogenesis is an important biochemical pathway, and as might be expected it is under rather strict metabolic control. Perhaps the most important control point is at the conversion of pyruvate to oxaloacetate. The enzyme involved, pyruvic carboxylase, has an absolute requirement for acetyl CoA as an allosteric effector. This is physiologically desirable, because an increase in acetyl CoA production from fatty-acid metabolism or pyruvate oxidation might provide more substrate than could be oxidized by the citric acid cycle and lead to an increase in its concentration. Activation of pyruvic carboxylase would then result in more oxaloacetate to condense with acetyl CoA and to shunt carbon toward gluconeogenesis. These same conditions lead to high NADH/NAD⁺ ratios, and this inhibits pyruvate kinase so that phosphoenol pyruvate is converted to triose phosphates rather than to pyruvate. A second major control point is at the level of fructose diphosphate. High levels of AMP, which would be associated with low ATP and which indicate a need for more energy, stimulate phosphofructokinase and inhibit fructose diphosphatase. This results in a flow of carbon toward pyruvate for energy production. When the ATP level increases and AMP concentration falls, this situation is reversed, and the flow of carbon toward glucose formation is enhanced.

GLYCOGEN SYNTHESIS

The synthesis of glycogen is a good example of a metabolic reaction series considered at one time to be the reverse of the degradative reaction. It has now been shown to occur by a completely different mechanism. The enzyme responsible for glycogen synthesis, *glycogen synthetase*, carries out the reaction indicated in Figure 15-4. The reaction involved is a transfer of

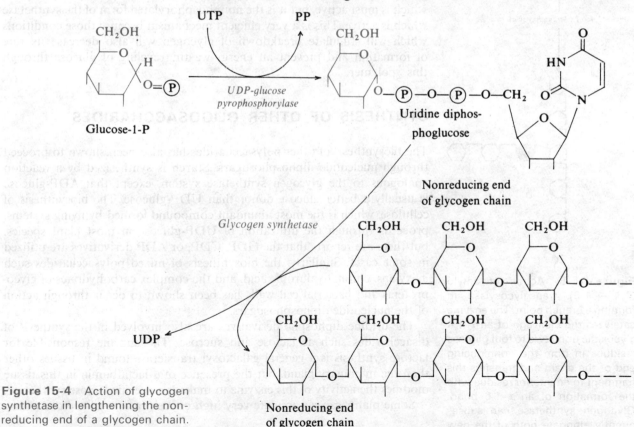

Figure 15-4 Action of glycogen synthetase in lengthening the non-reducing end of a glycogen chain.

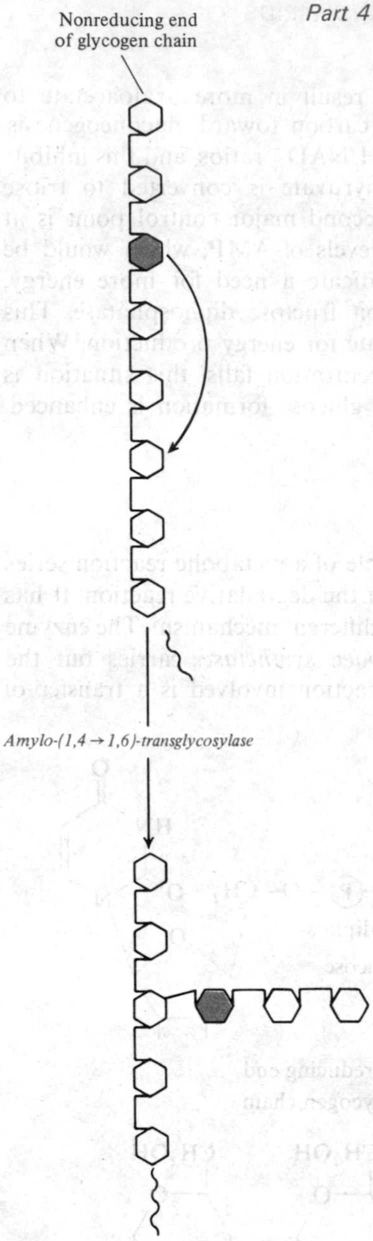

Nonreducing end
of glycogen chain

Amylo-(1,4 → 1,6)-transglycosylase

Figure 15-5 Action of amylo (1,4 → 1,6) transglycosylase in forming a branch point. The enzyme catalyzes the cleavage of an α-1,4 glycosidic bond two to four glucose residues in from the nonreducing end of the chain and transfers this fragment to an interior residue with the formation of an α-1,6 bond. Glycogen synthetase can subsequently elongate both of the new nonreducing ends.

glucose from UDP-glucose to the nonreducing end of a glycogen molecule. The UDP which is formed in the reaction can be reconverted to UTP by transfer of a phosphate from ATP; and in a reaction catalyzed by *UDP-glucose pyrophosphorylase*, UTP can be used to resynthesize UDP-glucose by reacting with glucose-1-P to liberate pyrophosphate.

Since the reaction catalyzed by glycogen synthetase is the addition of a glucose residue to a polysaccharide primer by an α-1,4 glycosidic bond, continuation of the process would yield only a linear nonbranched chain which would be analogous to amylose. Formation of the α-1,6 branch points of glycogen requires a second enzyme, *amylo (1,4 → 1,6) transglycosylase*, whose action is illustrated in Figure 15-5. This enzyme catalyzes the transfer of a short α-1,4-linked oligosaccharide fragment from the nonreducing end of the chain to form an α-1,6-linked branch point.

The enzyme responsible for glycogen formation, the synthetase, is controlled by the cellular concentration of cyclic AMP. A cAMP dependent kinase converts a more active form of the enzyme—the I, or independent, form—to a less active D, or dependent, form. This phosphorylated form of the enzyme is active only in the presence of glucose-6-P. The action of cAMP is, therefore, just the opposite of that which was discussed for phosphorylase in Chapter 10. It is the phosphorylated form of phosphorylase which is most active, but it is the nonphosphorylated form of the synthetase which is active. This is a very efficient mechanism because those conditions which will stimulate breakdown of glycogen will also decrease its rate of formation and prevent an energy-wasting cycling of glucose through this polymer.

SYNTHESIS OF OTHER OLIGOSACCHARIDES

The biosynthesis of other polysaccharides has also been shown to proceed through nucleotide diphosphosugars. Starch is synthesized by a reaction analogous to the glycogen synthetase system, except that ADP-glucose is usually a better glucose donor than UDP-glucose. The biosynthesis of cellulose, which is the most abundant compound formed by living systems, proceeds through the utilization of UDP-glucose in most plant species, but there are reports that the GDP, CDP, or ADP derivatives are utilized in some cases. Similarly, the biosynthesis of mixed polysaccharides such as xylans, chitin, hyaluronic acid, and the complex carbohydrates of glycoproteins and bacterial cell walls has been shown to occur through action of the nucleotide diphosphosugars.

The uridine diphospho derivatives are also involved in the synthesis of disaccharides such as lactose and sucrose. The enzyme responsible for lactose synthesis is a general galactosyl transferase found in tissues other than the mammary gland, but the presence of α-lactalbumin in this tissue modifies the activity of this enzyme to make it a specific lactose synthetase.

Some plants can accumulate very high concentrations of sucrose within

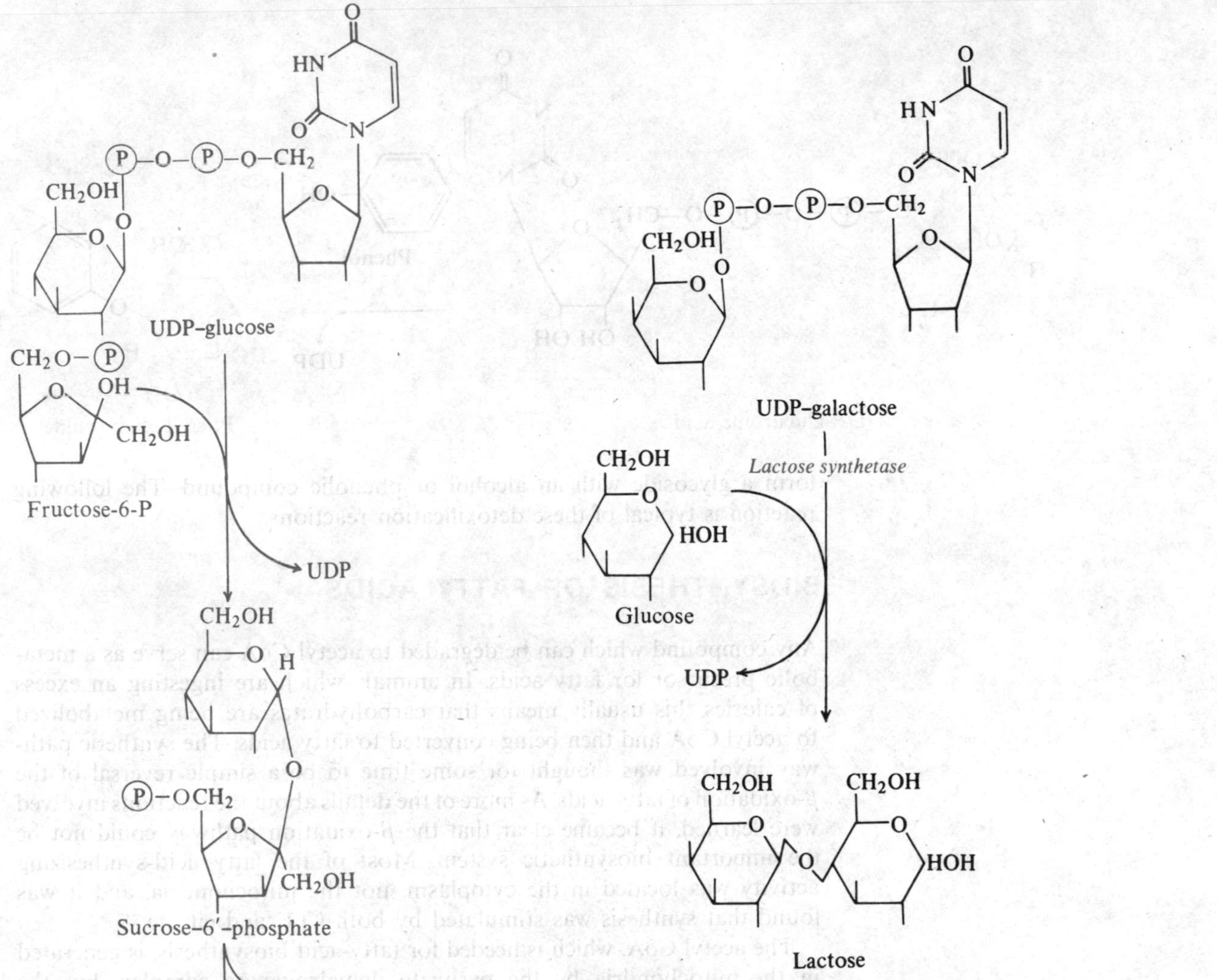

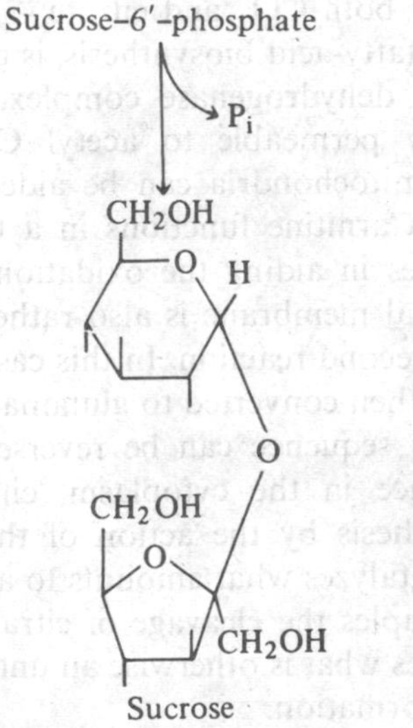

Figure 15-6 Enzymatic formation of lactose and sucrose.

the cell, and they do this by initially forming sucrose-6-P from UDP-glucose and fructose-6-P. The subsequent hydrolysis of the phosphate provides the driving force for the overall reaction. This allows the formation of a high concentration of sucrose even though the glycosidic bond formed in the reaction has a large negative ΔG^0 of hydrolysis (-6.6 kcal/mole). The reactions involved in lactose and sucrose formation are shown in Figure 15-6.

The oxidation of glucose to glucuronic acid also occurs as the UDPG derivative. Glucuronic acid is used to detoxify a large number of compounds, which are then excreted by the kidney. Glucuronic acid can be used either to form an ester with an acid which is to be detoxified, or to

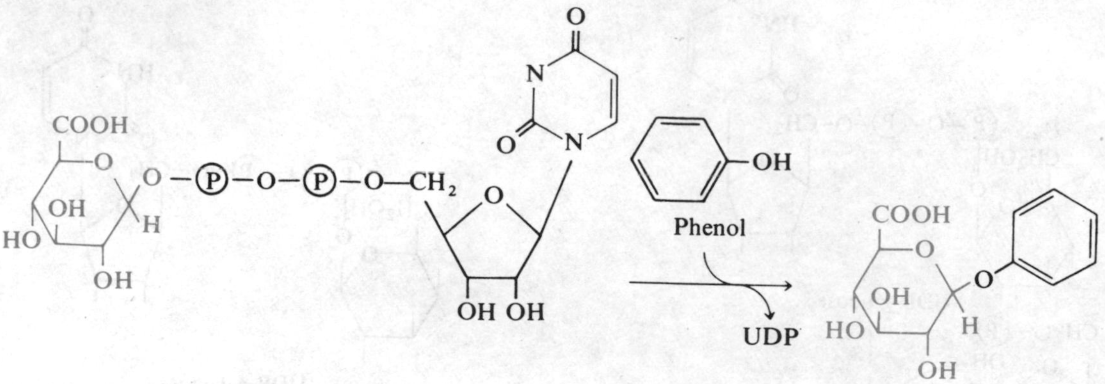

UDP-glucuronic acid

Phenol

UDP

Phenyl glucuronide

form a glycoside with an alcohol or phenolic compound. The following reaction is typical of these detoxification reactions.

BIOSYNTHESIS OF FATTY ACIDS

Any compound which can be degraded to acetyl CoA can serve as a metabolic precursor for fatty acids. In animals which are ingesting an excess of calories this usually means that carbohydrates are being metabolized to acetyl CoA and then being converted to fatty acids. The synthetic pathway involved was thought for some time to be a simple reversal of the β-oxidation of fatty acids. As more of the details about the reactions involved were learned, it became clear that the β-oxidation pathway could not be the important biosynthetic system. Most of the fatty acid-synthesizing activity was located in the cytoplasm, not the mitochondria, and it was found that synthesis was stimulated by both CO_2 and citrate.

The acetyl CoA, which is needed for fatty-acid biosynthesis, is generated in the mitochondria by the pyruvate dehydrogenase complex, but the mitochondrial membrane is not very permeable to acetyl CoA. The movement of acetyl CoA out of the mitochondria can be aided by the two pathways shown in Figure 15-7. Carnitine functions in a transport capacity in this system much as it does in aiding the oxidation of fatty acids. In some species the mitochondrial membrane is also rather impermeable to the citrate generated by the second reaction. In this case, citrate can be oxidized to α-ketoglutarate and then converted to glutamate, which diffuses out of the mitochondria. This sequence can be reversed in the cytoplasm to resynthesize citrate. Once in the cytoplasm, citrate can furnish acetyl CoA for fatty-acid synthesis by the action of the *citrate cleavage enzyme*. This enzyme, which catalyzes what amounts to a reversal of the condensing enzyme reaction, couples the cleavage of citrate to the breakdown of a mole of ATP. This drives what is otherwise an unfavorable reaction in the direction of acetyl CoA formation.

Enzymes of Fatty-Acid Synthesis

The key step in the fatty-acid biosynthetic pathway is the carboxylation of acetyl CoA by *acetyl CoA carboxylase* to form malonyl CoA. This is

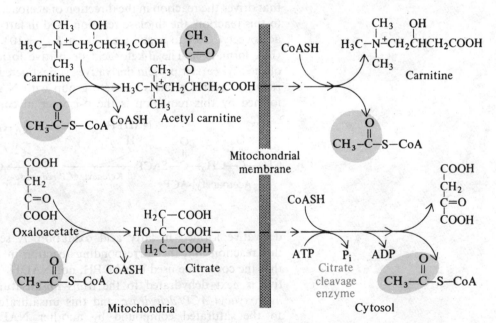

a typical two-step biotin-mediated carboxylation. The first step is an ATP-dependent carboxylation of an enzyme-bound biotin residue. The second step is a transfer of this activated CO_2 to acetyl CoA. The enzyme is strongly activated by citrate or isocitrate, either of which can promote the aggregation of inactive subunits of the enzyme into a large active complex. The inactive subunits are themselves composed of four polypeptide chains of about 100,000 molecular weight, only one of which contains one mole of bound biotin. In the presence of citrate, the protomers aggregate to a filamentus form which has a molecular weight of over 4 million. Before

Figure 15-7 Two mechanisms involved in the movement of acetyl CoA from the mitochondria to the cytosol for fatty-acid synthesis. Without some mechanism for aiding the transfer of acetyl CoA out of the mitochondria, fatty acid synthesis would be limited by its slow rate of diffusion.

341

being utilized further in fatty-acid biosynthesis, malonate is transferred from coenzyme A to an acyl carrier protein (ACP-SH) to form malonyl-ACP. Acyl carrier protein is a low molecular weight protein (about 10,000) which has an internal serine residue esterified to phosphopantetheine. This is the same modified pathothenic acid group that furnishes the free sulfhydryl group for CoA. A second transacylase forms acetyl-ACP from acetyl CoA, and by the action of a *β-ketoacyl-ACP synthetase*, the two acyl-carrier protein derivatives are combined to form acetoacetyl-ACP.

$$HO-\overset{\overset{\textstyle O}{\|}}{C}-CH_2-\overset{\overset{\textstyle O}{\|}}{C}-S-ACP$$

Malonyl-ACP

HS—ACP

CO_2

Condensing enzyme

$$CH_3-\overset{\overset{\textstyle O}{\|}}{C}-S-ACP$$

Acetyl-ACP

$$CH_3-\overset{\overset{\textstyle O}{\|}}{C}-CH_2-\overset{\overset{\textstyle O}{\|}}{C}-S-ACP$$

Acetoacetyl-ACP

This reaction proceeds with the elimination of the same CO_2 that was added to acetyl CoA to form malonyl CoA, and it is this decarboxylation that drives the reaction in the direction of acetoacetate synthesis. In contrast to this reaction, the thiolase reaction used in fatty-acid oxidation to cleave acetoacetyl CoA to acetyl CoA has a $K_{eq} = 10^5$ in the direction of acetyl CoA formation. The β-keto acid derivative formed by the condensation of the acyl carrier protein derivatives is next acted on by a *β-ketoacyl-ACP reductase* which reduces the keto group with NADPH. The hydroxy acid formed by this reduction is the D-isomer in contrast to the L-acyl CoA

$$CH_3-\overset{\overset{\textstyle O}{\|}}{C}-CH_2-\overset{\overset{\textstyle O}{\|}}{C}-SACP$$

Acetoacetyl-ACP

NADPH
H^+

β-Ketoacyl-ACP reductase

NADP⁺

$$CH_3-\overset{\overset{\textstyle H}{|}}{\underset{\underset{\textstyle OH}{|}}{C}}-CH_2-\overset{\overset{\textstyle O}{\|}}{C}-SACP$$

D-β-Hydroxybutyryl-ACP

derivative formed in fatty acid oxidation. A second difference between this reaction and the corresponding reaction in the oxidative pathway is that the coenzyme used is NADPH, not NADH. The hydroxy acid derivative is next dehydrated to the *trans-α,β*-unsaturated acyl-ACP by a *β-hydroxyacyl-ACP dehydrase*, and this unsaturated compound is reduced to the saturated compound by another NADPH-linked reaction, an *enoyl-ACP reductase*. In some species, this final enzyme contains a protein-bound flavin, but in others the transfer of electrons is directly from NADPH to the unsaturated acid. The saturated acyl-ACP derivative formed by the final reduction can next react with a second mole of malonyl ACP to

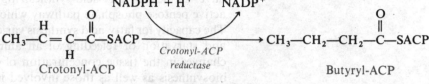

D-β-Hydroxybutyryl-ACP → (Enoyl-ACP dehydratase, H_2O) → Crotonyl-ACP

Crotonyl-ACP → (Crotonyl-ACP reductase, $NADPH + H^+$ → $NADP^+$) → Butyryl-ACP

Figure 15-8 Fatty acid biosynthesis. The individual reactions are detailed in the text. The reactions in boldface are the start of a second turn of the cycle. The β-keto acid formed at this point would be subjected to the same series of reduction, dehydration, and reduction as shown for the 4-carbon keto acid derivative. Note that only the two carbons at the methyl end of the completed fatty acid would be directly contributed by acetyl-ACP. The rest arise through the intermediate formation of malonate.

form a 6-carbon β-keto acyl-ACP and the entire sequence can be repeated. The overall pathway is indicated in Figure 15-8, and the net reaction for the synthesis of palmitate by this pathway is shown on next page.

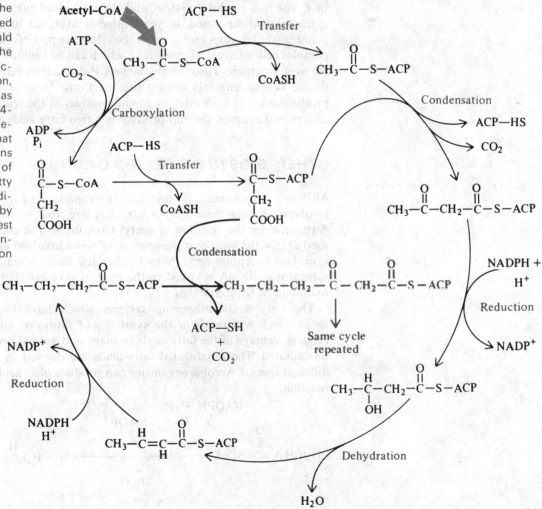

$$8 \text{ acetyl CoA} + 14(\text{NADPH} + \text{H}^+) + 7\text{ATP} \longrightarrow \text{Palmitate} + 8\text{CoASH}$$
$$+ 14\text{NADP}^+ + 7\text{P}_i + 6\text{H}_2\text{O}$$

The overall reaction for fatty acid synthesis requires a large amount of NADPH. Tissues such as liver, adipose tissue, and mammary gland, which have an active fatty-acid-synthesizing capacity, usually have a rather active pentose phosphate pathway which can furnish the NADPH needed. The capacity for fatty acid synthesis varies directly with the energy available, and starvation or refeeding of an animal can be shown to cause rapid changes in the tissue concentration of the enzymes involved ln fatty acid biosynthesis as well as those involved in NADPH generation.

Multienzyme Complexes

The fatty-acid-synthesizing system is reasonably similar in most species. In *E. coli* it is a soluble system, and the individual enzymes can be readily extracted and separated. In yeast, pigeon liver, rat liver, and mammary gland, all the enzymes except the carboxylase are part of a large multienzyme complex called *fatty acid synthetase*, which can be isolated as a high molecular weight particle. The intermediate ACP-derivatives remain firmly bound to the various enzymes during the reactions. *In vitro*, the yeast system produces an acyl CoA ester as an end product of the reaction; in bacterial and animal systems, the end product is a free fatty acid, usually palmitate.

OTHER BIOSYNTHETIC REACTIONS

Although mitochondria do not have the capacity to carry out the *de novo* synthesis of long-chain fatty acids, they are able to elongate acyl CoA derivatives by the addition of acetyl CoA units. The enzymatic reactions used are for the most part the reversal of those involved in fatty acid oxidation. One exception appears to be that the final reduction of the α,β-unsaturated acyl-CoA is linked to the oxidation of NADPH rather than the oxidation of reduced flavin.

The fatty acid-synthesizing systems which have been described are those which will result in the synthesis of saturated fatty acids only. A large percentage of the fatty acids in plant and animal lipids are, however, unsaturated. The unsaturated fatty acids are derived in different ways in different species. Aerobic organisms can produce oleic acid by the following reaction.

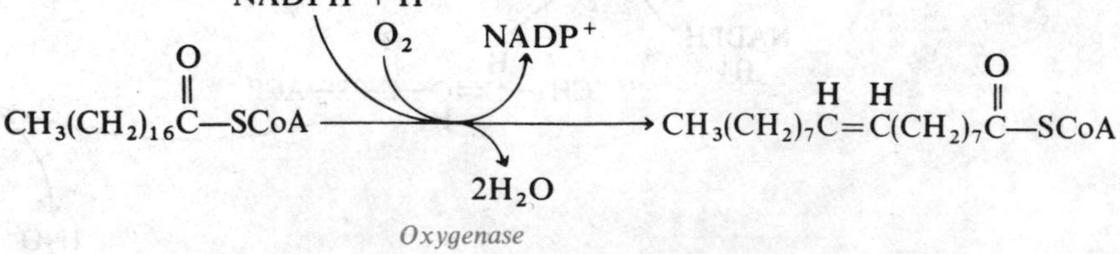

Oxygenase

This reaction, which occurs in the microsomes, is catalyzed by an enzyme belonging to the class of enzymes called *mixed-function oxidases*. In this reaction, an intermediate hydroxy acyl CoA derivative is formed, followed by a desaturation to produce the unsaturated fatty-acid derivative. In anaerobic bacteria, a different pathway is used—a β-hydroxy acyl-ACP derivative is dehydrated to form the β,γ-enoyl-ACP derivative rather than the α,β double-bond derivative usually formed in fatty-acid oxidation. The successive addition of 2-carbon units in the form of malonyl-ACP to this intermediate will result in the formation of the common monounsaturated fatty acids.

Plants can synthesize linoleic and linolenic acid from oleic acid by a reaction which again involves the mixed function oxidases. Mammals cannot synthesize these acids, but can elongate them to form a series of polyunsaturated fatty acids with up to 22-carbon atoms. Because mammals cannot synthesize linoleic or linolenic acids, they must be provided in the diet and are therefore called essential fatty acids.

SYNTHESIS OF TRIGLYCERIDES AND PHOSPHOLIPIDS

The key intermediate needed for triglyceride is phosphatidic acid. This compound, because it is a precursor for both triglyceride and phospholipid synthesis, is one of the most important intermediates in lipid biosynthesis. Phosphatidic acid is formed in a reaction which uses two moles of acyl CoA esters to acylate α-glycerol phosphate. Both glycerol and dihydroxyacetone phosphate can serve as a precursor for α-glyceryl phosphate formation by the reactions shown in Figure 15-9. The *glycerol kinase* is essentially absent in adipose tissue and intestinal mucosa, and in these tissues α-glycerol phosphate is formed by the reductive pathway. To complete the formation of a triglyceride, the phosphate of the phosphatidic acid is removed by a phosphatase, and the resulting 1,2-diglyceride is subjected to a further acylation by a third acyl CoA derivative. The overall pathway is indicated in Figure 15-9. This is the system which is active in most tissues. Intestinal mucosa has an active triglyceride-synthesizing system, and it appears that there is a pathway in this tissue which involves the direct acylation of a 2-monoglyceride to the 1,2-diglyceride and then to the triglyceride.

Phospholipid Synthesis

A number of different pathways lead to the synthesis of the phospholipids. As was the case with polysaccharide synthesis, they involve the participation of a nucleoside diphosphate, in this case the cytidine derivative. One of the most common phospholipids, phosphatidyl choline, can be formed by the reactions in Figure 15-9. A system analogous to this can form phosphatidyl ethanolamine from CDP-ethanolamine. These reactions

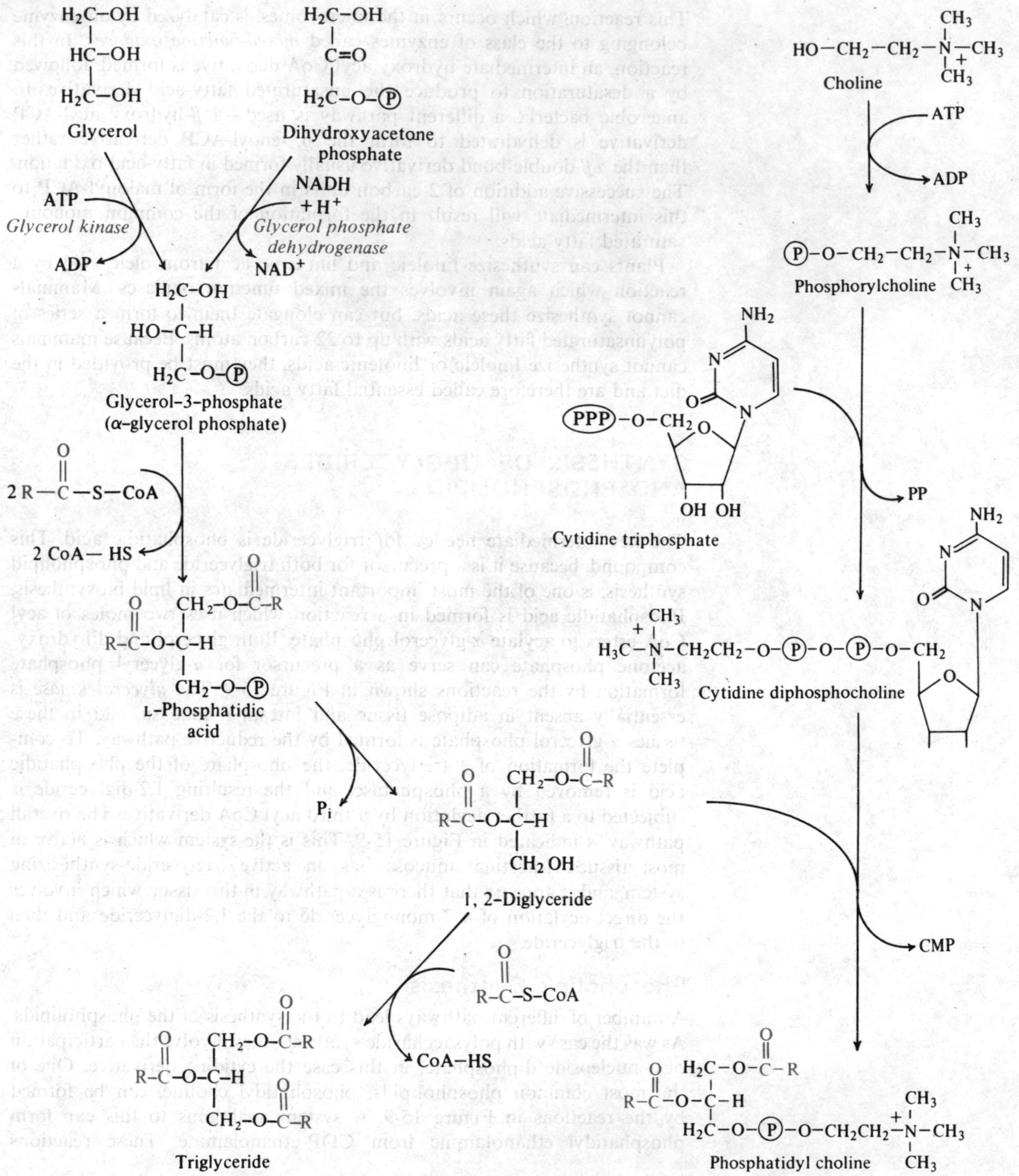

Glycerol

Dihydroxyacetone phosphate

ATP

Glycerol kinase

ADP

NADH +H⁺

Glycerol phosphate dehydrogenase

NAD⁺

Glycerol-3-phosphate (α-glycerol phosphate)

2 R—C—S—CoA

2 CoA—HS

L-Phosphatidic acid

Pᵢ

1, 2-Diglyceride

R—C—S—CoA

CoA—HS

Triglyceride

Choline

ATP

ADP

Phosphorylcholine

Cytidine triphosphate

PP

Cytidine diphosphocholine

CMP

Phosphatidyl choline

Figure 15-9 (*opposite*) Synthesis of triglycerides and phosphatidyl choline. Note that the 1,2 diglyceride is a common intermediate in the formation of these two types of lipids.

occur in smooth microsomal membranes and there is some specificity with regard to the 1,2-diglycerides which are used. Saturated fatty acids found in phospholipids are always in the 1 position, and unsaturated fatty acids are in the 2 position. An alternate, more general pathway for phospholipid synthesis which starts with the formation of CDP-diglyceride is shown in Figure 15-10. In this pathway the CDP-diglyceride transfers

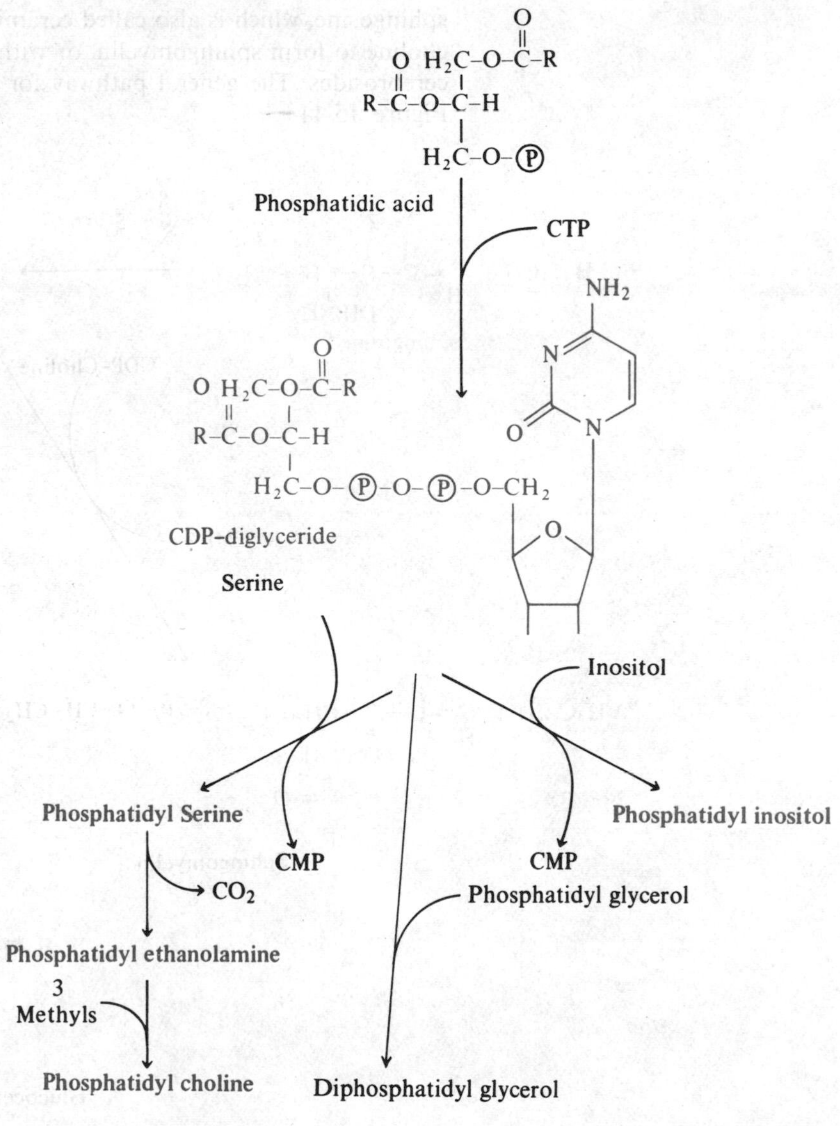

$$O \quad H_2C-O-\overset{\overset{O}{\|}}{C}-R$$
$$R-\overset{O}{\underset{\|}{C}}-O-\overset{}{C}-H$$
$$H_2C-O-\textcircled{P}$$

Phosphatidic acid

CTP

$$O \quad H_2C-O-\overset{\overset{O}{\|}}{C}-R$$
$$R-\overset{O}{\underset{\|}{C}}-O-\overset{}{C}-H$$
$$H_2C-O-\textcircled{P}-O-\textcircled{P}-O-CH_2$$

CDP–diglyceride

Serine

Inositol

Phosphatidyl Serine   **Phosphatidyl inositol**

CMP   CMP

CO_2 **Phosphatidyl glycerol**

Phosphatidyl ethanolamine

3 Methyls

Phosphatidyl choline   **Diphosphatidyl glycerol**

Figure 15-10 Alternate pathway for phosphatidyl choline synthesis. This pathway, through a CDP-diglyceride, also leads to the synthesis of the other common glycerol phospholipids.

a phosphatidic acid group to a number of different acceptors to form phosphatidyl inositol, phosphatidyl serine, or diphosphatidyl glycerol. There are then pathways to convert phosphatidyl serine to phosphatidyl ethanolamine and phosphatidyl choline.

The other main class of complex lipids contains the fatty alcohol sphingosine rather than glycerol, and these compounds are called sphingolipids. Sphingosine is synthesized from palmitic acid and serine, and it reacts with an acyl CoA molecule to form the important intermediate *N*-acylsphingosine, which is also called ceramide. Ceramide can react with CDP-choline to form sphingomyelin, or with uridine diphospho sugars to form cerebrosides. The general pathway for sphingolipid synthesis is shown in Figure 15-11.

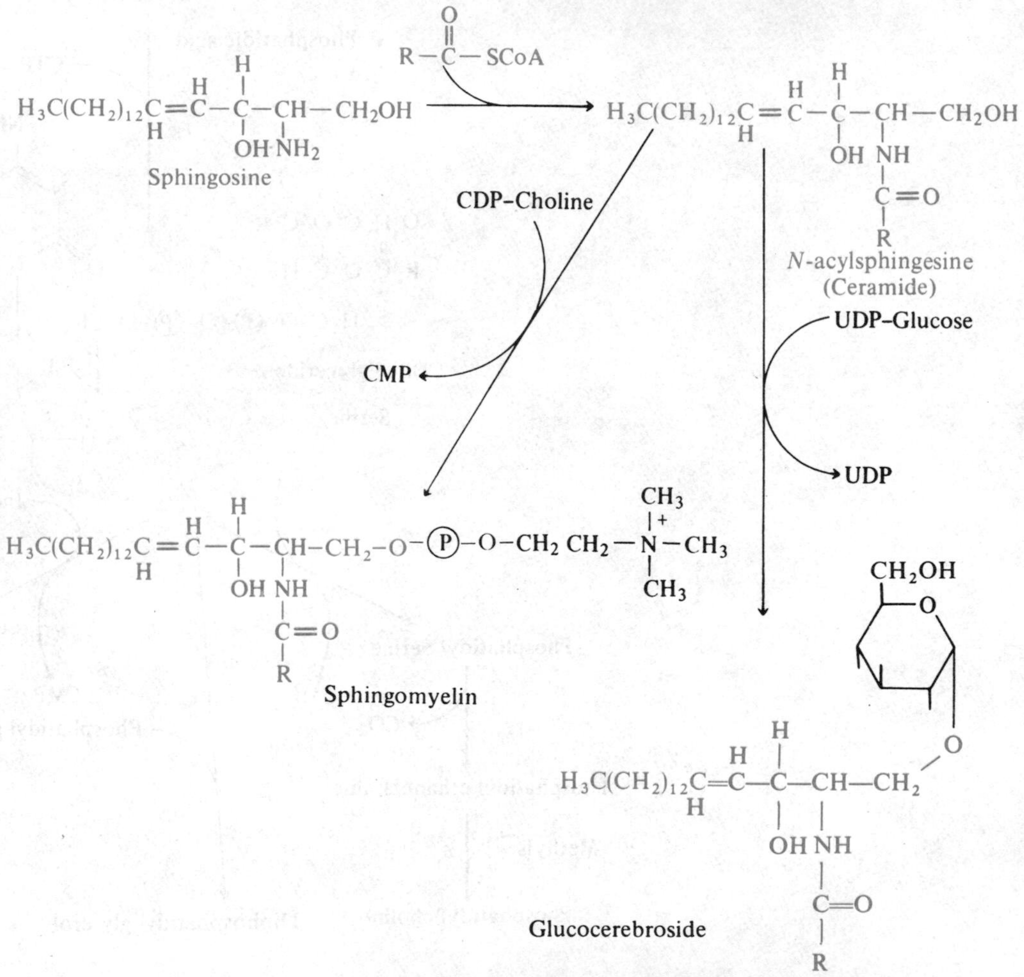

Figure 15-11 Synthesis of sphingolipids. Ceramide serves as the acceptor for groups transferred from nucleotide diphosphates.

ONE-CARBON METABOLISM

The methyl groups shown in Figure 15-10 as being transferred to phosphatidyl ethanolamine to form phosphatidyl choline represent a new type of biochemical reaction, the transfer of one-carbon units. In this particular case, the transfer is at the oxidation state of a methyl group, and the methyl donor is *S*-adenosyl methionine (SAM). One-carbon transfers can also occur at oxidation states which would be equivalent to those of methanol, formaldehyde, formate, or carbon dioxide. It has previously been shown that transfers of carbon at the oxidation state of CO_2 occur as a derivative of the vitamin biotin (Chapter 11), whereas interconversions of the other one-carbon compounds are associated with the action of another vitamin, folic acid.

> *Folic Acid (Pteroylglutamic Acid)* A large number of factors that were necessary in the nutrition of various species to prevent specific symptoms as well as an unidentified bacterial growth factor were shown in the early 1940s to be similar compounds. The vitamin is a complex molecule composed of a pteridine ring, *p*-aminobenzoic acid, and glutamic acid. In some species the active factor has a number of glutamic acid residues joined in peptide linkages. The reduced form of the vitamin, tetrahydrofolic acid, is involved in the metabolism of one-carbon units in a large number of reactions. A deficiency of the vitamin in the human diet causes macrocytic and megaloblastic anemias, sprue, and some malabsorption syndrome.
>
> The vitamin is present in high concentrations in liver, and is widely distributed in green leafy vegetables. The recommended dietary allowance is 0.4 mg per day.

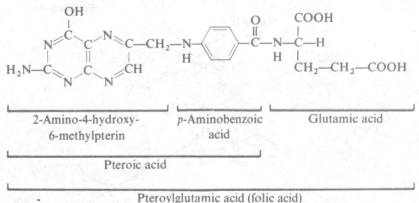

Folic acid can exist in a number of forms, and it is the tetrahydro derivative, "H_4 folate," formed by the enzymatic reduction of folate which is the coenzyme-active form. One-carbon compounds are oxidized or reduced as derivatives of the N-5 or N-10 positions of H_4 folate, and these compounds are shown in Figure 15-12. There are some metabolic reactions which release formate or formaldehyde and there are other reactions which directly transfer single carbons from a metabolic intermediate to a folic acid derivative. The major contributors to the "1-carbon pool" are choline and the amino acids serine, glycine, and histidine. At whatever oxidation state these compounds enter the pool, the reactions in Figure 15-12 can interconvert them to the other forms. There are very few reactions where a methyl group is transferred directly from N^5-Me-H_4 folate to an acceptor

Figure 15-12 Structure of tetrahydrofolic acid. The important metabolic interconversions of one-carbon derivatives of H_4folate are shown.

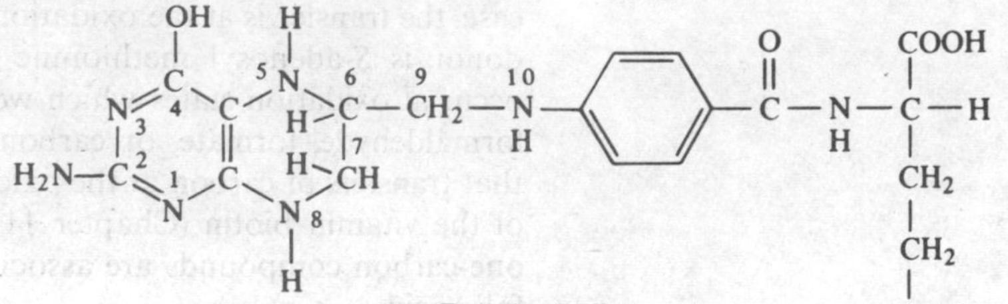

Tetrahydrofolic acid

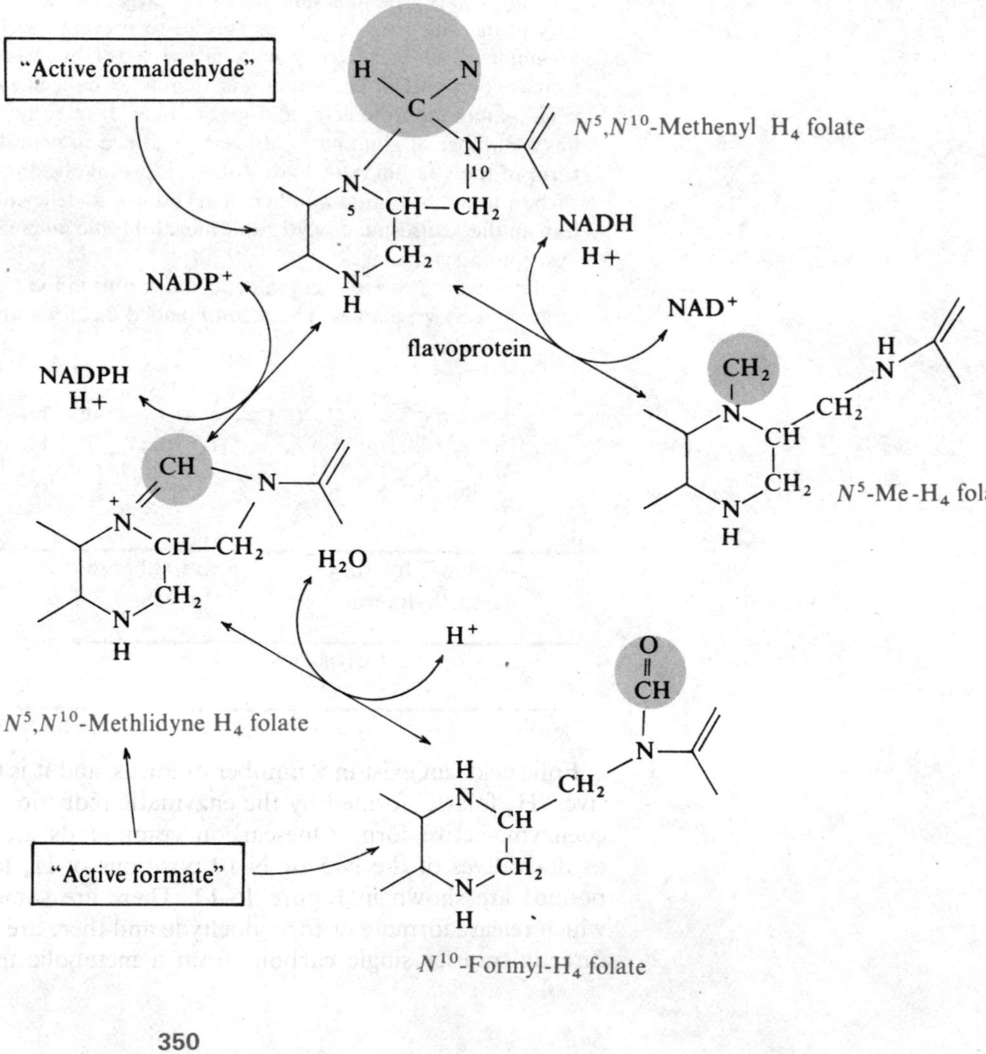

"Active formaldehyde"

N^5,N^{10}-Methenyl H_4 folate

NADH
H+

NADP+

NAD+

NADPH
H+

flavoprotein

N^5-Me-H_4 folate

N^5,N^{10}-Methlidyne H_4 folate

H_2O

H+

"Active formate"

N^{10}-Formyl-H_4 folate

molecule. Rather, most of these transfers occur through the participation of S-adenosyl methionine, and this methyl carrier is regenerated from the folate pool by a vitamin B_{12}-mediated reaction. These interconversions are shown in Figure 15-13.

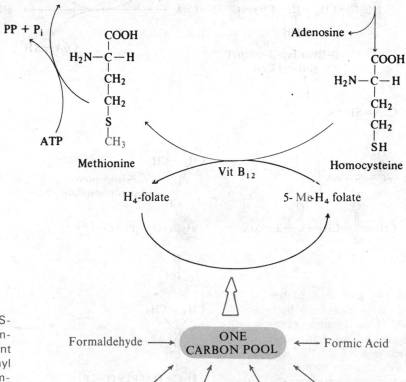

Figure 15-13 Metabolism of S-adenosylmethionine. This compound serves as an important methyl donor, and the methyl group is regenerated from a common "one carbon" pool.

ISOPRENOID BIOSYNTHESIS

Sterols and other isoprenoid compounds are synthesized from acetyl CoA by a complex series of reactions which involve a number of intermediates distinctively different from those found in other metabolic pathways. The initial reactions of this pathway involve the conversion of acetate into a 5-carbon intermediate which is the precursor of all the isoprenoid compounds. The first reaction in the series leading to this 5-carbon compound is the condensation of acetyl-CoA with acetoacetyl-CoA to form β-hydroxy-β-methyl-glutaryl-CoA. This is the same intermediate which was important in ketone body formation. The carboxyl group which is esterified to CoA is reduced by NADPH with the liberation of free coenzyme A and the formation of mevalonic acid. The discovery of this 6-carbon dicarboxylic acid was the key to the elucidation of pathway for the conversion of acetate to the sterols. Mevalonic acid can be further decarboxylated and phosphorylated to form two isomeric pyrophosphates. These initial reactions in the pathway to isoprenoid synthesis are indicated in Figure 15-14.

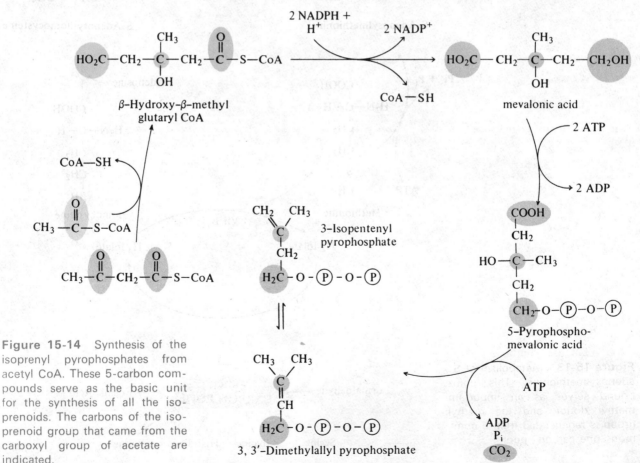

Figure 15-14 Synthesis of the isoprenyl pyrophosphates from acetyl CoA. These 5-carbon compounds serve as the basic unit for the synthesis of all the isoprenoids. The carbons of the isoprenoid group that came from the carboxyl group of acetate are indicated.

Squalene and Cholesterol Formation

The pathway followed to convert these 5-carbon intermediates to sterols involves successive condensations of 5-carbon units to form geranyl pyrophosphate (10-carbons) and farnesyl pyrophosphate (15-carbons). Both of these are what are called "head to tail" condensations. The carbon that carries the pyrophosphate group attacks one of the methyl carbons of the second reactant. Farnesyl pyrophosphate can condense with its isomer, nerolidol pyrophosphate, in an NADPH-requiring reaction to form the 30-carbon isoprenoid, squalene (Figure 15-15). Squalene is a hydrocarbon found in relatively high concentrations in shark liver and in at least trace concentrations in livers of other animals. It is the direct precursor of the sterols. When the squalene structure is rewritten (Figure 15-16), it can be seen that a shift in a small number of carbon-carbon bonds will result in closure of the rings to form the basic sterol ring system. The reactions involved utilize molecular oxygen, NADPH, and a microsomal mixed function oxidase to form the 2,3-epoxide, which then cyclizes to form the first sterol intermediate, lanosterol. This condensation requires a number of shifts of both hydrogen atoms and methyl groups. Lanosterol can be converted to cholesterol by a series of reactions which remove three of the five methyl groups on the rings, shift the double bond in the ring system, and saturate the side-chain double bond. These reactions are catalyzed by a multienzyme system that is tightly bound to the microsomal membrane. The overall rate of cholesterol synthesis can be regulated by a number of factors. The major control point seems to be at the level of the β-hydroxy-β-methylglutaryl coenzyme A reductase, which produces mevalonic acid. Cholesterol will inhibit this reaction and will therefore limit its own production. The enzyme shows a distinct diurnal variation in activity, and cholesterol synthesis in experimental animals is high at night and depressed during the day. It is probable that both the activity of the enzyme and the continued synthesis of the enzyme can be altered by increased cholesterol concentrations. Some control of the enzymes involved in converting squalene to cholesterol may also be exerted by variations in tissue cholesterol content.

Other Isoprenoids

Cholesterol can serve as a precursor for the synthesis of other physiologically important steroids such as the bile acids and the steroid hormones. A large number of other important, nonsteroid, compounds are also synthesized from mevalonate. Natural rubber is a high molecular weight poly-*cis*-isoprene formed by the condensation of isopentenyl pyrophosphate units, and monoterpenes such as camphor and menthol are formed from the 10-carbon intermediate geranyl pyrophosphate. Carotinoids, including β-carotene, the precursor to vitamin A, are 40-carbon compounds derived from the condensation of two geranylgeranyl pyrophosphates. Many other compounds which have important biochemical functions, including coenzyme Q or ubiquinone, are synthesized from these 5-carbon precursors.

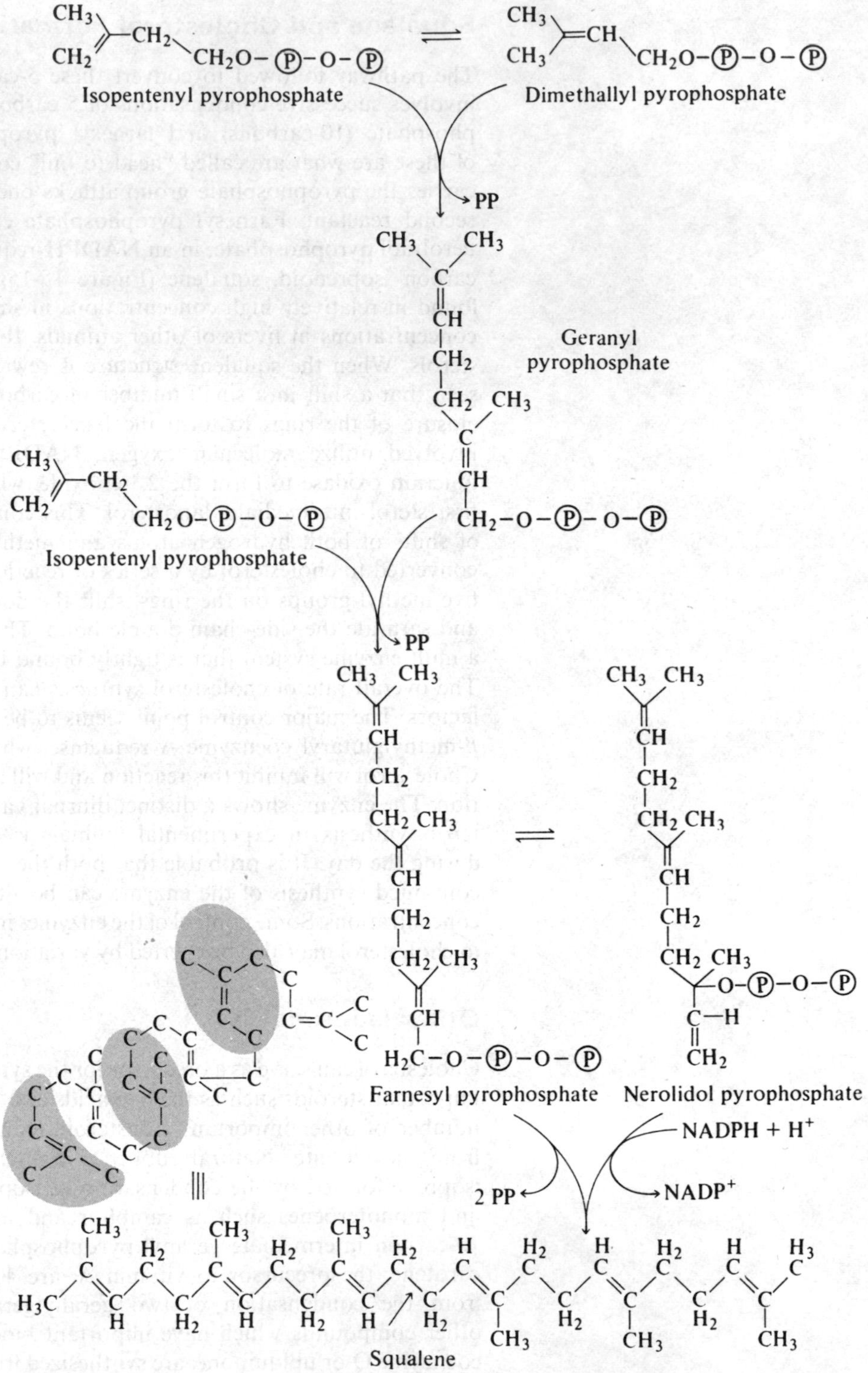

Figure 15-15 Conversion of isoprenyl pyrophosphates to squaline. The arrow indicates the point where there has been a "head to head" condensation of two 15-carbon compounds, and the molecule is symmetrical about this point. When the molecule is written in a different shape, it is easy to see its relationships to the sterol ring. Alternate isoprene units are shaded.

Isopentenyl pyrophosphate

Dimethallyl pyrophosphate

Geranyl pyrophosphate

Isopentenyl pyrophosphate

Farnesyl pyrophosphate

Nerolidol pyrophosphate

Squalene

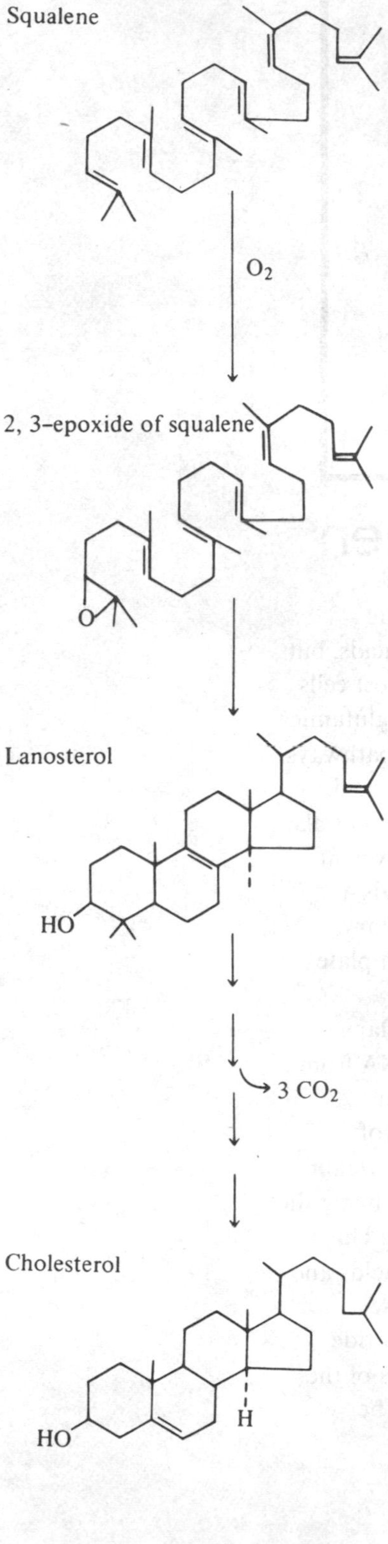

Squalene

O₂

2, 3–epoxide of squalene

Lanosterol

HO

3 CO₂

Cholesterol

HO   H

Figure 15-16 Conversion of squaline to cholesterol. A mixed function oxidase creates an epoxide of squalene which then cyclizes to form lanosterol. The conversion of lanosterol to cholesterol involves a number of intermediates which have been omitted from this diagram.

PROBLEMS

1. Alanine is a good source of carbon for gluconeogenesis. Where would the label from methyl-^{14}C-labeled alanine appear in glucose?

2. Synthesis of glycogen through the glycogen synthetase system requires a considerable amount of energy in the formation of UDPG. Why would this system have evolved rather than the synthesis of glycogen through a reversal of phosphorylase?

3. The first step in fatty-acid synthesis attaches CO_2 to acetyl CoA, yet if an *in vitro* fatty-acid synthesizing system is incubated with $^{14}CO_2$, no radioactivity is found in the fatty acids. Why?

4. How can adipose tissue, which has essentially no glycerol kinase activity, form triglycerides through a pathway that requires α-glycerol phosphate as an intermediate?

5. Geranol is a 10-carbon isoprenoid. Which of its carbons are derived from the methyl carbin (M) and which from the carboxyl carbon (C) of acetate?

6. Which coenzymes are associated with the transfer of one-carbon compounds to an acceptor molecule at the oxidation level of CO_2, formate, and methyl groups?

Suggested Readings

Articles and Reviews

Bloch, K. The Biological Synthesis of Cholesterol. *Science* **150**:19–28 (1965).

Hassid, W. Z. Biosynthesis of Oligosaccharides and Polysaccharides in Plants. *Science* **165**:137–144 (1969).

Kennedy, E. P. The Metabolism and Function of Complex Lipids. *Harvey Lect.* **57**:143–171 (1962).

Larner, J., C. Villar-Palasi, N. D. Goldberg, J. S. Bishop, F. Huijing, J. I. Wenger, H. Sasko, and N. B. Brown. Hormonal and Non-Hormonal Control of Glycogen Synthesis. *Adv. Enzyme Regulation* **6**:409 (1968).

Lynen, F. The Role of Biotin-Dependent Carboxylations in Biosynthetic Reactions. *Biochem. J.* **102**:381–400 (1967).

Oesterhelt, D., H. Bauer, and F. Lynen. Crystallization of a Multienzyme Complex: Fatty Acid Synthetase from Yeast. *Proc. Nat. Acad. Sci. USA* **63**:1377–1382 (1969).

Pontremoli, S., and E. Grazi. Gluconeogenesis, in *Carbohydrate Metabolism and Its Disorders*, Dickens, F., P. J. Randle, and W. J. Whelan (eds.), vol. 1, pp. 259–295. Academic Press, New York (1968).

biosynthesis of nitrogenous compounds

chapter sixteen

Most organisms need some form of organic nitrogen compound to synthesize amino acids, but only a few can reduce N_2 to ammonia. Most cells can fix NH_3 into α-ketoglutarate to form glutamic acid, and there are specialized metabolic pathways for the synthesis of each individual amino acid. Pyrimidines and purines are not dietary essentials, and they are formed from low molecular weight nitrogenous precursors. Enzymes called DNA polymerases synthesize a new complementary strand of DNA upon an existing DNA template by the condensation of the four deoxyribonucleotidetriphosphates. A similar enzyme, RNA polymerase, synthesizes RNA from a DNA template and the ribonucleosidetri-phosphates. This DNA-directed synthesis of messenger RNA molecules is called *transcription*, and the formation of a specific protein, utilizing the mRNA as a template, is called *translation*. This is accomplished by activating the amino acids and transferring them to a specific transfer RNA molecule that can recognize a three-nucleotide codon on the mRNA. The actual synthesis of the polypeptide chain occurs on the surface of the ribosome.

NITROGEN FIXATION

All living organisms need some kind of metabolizable nitrogen compounds to synthesize the amino acids, nucleotides, and other nitrogenous compounds required for life. Plants and many microorganisms can utilize ammonia, whereas various animal species need a dietary source of at least some of the amino acids. The ability of an organism to convert atmospheric nitrogen, N_2, to ammonia is the process called nitrogen fixation and is a property restricted to a relatively small number of organisms. Although the nodule-forming *Rhizobia*, which grow in a symbiotic relationship with leguminous plants, are commonly thought of as the important nitrogen-fixing bacteria, a large number of free living forms have this capacity. The blue-green algae, the aerobic bacteria *Azotobacter*, and both nonphotosynthetic and photosynthetic anaerobic bacteria can carry out active nitrogen fixation. These anaerobes have proved to be the most useful organisms for studying the biochemical details of the system.

Although other products were considered likely at one time, it is now clear that the initial product of nitrogen fixation is NH_3. The conversion of N_2 to NH_3 involves a change in the oxidation state of nitrogen from 0 to -3, which means that a total of 6 electrons is required for each molecule of N_2 reduced. Although numerous attempts have been made to identify intermediates in the pathway of nitrogen reduction, none have been successful; the first detectable product is NH_3. It is therefore assumed that the intermediate reduction products remain firmly bound to the enzyme.

Nitrogenase

In the early 1960s cell-free systems which were able to convert a small amount of N_2 to NH_3 were developed by a number of investigators. These systems required high concentrations of pyruvate and liberated both CO_2 and H_2. The breakdown of pyruvate provided ATP to the system, and the formation of hydrogen is now known to be a side reaction that competes with the reduction of N_2 for the electrons released by pyruvate oxidation. The enzyme complex that is able to fix N_2 is now called *nitrogenase*. It consists of an iron-containing protein, *Fe protein*, and a protein containing both iron and molybdenum, the *MoFe protein*. The iron in both proteins is bound to acid labile sulfur. Neither of these proteins has catalytic activity by itself, but when mixed with ATP and a source of electrons they constitute the active nitrogenase complex. The reaction probably requires 4 ATPs per pair of electrons transferred, or twelve ATPs per mole of N_2 reduced. The physiological reductant in the fixation process is a ferredoxin, the low redox potential iron protein also involved in photosynthesis.

There is now sufficient data available at least to postulate a molecular mechanism for the fixation process (Figure 16-1). The Fe protein can be reduced by ferredoxin, and this reduced protein will bind ATP. The binding of ATP to this protein lowers its redox potential, and it is able to transfer

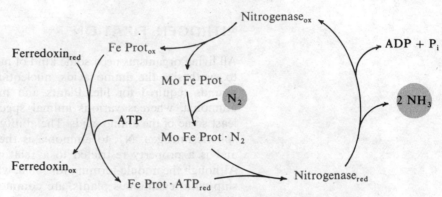

Figure 16-1 Nitrogenase complex. Experimentally the complex is usually studied with $S_2O_4^{2-}$ as a source of electrons rather than reduced ferredoxin. The system will reduce other compounds, and C_2H_2 is often used as a substrate because of the ease of detection of the product C_2H_4. The point at which ADP is released from the system may not be until after the nitrogenase complex is split and the Fe protein has been reduced.

electrons to the MoFe protein and cause its reduction. The cleavage of ATP to ADP and Pi probably occurs at this point, but may occur at a later step. The reduced MoFe protein binds N_2, reduces it to NH_3, and in the process it becomes reoxidized and the cycle repeats itself. It is not known if the reduction occurs through a series of enzyme-bound distinct intermediates, or if there is an essentially instantaneous reduction of N_2 to NH_3 on the enzyme surface.

METABOLISM OF NH₃

Cells fixing N_2 do not build up high concentrations of ammonia, since it is rapidly converted into other compounds. In most cells a very active *glutamic dehydrogenase* forms glutamic acid from α-ketoglutarate, and this amino acid can participate in a number of transaminase reactions. Much of the free NH_3 produced is also taken up by *glutamine synthetase*, and the amide nitrogen formed in this reaction can be transferred to other compounds.

COOH   NADH+
| H⁺   NAD⁺
C=O   COOH   ADP   COOH
| NH₂—C—H   ATP   Pᵢ   NH₂—C—H
CH₂   *Glutamic* | |
| *dehydrogenase* CH₂   CH₂
CH₂   NH₃   CH₂   *Glutamine* |
| | *synthetase*   CH₂
COOH   COOH   NH₃ |
                                                                C—NH₂
                                                                ‖
                                                                O

α-Ketoglutaric   Glutamic   Glutamine
acid   acid

When living organisms die and decay, the nitrogenous compounds are degraded to NH_3, which has a relatively short life in the soil. It is utilized by the autotropic bacteria *Nitrosomonas* and *Nitrobacter*. These organisms obtain their energy by oxidizing NH_3 to NO_2^- and then to NO_3^-. The major portion of the nitrogen in the soil is thus converted to NO_3^-, which is absorbed by plants and reduced to NH_3 for the synthesis of nitrogen containing compounds.

AMINO ACID SYNTHESIS

As indicated in Chapter 12, animals can synthesize some amino acids, but need others preformed in the diet. Transaminases are present in tissues which will act on almost any keto acid; therefore, the inability of an animal to form a particular amino acid is really a failure to synthesize the corresponding keto acid. The amino acids that are essential in the diet of animals are, however, synthesized in plants and microorganisms by pathways lacking in animal tissues.

Nonessential Amino Acids

The amino acids *alanine, aspartic acid,* and *glutamic acid* can be formed by the direct transamination of the corresponding keto acids, pyruvate, oxaloacetate, and α-keto glutarate. Glutamic acid can also be formed by the reductive amination of α-keto glutarate, the reverse of the reaction previously described in Chapter 12 for the degradation of glutamate. *Proline* is formed from glutamic acid through its reduction to glutamic acid γ-semialdehyde by an NADH-linked enzyme. The pyroline-5-carboxylic

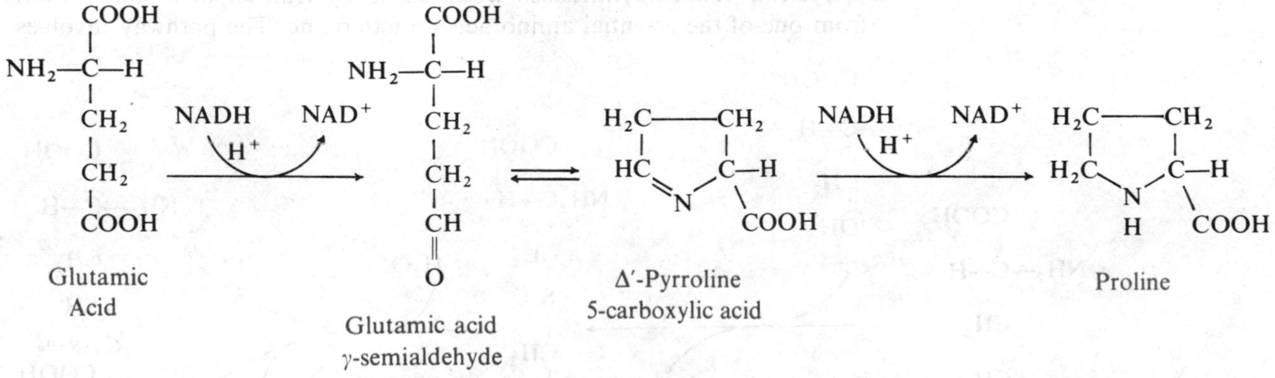

Glutamic Acid

Glutamic acid γ-semialdehyde

Δ′-Pyrroline 5-carboxylic acid

Proline

acid, which is in equilibrium with the glutamic acid derivative, can then be reduced to proline.

Tyrosine is synthesized from phenylalanine through the action of a complex mixed function oxidase called phenylalanine hydroxylase, and *serine* can be formed from the glycolytic intermediate, 3-phosphoglyceric acid. In this pathway, 3-PGA is oxidized to 3-phospho-hydroxypyruvic

COOH   NAD⁺   NADH H⁺   COOH   COOH   H₂O   COOH

3-Phosphoglyceric acid   3-Phosphohydroxy-pyruvic acid   3-Phospho-serine   Serine

acid, which can be transaminated to 3-phosphoserine. There is a phosphatase which hydrolyzes this to yield serine.

Although a number of pathways contribute to *glycine* formation in mammalian tissue, the most important one is direct conversion from serine. This is a pyridoxal phosphate-mediated reaction where the hydroxy methyl group of serine is transferred to tetrahydrofolic acid.

H₄ folate   5,10-methylene-H₄ folate

Serine   Pyridoxal phosphate   Glycine

Cysteine can be synthesized from serine by transfer of a sulfur atom from one of the essential amino acids, methionine. The pathway involves

Homocysteine   Serine   H₂O   Cystathionine   H₂O   NH₃   Cysteine   α-Ketobutyrate the formation of homocysteine from methionine through the intermediate 5-adenosylmethionine by reactions discussed in Chapter 15. Homocysteine then reacts with serine to form cystathionine, which breaks down to yield cysteine. The carbon skeleton of cysteine is derived from serine, and all that is contributed by the methionine is the sulfhydryl group.

Essential Amino Acids

Arginine for tissue protein synthesis can be formed by the reactions of the urea cycle (Figure 12-4) but these reactions would deplete the ornithine pool unless there is some way of replenishing it. It is the enzymes needed to synthesize ornithine which are absent in mammalian tissue. Ornithine is formed by reduction of the γ-carboxyl group of glutamic acid to an aldehyde, followed by a transamination at this position. These reactions occur as the *N*-acetyl derivative of glutamic acid.

Three of the essential amino acids, *threonine*, *methionine*, and *lysine*, can be derived from aspartic acid. The key intermediates in these conversions are aspartic-β-semialdehyde, which can be converted to diaminopimelic acid and then to lysine, and homoserine, which can be formed from aspartic-β-semialdehyde. Homoserine can be converted to threonine, or, through cystathionine as an intermediate, it can be converted to methionine. During this conversion the sulfur comes from cysteine. These general pathways are illustrated in Figure 16-2. There is also a pathway for lysine synthesis which starts with α-ketoglutarate and goes through aminoadipic acid as an intermediate.

Three of the common amino acids, *phenylalanine*, *tryptophan*, and *tyrosine* require the formation of an aromatic ring and present an additional biosynthetic problem. Tyrosine can be formed from phenylalanine in mammalian tissue, but the other two amino acids can be synthesized only in plants and microorganisms. The biosynthesis proceeds through the formation of shikimic acid from intermediates of carbohydrate metabolism. Although shikimic acid is not an aromatic compound, it does contain a six-member ring, and is the precursor for the aromatic ring of both phenylalanine and tryptophan (Figure 16-3).

The branched-chain amino acids, *valine*, *leucine*, and *isoleucine*, are formed from α-keto acids that are intermediates in other metabolic pathways. The conversion of pyruvate to valine is illustrated in Figure 16-4, and the formation of isoleucine from α-ketobutyric acid proceeds through a very similar series of reactions. The precursors of leucine are acetyl CoA and α-ketoisovaleric acid, which is an intermediate in valine synthesis.

Histidine is formed by an involved series of reactions starting with ribose and ATP. All five carbons of ribose, along with a carbon and nitrogen from the adenine ring of ATP, and a nitrogen from the amide group of glutamine, are retained in histidine, the end product.

This brief description should serve as a general indication of the pathways by which the various amino acids can be synthesized. In many cases the details of the biosynthetic pathway for the formation of an essential amino

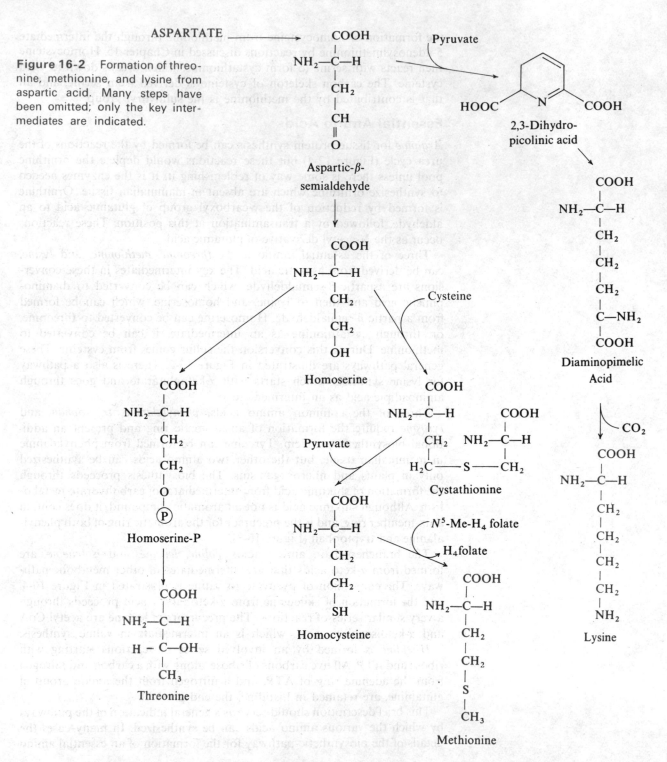

Figure 16-2 Formation of threonine, methionine, and lysine from aspartic acid. Many steps have been omitted; only the key intermediates are indicated.

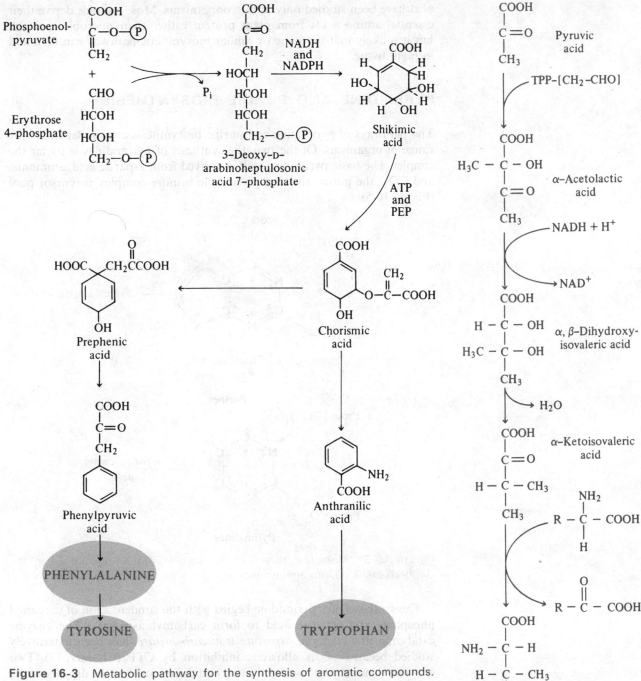

Figure 16-3 Metabolic pathway for the synthesis of aromatic compounds. A number of steps have been omitted; only key intermediates are shown.

Figure 16-4 Metabolic pathway for the synthesis of valine.

acid have been studied only in microorganisms. Most animals derive their essential amino acids from plant protein rather than microbial protein, but it is likely that the same or similar biosynthetic pathway can be found in plant tissue.

PYRIMIDINE AND PURINE BIOSYNTHESIS

The pathways of pyrimidine and purine biosynthesis are similar in a wide range of organisms. Of the two, the synthesis of pyrimidines is by far the simpler. The basic pyrimidine ring is derived from aspartic acid, ammonia, and CO_2; the purine ring is derived from a more complex precursor pool (Figure 16-5).

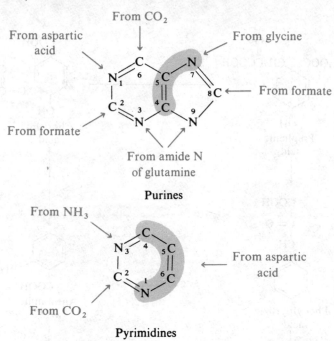

Figure 16-5 Metabolic source of the carbon and nitrogen used in the synthesis of the purines and pyrimidines.

The synthesis of a pyrimidine begins with the condensation of carbamyl phosphate and aspartic acid to form carbamyl aspartate. The enzyme catalyzing this reaction, *aspartate transcarbamylase*, has been extensively studied because of its allosteric inhibition by CTP (Chapter 17). Two enzymes then act, first to close the ring and then to oxidize the dihydroorotic acid to orotic acid. The intact pyrimidine formed by these reactions is condensed with 5-phosphoribosyl-1-pyrophosphate (PRPP) to form orotidine-5′-phosphate. This nucleotide monophosphate is then decarboxylated

364

to form uridine monophosphate, which, after conversion to UDP and then to UTP, is aminated to form CTP. This rather straightforward pathway for pyrimidine biosynthesis is summarized in Figure 16-6.

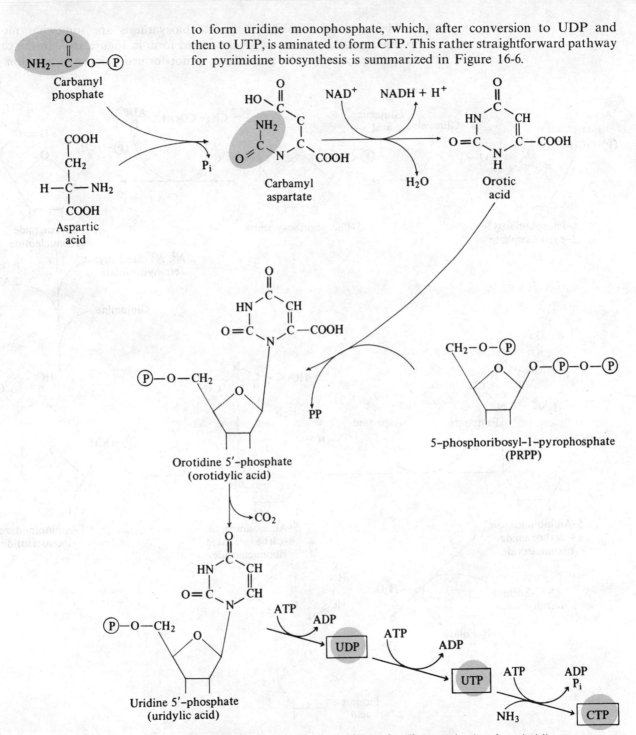

Figure 16-6 Metabolic pathway for the synthesis of pyrimidines.

The reactions involved in purine biosynthesis are somewhat more complex, and are presented in condensed form in Figure 16-7. In the case of purine synthesis, the purine ring is not formed before the addition of

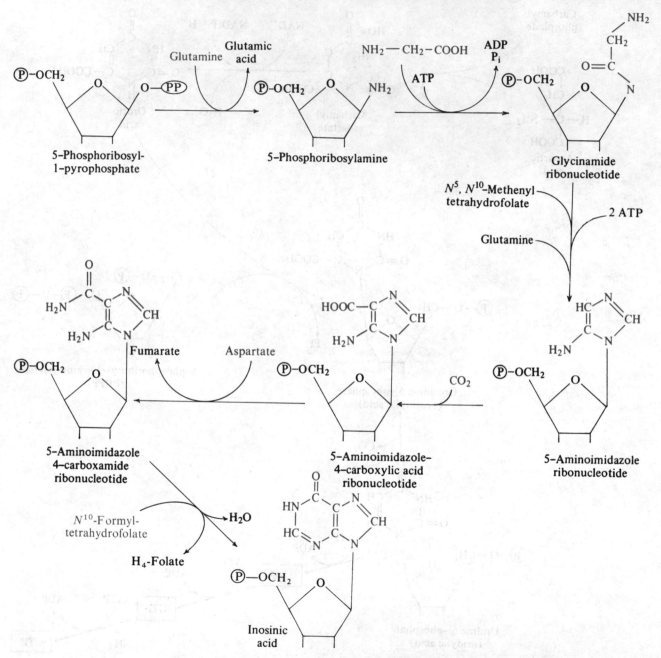

Figure 16-7 Metabolic pathway for the synthesis of purines.

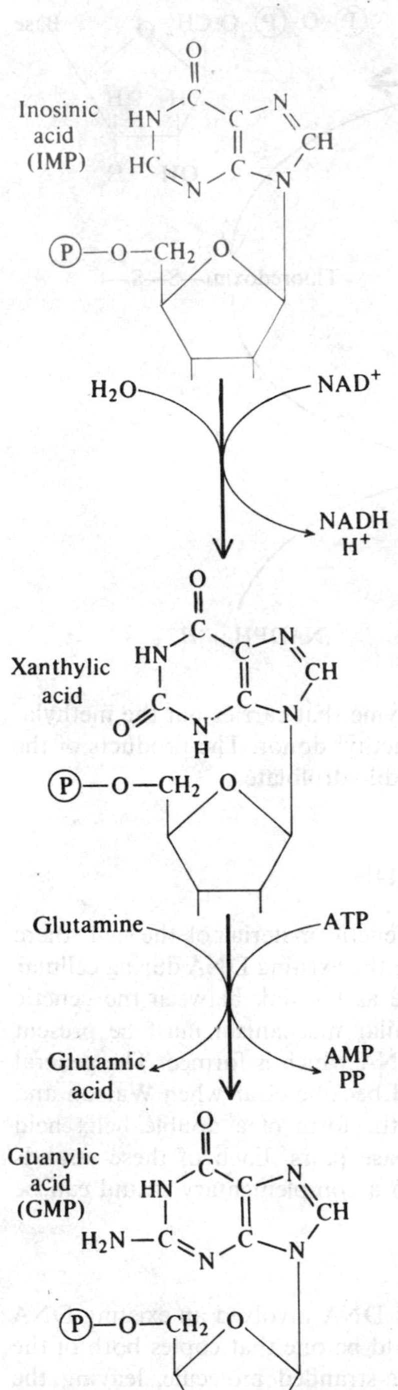

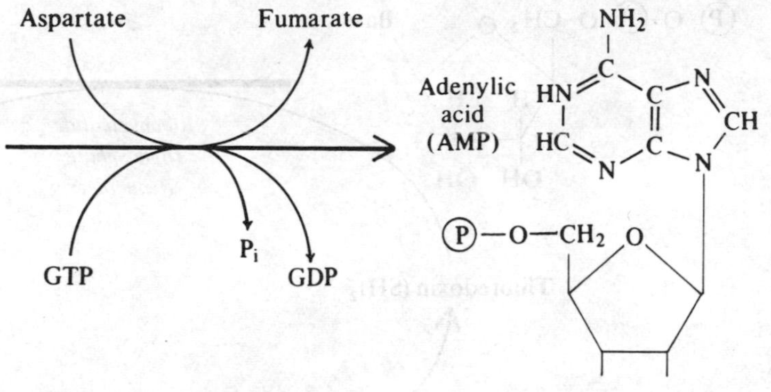

Figure 16-8 Metabolic pathway for the conversion of inosinic acid into the common purine nucleotides.

ribose; rather, the first reaction involves the interaction of PRPP with the amide nitrogen of glutamine to form 5-phosphoribosylamine. This compound is then condensed with glycine in a reaction which requires ATP. After the donation of a formate group from a tetrahydrofolate derivative and a second amide nitrogen from glutamine, the ring is closed by another ATP-requiring reaction. The compound formed, 5′-amino-imidazole ribonucleotide, is carboxylated to give a carboxylic acid group at what will be the carbon 6 of the purine ring. The remaining nitrogen is added by a reaction which involves the condensation of aspartic acid with this carboxyl group and the subsequent loss of fumarate. The last carbon added to the ring comes from a second participation of a folic acid derivative. Inosinic acid, which is the product of this reaction, is not one of the purines commonly found in polynucleotides, but is converted to adenylic acid and guanylic acid by the reactions shown in Figure 16-8.

The deoxyribonucleotides required for DNA synthesis are not derived through a pathway that starts with 2-deoxyribose. Rather, the corresponding ribonucleotides are synthesized and are reduced to deoxysugar derivatives at the level of the nucleotide di- or triphosphates. The reactions involved have been demonstrated in mammalian tissue, but the most extensively studied pathway is that found in *E. coli*. In this species the reduction is at the level of the diphosphates and involves a sulfhydryl containing protein, *thioredoxin*. The reduced form of this enzyme [thioredoxin-$(SH)_2$] can donate a pair of electrons and hydrogen to form 2-deoxyribose and thio-redoxin-S_2. The oxidized form of the protein can be reduced by NADPH (Figure 16-9). The pathway in some other bacterial species has been shown to differ significantly. In some species the reduction is at the level of the triphosphates, and it is dependent on a coenzyme form of vitamin B_{12}. What pathway operates in mammalian tissue is not clear, but it appears to be similar to that found in *E. coli*.

DNA contains thymidylic acid rather than deoxyuridylic acid, and this conversion is made at the level of the deoxyribonucleotide. In this reaction,

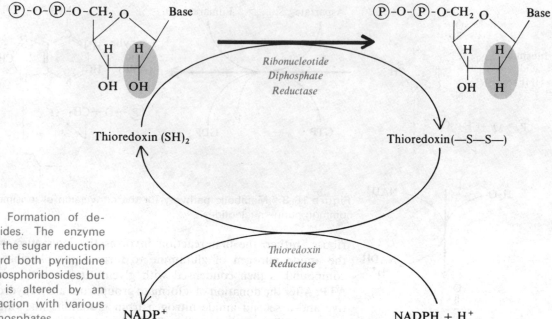

Figure 16-9 Formation of deoxyribonucleotides. The enzyme responsible for the sugar reduction is active toward both pyrimidine and purine diphosphoribosides, but its specificity is altered by an allosteric interaction with various nucleotide triphosphates.

dUMP serves as the substrate for an enzyme that carries out the methylation by using N^5, N^{10}-methylene FH_4 methyl donor. The products of the reaction are thymidylic acid, TMP, and dihydrofolate.

SYNTHESIS OF NUCLEIC ACIDS

Because DNA molecules serve as the genetic material of the cell, there must be a means of accurately duplicating the existing DNA during cellular division. If an RNA molecule is to serve as the link between the genetic information and cellular proteins, a similar mechanism must be present to assure accurate transcription of the RNA which is formed. The general way in which this could be accomplished became clear when Watson and Crick postulated that DNA existed in the form of a double helix held together by hydrogen bonds between base pairs. Each of these strands can then serve as a template upon which a complementary strand can be synthesized.

DNA Synthesis

If it were assumed that the replication of DNA involved an existing DNA strand as a template, the mechanism could be one that copies both of the strands to form an entirely new double-stranded molecule, leaving the old one intact; or one of the new strands could be associated with each of the old strands. The first possibility would be called a *conservative* mechanism, and the latter a *semiconservative* mechanism, of duplication.

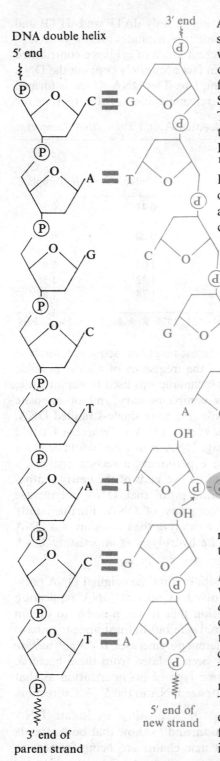

DNA double helix
5′ end

3′ end

C ≡ G

A ≡ T

G   C

C   G

T   A

A ≡ T

OH

OH

5′ end

C ≡ G

T ≡ A

5′ end of
new strand

3′ end of
parent strand

A third possibility, which is at least theoretically conceivable, is that each strand of DNA contains segments which are newly synthesized, alternating with segments from the old strand. This method of duplication would be called a *dispersive mechanism.* The proof that normal DNA replication follows a semiconservative mechanism was obtained by Meselson and Stahl before there was much knowledge of the enzymatic process involved. These investigators were able to set up conditions where the old DNA of the cell was labeled with heavy nitrogen, ^{15}N, and the newly synthesized DNA contained the normal isotope of nitrogen, ^{14}N. The presence of the ^{15}N causes a density change in the DNA sufficient to allow separation of DNA containing ^{15}N from DNA containing ^{14}N by density gradient centrifugation. Cells were grown in a ^{15}N medium and then switched to a ^{14}N medium. It was shown that all the DNA produced during the first cycle of cell division following transfer to the new medium was of a density

Figure 16-10   Action of DNA polymerase in the synthesis of DNA. Note that the two strands of the DNA double helix must separate for the new strands to be synthesized on them.

midway between that of the old and the new DNA. It was concluded from this that the cells must have contained one chain of each.

DNA Polymerase I   The first enzyme to be implicated in DNA synthesis was isolated from *E. coli* by Kornberg in 1956 and was called *DNA polymerase.* This enzyme, now called *DNA polymerase I,* has a single polypeptide chain and a molecular weight of 109,000. Although it is now clear that this enzyme is not the one responsible for DNA replication in the bacterial cell, it catalyzes a similar reaction and has been very extensively studied. To synthesize a new DNA strand, the enzyme requires the deoxyribonucleoside triphosphates as substrates and a source of DNA as a template. The enzyme catalyzes the addition of a deoxynucleotide to the free 3′ end of a growing polynucleotide strand and the liberation of pyrophosphate from the deoxynucleoside triphosphate which is added. This action is illustrated in Figure 16-10. If given a synthetic DNA such as poly (dA-dT), the enzyme will catalyze the incorporation of only dATP and dTTP and will not use the other two deoxynucleoside triphosphates (dGTP or dCTP) if they are also added. A number of different kinds of evidence contributed to the realization that this enzyme was in fact accurately copying the DNA which was added to the system as a template. The DNA which is formed is double stranded and has the same base composition (Table 16-1) as the

Table 16-1 Comparison of the Base Composition of Enzymatically Synthesized DNA and Their DNA Templates

Source of DNA Template	Base Composition of the Enzymatic Product				$\dfrac{A + T}{G + C}$ in Product	$\dfrac{A + T}{G + C}$ in Template
	Adenine	Thymine	Guanine	Cytosine		
Micrococcus lysodeiticus (a bacterium)	0.15	0.15	0.35	0.35	0.41	0.39
Aerobacter acrogenes (a bacterium)	0.22	0.22	0.28	0.28	0.80	0.82
Escherichia coli	0.25	0.25	0.25	0.25	1.00	0.97
Calf thymus	0.29	0.28	0.21	0.22	1.32	1.35
Phage T2	0.32	0.32	0.18	0.18	1.78	1.84

Source: Reprinted by permission from James D. Watson, *Molecular Biology of the Gene*, 2nd ed., © 1970, W. A. Benjamin, Menlo Park, California.

template added. The techniques of nearest-neighbor sequence analysis (Figure 6-13) allows a determination of the frequency of all the possible dinucleotides in a nucleic acid, and this technique was used to demonstrate that the daughter strand formed was complementary and of opposite polarity to the template. The enzyme will also copy single-stranded DNA, and Kornberg's group was able to show that DNA polymerase I could copy a small single-stranded circular phage DNA *in vitro* and could produce a biologically active molecule. In these experiments, a second enzyme, a *polynucleotide ligase*, was needed to join the two ends of the newly synthesized DNA. This experiment was definite proof that DNA polymerase had the ability to produce an error-free copy of DNA. Further study indicated that not only could the enzyme catalyze the extension of a DNA chain, but that it could also catalyze the hydrolysis of an existing DNA chain.

DNA Polymerase II and III   The possibility that the original DNA polymerase might not be the enzyme involved in normal DNA replication was strongly suggested by the observation that it was possible to obtain bacterial mutants which grew at a normal rate but did not appear to have any of the polymerase I activity. Two different polymerases, *DNA polymerase II* and *DNA polymerase III*, have now been isolated from these bacteria. Both of these enzymes catalyze the same type of polymerization as that shown by DNA polymerase I, and synthesize DNA in the $5' \rightarrow 3'$ direction.

Replication of Double-Stranded DNA   It is possible to isolate DNA molecules during the process of replication, and to show that both strands are replicated simultaneously and that new chains are being formed on

Figure 16-11 DNA replication. A series of short $5' \rightarrow 3'$ chains can be formed along both strands of the parent molecule, and these short segments are joined by a second enzyme.

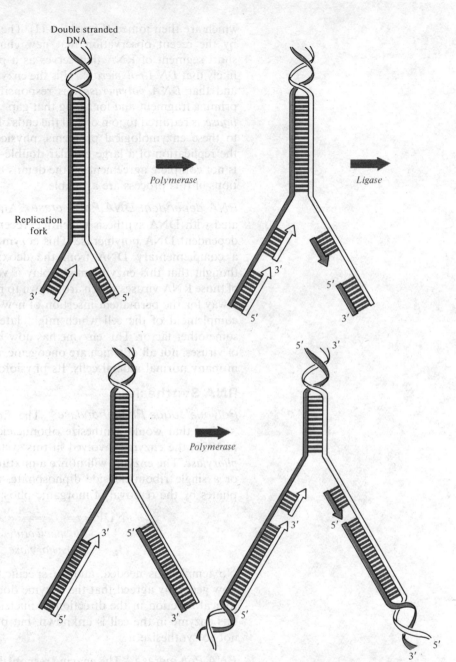

both parent strands as they separate to form a replication fork. As the two parent strands are of opposite polarity, this would suggest that replication in one strand proceeds in the $3' \rightarrow 5'$ direction, and on the other from the $5' \rightarrow 3'$ direction. No enzymes which will polymerize DNA in the $3' \rightarrow 5'$ direction have been found, but there are alternate explanations. The most likely is that both strands are synthesized in short $5' \rightarrow 3'$ segments, which are then joined (Figure 16-11). The synthetic process is complicated by the recent observation that new chains are probably initiated by a short segment of RNA that serves as a primer for DNA elongation. It is likely that *DNA polymerase III* is the enzyme responsible for the elongation and that *DNA polymerase I* is responsible for the excision of the RNA priming fragment and for filling that gap with DNA. A third enzyme, *DNA ligase*, is required to join or seal the ends of two short fragments. In addition to these enzymological problems, physical problems are associated with the replication of a large circular double-helical molecule. Although there is not complete agreement on the details of this process, plausible explanations of this process are available.

RNA-dependent DNA Polymerase   Another enzymatic activity associated with DNA synthesis which has recently been discovered is an RNA-dependent DNA polymerase. This enzyme uses an RNA template to form a complementary DNA from the deoxynucleoside triphosphates. It is thought that this enzyme might play a very significant role in the action of those RNA viruses which are known to produce malignancies. It provides a way for the permanent insertion of new genetic material into the genetic complement of the cell which might later be expressed by the action of some other factor. This enzyme has now been identified in a large number of viruses, not all of which are oncogenic, and in fact may be demonstrated in many normal animal cells. Its physiological function is not yet known.

RNA Synthesis

Polynucleotide Phosphorylase   The first enzymatic system to be described that would synthesize ribonucleic acid was discovered by Ochoa in 1955; the enzyme involved in this system is called *polynucleotide phosphorylase*. The enzyme will utilize a mixture of the nucleoside diphosphates or a single ribonucleoside diphosphate, and will polymerize the diphosphates by the removal of inorganic phosphate to form a ribonucleic acid.

$$n\text{NDP} \underset{\substack{Polynucleotide \\ phosphorylase}}{\rightleftharpoons} (\text{NMP})_n + n\text{P}_i$$

No template is needed, and no specificity of sequence is involved. It is now generally agreed that the enzyme does not have an important physiological function in the direction of nucleic acid synthesis. The function of this enzyme in the cell is unknown, but presumably it is to degrade RNA, not to synthesize it.

RNA Polymerase   The enzyme responsible for most of the RNA synthesis in the cell is called *RNA polymerase*. This enzyme has many of the properties of DNA polymerase. It uses DNA as a template, and the nucleoside triphosphates as substrates to synthesize an RNA molecule which has the same $(A + U)/(G + C)$ ratio as the $(A + T)/(G + C)$ ratio of the double-stranded DNA used as a primer (Table 16-2). The action catalyzed by the enzyme is the attack of the free $3'$ OH of the growing chain on the

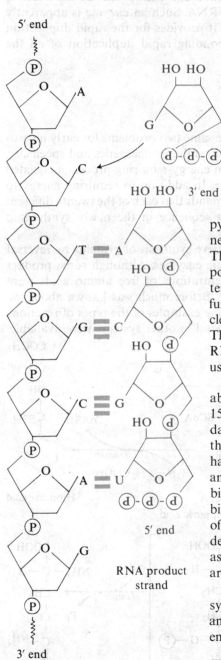

5' end

HO HO

G

d-d-d

HO HO 3' end

T ≡ A

HO d

G ≡ C

HO d

C ≡ G

HO d

A ≡ U

d-d-d

5' end

RNA product
strand

3' end

DNA template
strand

Table 16-2 Comparison of the Base Composition of Enzymatically Synthesized RNAs with the Base Composition of Their Double-Helical DNA Templates

Source of DNA Template	Composition of the RNA Bases				$\dfrac{A + U}{G + C}$ Observed	$\dfrac{A + T}{G + C}$ in DNA
	A	U	G	C		
T2	0.31	0.34	0.18	0.17	1.86	1.84
Calf thymus	0.31	0.29	0.19	0.21	1.50	1.35
E. coli	0.24	0.24	0.26	0.26	0.92	0.97
Micrococcus lysodeikticus (a bacterium)	0.17	0.16	0.33	0.34	0.49	0.39

Source: Reprinted by permission from James D. Watson, *Molecular Biology of the Gene*, 2nd ed., © 1970, W. A. Benjamin, Menlo Park, California.

pyrophosphate bond of the incoming nucleoside triphosphate to form a new phosphodiester bond and to liberate pyrophosphate (Figure 16-12). The chain elongation therefore proceeded from the 5' to the 3' end. RNA polymerase acts on only a single strand of the double-stranded DNA template, so a short segment of the DNA must unwind for the enzyme to function. How the correct strand is chosen *in vivo* is not known, but it is clear that it is the same strand for the entire length of the DNA chain. The enzyme will also act on a single-stranded viral DNA and produce an RNA which in this case is complementary to the composition of the DNA used as a primer.

RNA polymerase is a multisubunit protein with a molecular weight of about 500,000. It is composed of 2 α chains, 40,000 daltons each; a β chain, 150,000 daltons; a β' chain, 160,000 daltons; and a σ-factor chain, 90,000 daltons. The holoenzyme can, therefore, be represented as $\alpha_2\beta\beta'\sigma$. One of these subunits, the sigma factor, is less tightly bound than the other and has the role of recognizing a specific nucleotide sequence in a DNA molecule and initiating RNA synthesis at this location. The antibiotic rifampicin binds to the β subunit and prevents RNA synthesis, whereas actinomycin D binds to guanine residues on the DNA template and prevents movement of the enzyme along it. In addition to the necessity of the sigma factor to determine where the polymerase starts to copy DNA, other proteins not associated with the holoenzyme complex, the rho and the kappa factors, are involved in termination of transcription.

RNA polymerase would seem to be the enzyme responsible for the synthesis of all the major classes of RNA in the cell—transfer, ribosomal, and messenger. However, under some conditions, it is not the only cellular enzyme which can carry out this function. Considerable work has been

Figure 16-12 Action of RNA polymerase in the synthesis of RNA on a DNA template. The action is analogous to that of DNA polymerase, except that it utilizes the ribonucleotide triphosphates rather than the deoxyribonucleotide triphosphates.

done on an enzyme called *RNA synthetase* or *RNA replicase*. This enzyme is induced in some cells by viral infection and then promotes the infective process by duplicating only the viral RNA. Such an enzyme is apparently necessary for efficient viral infection. It provides for the rapid duplication of the viral RNA without a corresponding rapid duplication of all the RNA in the cell.

SYNTHESIS OF PROTEINS

The biosynthesis of proteins posed the same two problems for early investigators as were involved in nucleic acid synthesis; energetics and specificity. The formation of a peptide bond is an energy-requiring process associated with a ΔG^0 of about $+3000$ cal/mole. In addition to requiring energy to drive this reaction, the process also demands that each of the twenty different amino acids be placed in its correct sequence in the newly synthesized polypeptide chain.

Early biochemists attempted to achieve synthesis of proteins by reversal of the reaction catalyzed by proteolytic enzymes. Although some product could be demonstrated if high concentrations of free amino acids were used, it has no specificity of sequence. Before much was known about the requirements for protein synthesis, some examples of the types of reactions used to form an amide or a peptide bond in other systems were available.

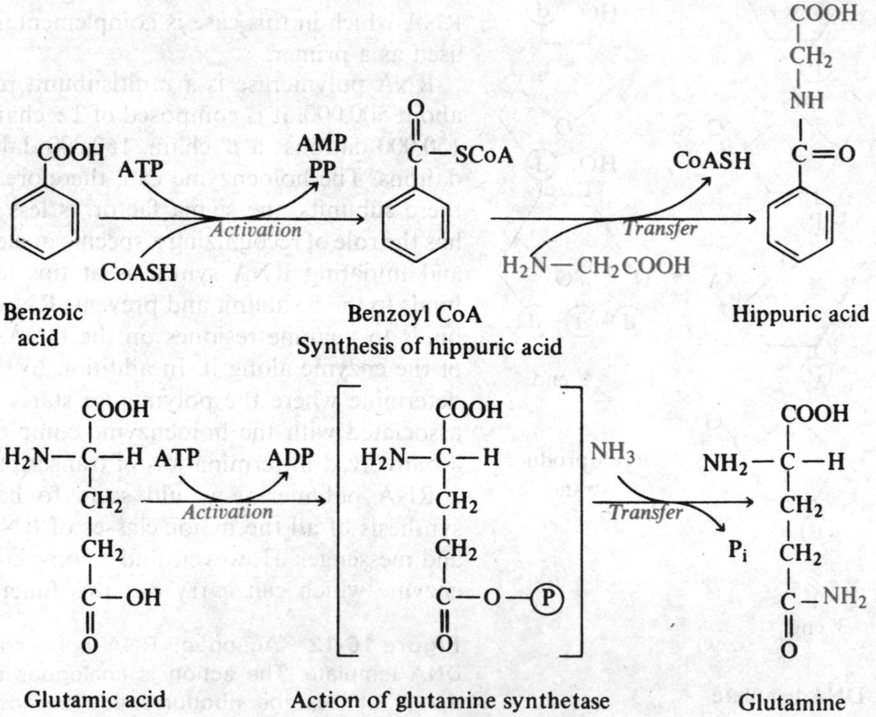

Figure 16-13 Synthesis of non-protein amide bonds. In both cases, the carboxyl group has been activated, and the activating group leaves as the amide bond is formed. The γ-glutamyl phosphate, which is an intermediate in the second reaction, remains tightly bound to the enzyme.

The reactions involved in hippuric acid and glutamine synthesis are shown in Figure 16-13. In both cases, the carboxyl group was activated and the amide then formed.

Some information about the process of protein synthesis was also learned by administering radioactive amino acids to animals and following their incorporation into various tissues. Those tissues which most rapidly incorporated radioactive amino acids were the pancreas, liver, and intestinal mucosa—all tissues which are rapidly dividing or which are producing proteins for export from the cell. When subcellular fractions were obtained from tissues (Chapter 9) after administering radioactive amino acids to an animal, a very uneven distribution of the label was found (Figure 16-14).

Figure 16-14 Incorporation of radioactive amino acids into the protein of various subcellular fractions of the pancreas. A single dose of radioactive amino acids was given to the animal at 0 time.

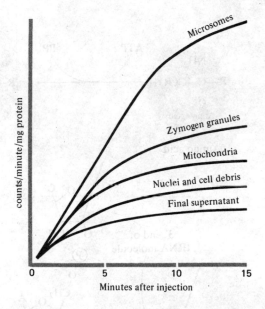

The fraction most rapidly labeled was the microsomal fraction, whereas the soluble proteins of the cell accumulated radioactive protein slowly. In general, protein synthesis was found to be associated with cells and cell fractions that were rich in RNA. It was also clear by the time these investigations were being carried out that the chemical entity responsible for the transfer of genetic information was a segment of a DNA molecule. DNA, however, did not appear to be directly involved in protein biosynthesis. In nucleated cells, protein synthesis was found to be occurring in the cytoplasm, not in the nucleus. There are also some cells without a nucleus, such as the reticulocyte, which do carry on an active synthesis of proteins. It was therefore generally accepted that the flow of genetic information was from DNA to RNA, a *transcriptional* or copying, process, and then from RNA to protein, a *translational* process.

$$\text{DNA} \xrightarrow{\text{transcription}} \text{RNA} \xrightarrow{\text{translation}} \text{Protein}$$

Amino Acid Activation

The amino acids themselves do not interact with the RNA molecules directing protein synthesis, and the first step in protein biosynthesis accomplishes both the activation of the amino acid and its attachment to a tRNA molecule. Both reactions are catalyzed by an enzyme called an *amino acyl-tRNA synthetase*. There is a different enzyme for each of the amino acids, and each is very specific. The enzyme must recognize both the amino acid and the correct "uncharged" tRNA. The reaction catalyzed is shown in Figure 16-15. The amino-acyl-AMP which is formed is not released as a product of the reaction, but remains tightly bound to the

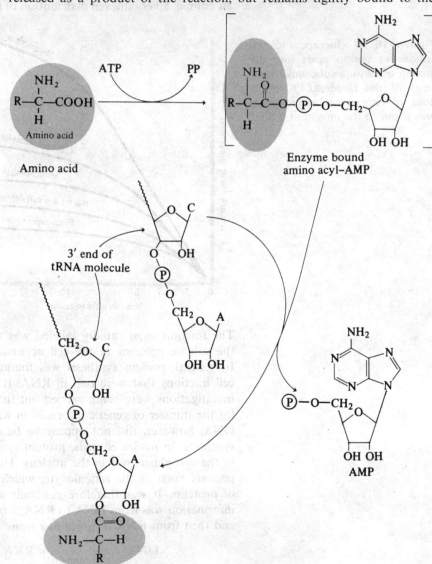

Figure 16-15   Action of the amino acyl tRNA synthetase. Both the activation of the amino acid and its transfer to the -C-C-A end of the correct tRNA are catalyzed by the same enzyme.

enzyme surface. The enzyme then interacts with a tRNA and transfers the amino acid to the free 3′-hydroxyl of the terminal adenosine at the C-C-A end of the tRNA. The bond formed is an ester, and because of the free hydroxyl on the 2′ position of the ribose, this particular type of ester is a high-energy bond. By the action of the synthetase, the amino acid has been both activated and transferred to a molecule which can recognize a specific sequence on a second nucleic acid.

Ribosomes

The actual polymerization of the activated amino acids into polypeptides occurs on the surface of the ribosome. Ribosomes are ribonucleoprotein particles present in all cells, but those from bacterial cell preparations have been most extensively characterized. The general composition and morphology of the ribosome is illustrated in Figure 16-16.

Figure 16-16 Protein and nucleic acid composition of the bacterial ribosome.

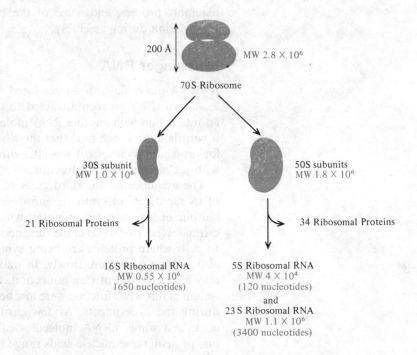

200 Å   MW 2.8 × 10⁶

70S Ribosome

30S subunit
MW 1.0 × 10⁶

50S subunits
MW 1.8 × 10⁶

21 Ribosomal Proteins

34 Ribosomal Proteins

16S Ribosomal RNA
MW 0.55 × 10⁶
1650 nucleotides)

5S Ribosomal RNA
MW 4 × 10⁴
(120 nucleotides)
and
23 S Ribosomal RNA
MW 1.1 × 10⁶
(3400 nucleotides)

Particles this size are not usually referred to in terms of a molecular weight, but rather in terms of their sedimentation coefficient expressed in Svedberg units, S. Sedimentation coefficients are calculated in reciprocal seconds and $1S = 10^{-13}$ sec. The characteristics of the ribosome vary somewhat in different organisms, and the ribosome from higher animals consists of a 40S and a 60S subunit, which together form an intact 80S particle. The 70S particle shown in Figure 16-16 is typical of a bacterial ribosome. A total of twenty-one different proteins has been isolated from the bacterial 30S subunit and characterized. Most of these occur once per subunit, but some are present in fractional amounts, indicating that there may be some heterogeneity. Some of these proteins have catalytic activity; others are required only for correct assembly of the particle. It is known that only four or five of these proteins are actually in contact with the 16S RNA. Largely through the efforts of Nomura and his group, it has been shown that the twenty-one isolated ribosomal proteins can combine spontaneously with the 16S RNA from the 30S ribosome to form a functional 30S ribosome. This indicates that no special enzymic mechanism is needed to form the ribosome, but that the interactions between the various proteins and the RNA are sufficient to result in the assumption of the correct tertiary structure.

Some success has also been achieved in reconstituting the large, more complex 50S particle. The assembly of ribosomes *in vivo* is known to differ somewhat from the process studied *in vitro*. The ribosomal RNA is synthesized as a higher molecular weight precursor which is cleaved during the assembly process, and some of the bases in the RNA are subjected to methylation during assembly.

· Messenger RNA

A tRNA molecule with an activated amino acid attached, referred to as a *charged* tRNA, is then attracted to the surface of the ribosome through an interaction with another RNA molecule, the messenger RNA. Although it was thought at one time that the ribosomal RNA served as the template for protein biosynthesis, it was later shown that this function was associated with the messenger RNA fraction.

The existence of this third class of RNA was first recognized because of its rapid rate of synthesis and degradation. In bacteria, it may have a half-life of less than two minutes, which is still a long time in relation to estimates (*ca.* 10–20 sec) of the time needed to synthesize a complete protein. In cells where proteins are being synthesized at a slower rate, mRNA is also turning over more slowly. In mammalian cells many of the mRNAs may have a half-life of a few hours or days. Early estimations of the molecular weight of this RNA fraction were low because of degradation of the molecule during the experiments. As few proteins contain less than 100 amino acids and some mRNA molecules carry the information for more than one protein, these nucleic acids range in size from 300 to 3000 nucleotides. The molecular weight of some mRNA molecules may therefore be over one million. As a heterogeneous class, the mRNA has the same ratio of $(A + U)/(G + C)$ as the $(A + T)/(G + C)$ ratio of the DNA of the organism. An individual mRNA molecule will, however, not have this property, as it represents the product of only a short segment of the chromosomal DNA which might have a base composition much different from that of the entire molecule. This is also true of the ribosomal RNA and tRNA molecules; they are synthesized by RNA polymerase using DNA as a template, but represent only a short segment of it. It is the mRNA fraction that carries a sequence of three nucleotides, called the *codon*, which can base pair to a complementary sequence on a tRNA, the *anticodon*, to direct the synthesis of proteins of specific sequence.

Peptide Bond Formation

Although there are free ribosomes and ribosomal subunits in the cytoplasm of cells, the ribosomes that are active in protein synthesis are those which are bound to a mRNA by a site on the 30S subunit to form a beadlike chain called a *polysome* (Figure 16-17). In eucaryotes, these mRNA-ribosome

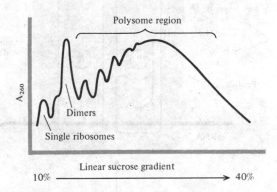

Figure 16-17 Photograph of a polysome preparation from rabbit reticulocytes. (Courtesy of A. Rich. Copyright © 1962 by the American Association for the Advancement of Science.) Most of the polysomes contain five ribosomes. The curve shows the separation of rat liver polysomes by density gradient centrifugation. The heavier polysomes have moved further down the centrifuge tubes during a given period. Polysomes containing up to seven ribosomes can be clearly resolved when the contents are removed and the amount of material in all parts of the gradient examined.

complexes are often bound to membranes. The amino acids are polymerized on the ribosome by the process diagrammatically illustrated in Figure 16-18. Each ribosome has two sites which can bind to a tRNA molecule. One site, called the P, or *peptidyl*, site binds at tRNA molecule to which is attached the amino terminal end of the growing peptide chain. A second site, the A, or *amino acyl*, site binds a tRNA which carries the next amino acid to be inserted into the growing chain. The peptide bond is formed by an attack on the activated carboxyl group of the growing peptide chain by the free amino group of the amino acid bound by its tRNA to the A site. As the new peptide bond is formed, the tRNA which previously held the growing peptide chain is released from the ribosome, and the nascent peptide is left attached to the tRNA bound to the A site. Before another amino acid can be added, the tRNA which carries the growing chain must be transferred from the A to the P site. This movement is called *translocation* and is accompanied by a movement of the mRNA relative to the ribosome. This movement shifts the mRNA by one codon so that the A site has a

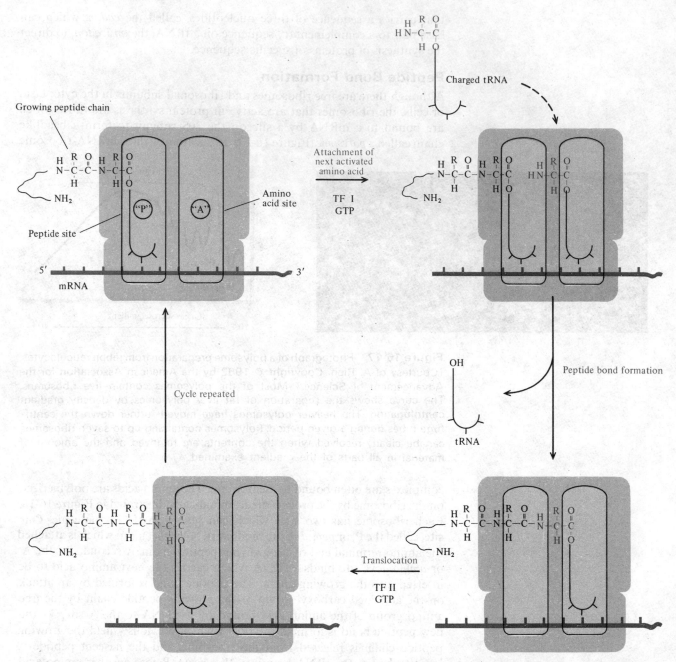

Figure 16-18 Diagrammatic representation of the steps involved in peptide synthesis. The sequential steps involved are outlined in the text.

free codon which specifies the tRNA carrying the next amino acid. It should be clear that an individual mRNA molecule can interact with more than one ribosome at a time and that there will be a series of partially completed peptides on the ribosomes forming a polysome.

In addition to the ribosomes, charged tRNAs, and mRNA molecules, other protein factors and GTP are required for peptide chain elongation. A protein called *transfer factor I* (TF I) interacts with the charged tRNA and GTP to place the tRNA at the correct site on the ribosome. Simultaneous with this attachment GTP is broken down to GDP and inorganic phosphate. The actual formation of the peptide bond is catalyzed by a protein called *peptidyl transferase*, which is one of the proteins of the 50S ribosomal subunit. A second protein, *transfer factor II* (TF II), is involved in the translocation process, and its function is also associated with the hydrolysis of GTP to GDP and inorganic phosphate.

Initiation

The initiation of a new polypeptide chain on a ribosome is a very complex reaction, and it has been described in detail only in bacterial systems. In these systems, the amino acid which is at the N-terminal position of the growing chain is always N-formyl methionine. This amino acid derivative can be inserted only at this position in the growing peptide, because it has no free amino group to participate in peptide bond formation. The *N*-formyl methionine is formed from one particular species of methionine tRNA, $tRNA^{fMet}$ by the action of an enzyme requiring N^{10}-formyltetrahydrofolate on the methionine charged $tRNA^{fMet}$ (Figure 16-19) to form $fMet-tRNA^{fMet}$. There are other species of methionine tRNA that cannot be formylated and these tRNAs are used to insert methionine into the interior of peptide chains. Peptide chains are initiated when the 30S subunit binds to a specific region of the mRNA which includes the fMet codon AUG. This binding requires a protein called initiation factor F3. This complex

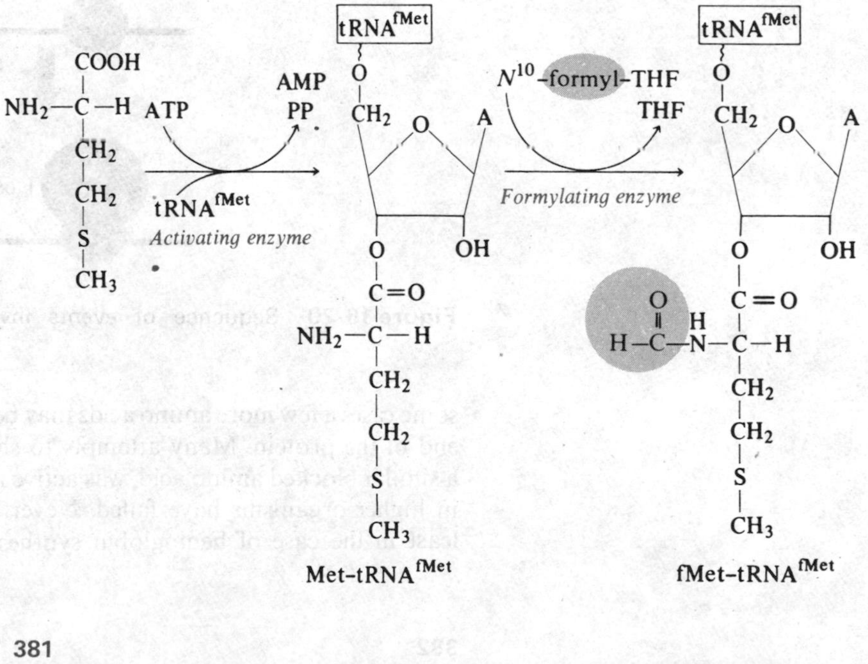

Figure 16-19 Formation of the formyl methionine-tRNA. The action of the amino acyl tRNA synthetase is followed by the formylating enzyme.

then binds to fMet-tRNA^{fMet} in a reaction that requires initiation factors F1 and F2 and GTP to form the 30S initiation complex. The 50S subunit interacts with this complex to place the charged tRNA^{fMet} at the P site on the 50S subunit. This interaction is coupled with the hydrolysis of GTP and the release of the three initiation factor proteins (Figure 16-20).

The formyl group does not remain on the terminal amino acid after the protein chain has been completed. Rather, it is cleaved off by a specific enzyme. In many cases the terminal methionine is also cleaved, and in

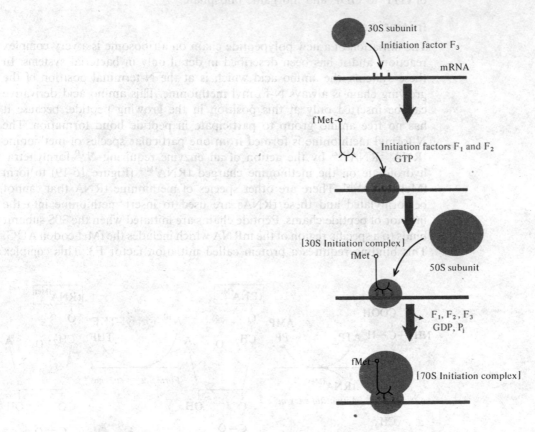

Figure 16-20 Sequence of events involved in peptide chain initiation.

some cases a few more amino acids may be removed from the amino terminal end of the protein. Many attempts to show that *N*-formyl methionine, or a similar blocked amino acid, was active in initiation of protein biosynthesis in higher organisms have failed. Nevertheless, it has been shown that, at least in the case of hemoglobin synthesis, the first amino acid added to the N-terminal end is methionine. Methionine is, however, removed from the growing chain after about twenty amino acids have been added, and is never found when the growing peptide exceeds this length.

Termination

After the final amino acid is added to the growing peptide, the protein is still attached to the ribosome by the tRNA that brought in the last amino acid. The ester bond between the carboxyl group of the carboxyl terminal amino acid and the hydroxyl group on the adenosine residue of the tRNA must be hydrolyzed. The entire system will then fall apart. Specific codons in the messenger RNA molecule signal the completion of a protein. It has been shown that the codons UAG, UAA, and UGA will signal chain termination if they arise by a mutation in the middle of a chain; the normal termination signal may be more complex. These termination sequences somehow interact with specific protein factors to cause termination, but the detailed mechanism of this action is not understood.

Finalization of Protein Structure

The messenger RNA can direct the synthesis of a linear polypeptide of a determined sequence, but the active protein may have a prosthetic group attached to it. It also has a definite three-dimensional configuration. Most proteins have disulfide bonds which link different parts of the molecule. The present view is that each polypeptide has one conformation of maximum thermodynamic stability, and that this will form without the requirement for an additional enzyme. Once formed, this tertiary structure can be stabilized by the formation of disulfide bonds from interacting sulfhydryl groups. Although this oxidation will go spontaneously, there is some evidence that an enzyme speeds up the rate of this process. There is also evidence that at least in some cases the amino terminal end of the growing protein begins to assume its final configuration before it is removed from the ribosome.

Many proteins contain a prosthetic group and these may be attached to the polypeptide chain by a number of mechanisms. Heme groups and flavin coenzymes, for example, may spontaneously interact with the correct site on the polypeptide to form the complete protein. Glycoproteins are formed in the smooth membranes of the cell by a series of uridinediphospho-nucleotide-sugar-mediated glycosylations of the polypeptide chain.

There may be extensive alterations and modifications of peptides after their synthesis. It is known that many of the proteins formed by some viruses are all synthesized as a single long polypeptide chain and later cleaved to the small proteins. It has also been shown that the protein insulin, which contains two polypeptide chains held together by disulfide bonds, is really synthesized as a single polypeptide chain, proinsulin, and a segment is removed by a hydrolytic cleavage (Figure 16-21). More recently evidence has been accumulating which would suggest that a large number

Figure 16-21 Structure of pro-insulin and insulin. The hormone is synthesized as a single chain which is later cleaved to insulin, a fragment called the C-peptide, and four free amino acids. The correct disulfide bonds are formed when the hormone is in the proinsulin form.

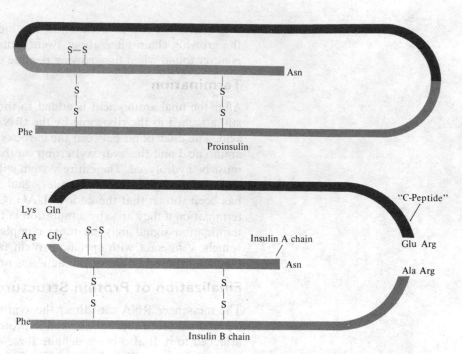

of mammalian proteins produced for export from the cell may be initially synthesized as higher molecular weight precursor forms.

IN VITRO PREPARATIONS USED

Early evidence pertaining to the mechanism of protein synthesis was obtained from a study of two different ribosomal systems: the system from *E. coli* and that obtained from the immature red blood cell or reticulocyte. Most of the information has been derived by studying the incorporation of radioactive amino acids into the material which can be precipitated by trichloroacetic or perchloric acid. For the most part, the incorporation represents the completion of peptide chains which were on the ribosomes when they were isolated. In general, the initiation process is very inefficient in these *in vitro* systems. In some systems, however, the complete synthesis of a protein has been obtained in an *in vitro* system. The ribosomes isolated from reticulocytes almost exclusively synthesize hemoglobin. A net synthesis of this protein can be demonstrated in this preparation. The net synthesis of reasonable amounts of a particular type of lysozyme coded for by a bacterial virus was also one of the early demonstrations of the adequacy of *in vitro* systems. It is now possible to show the synthesis of a number of proteins by these *in vitro* systems, and the *in vitro* translation of specific mRNA's by a ribosomal preparation from a different source is becoming a standard biochemical tool.

THE GENETIC CODE

Early in the studies of protein biosynthesis it was realized that a sequence of bases (the codon) in a mRNA molecule was acting as a template for protein synthesis by base pairing with a sequence (the anticodon) in each of the appropriate charged tRNA molecules. It was, however, some time before anything was known about the details of the process. As there are only four nucleotides, the use of two adjacent bases to form each codon would result in only 4^2 or 16 possible two-letter codons. As more than 16 amino acids are to be coded for, this is clearly not sufficient. There are 4^3 or 64 possible unique three-letter codons, and a triplet code would therefore provide sufficient information. There is no reason why a four- or five-letter code could not be used. The number of nucleotides per codon was a controversial question for some time; it was finally determined to be three by experiments which were largely genetic in nature.

Identification of Codons

The first indication that it might be possible to determine which bases were acting as a code for individual amino acids came in 1961, when Nirenberg incubated a synthetic RNA, polyuridylic acid, with a ribosomal protein-synthesizing system. He found that it resulted in the formation of a polypeptide chain that contained only phenylalanine. This observation established that UUU was a codon for phenylalanine. Similar experiments were done with the homopolymers, polyadenylic acid and polycytidylic acid, to establish that AAA and CCC were codons for lysine and proline respectively. Further information was gained by synthesizing copolymers containing two different nucleotides and determining which amino acids would be incorporated into polypeptide chains when these were used as synthetic messengers. The frequency of the different possible three-letter codons that can result from two bases will depend on the ratio of the total amounts of the two different bases in the polynucleotide. By determining which amino acids were most efficiently incorporated into polypeptides under the influence of these polynucleotides, some indication of possible codons was obtained. The results of these experiments were, however, rather ambiguous. For example, there was no way of knowing whether the code for an amino acid whose incorporation was directed by a codon containing two A's and one C, would be CAA, AAC, or ACA.

A second approach was based on the observation that complexes composed of ribosomes and natural or synthetic mRNAs will bind tRNA molecules even if protein synthesis is not proceeding. It was then shown that an oligonucleotide as short as a trinucleotide would bind to ribosomes and promote the binding of the specific tRNA for which it was a code. Eventually all the 64 possible trinucleotides were chemically synthesized and then tested in a ribosomal system to see which charged tRNA would be bound in their presence. A large number of correct codons were assigned on this basis. Certain of the trinucleotides, however, are not very efficiently bound to the ribosome; the data from these involve a great deal of uncertainty.

The method which finally yielded the most definitive data on the assignment of specific codons was that developed largely by Khorana. He prepared short, synthetic, deoxyribo-oligonucleotides and used them as a template for RNA polymerase to produce a series of polyribonucleotides of known sequence. The first systems tested were copolymers of regular repeating sequence. For example, UGUGUGUGU was tested and it was found that it directed the formation of a polypeptide containing only valine and cysteine in a repeating sequence. This was unambiguous proof that the two possible codons in this synthetic message, UGU and GUG, were the codons for valine and cysteine. Which of the two coded for each could be determined in subsequent experiments? The use of a large number of such copolymers using two, three, and even four bases of a known sequence eventually led to the determination of the codons for all the amino acids. These data confirmed much of the data from the nucleotide-binding studies and cleared up the more ambiguous cases. The information from these two methods has now resulted in the generally accepted assignment of codons shown in Table 16-3.

Table 16-3 The Genetic Code

First Position[a]	Second Position				Third Position
	U	C	A	G	
U	Phe	Ser	Tyr	Cys	U
	Phe	Ser	Tyr	Cys	C
	Leu	Ser	Term[b]	Term	A
	Leu	Ser	Term	Trp	G
C	Leu	Pro	His	Arg	U
	Leu	Pro	His	Arg	C
	Leu	Pro	Gln	Arg	A
	Leu	Pro	Gln	Arg	G
A	Ileu	Thr	Asn	Ser	U
	Ileu	Thr	Asn	Ser	C
	Ileu	Thr	Lys	Arg	A
	Meth[c]	Thr	Lys	Arg	G
G	Val	Ala	Asp	Gly	U
	Val	Ala	Asp	Gly	C
	Val	Ala	Glu	Gly	A
	Val[c]	Ala	Glu	Gly	G

[a] The first position is assumed to be the 5′ end of the triplet.
[b] Chain terminating.
[c] These codons also code for tRNA[fMet] in chain initiation.

Properties of the Genetic Code

Inspection of the code indicates that it is *degenerate*. That is, there is more than one codon for most of the amino acids. For the most part, the degeneracy is restricted to cases where the first two bases are the same and the third base differs, but there are exceptions to this generalization. This degeneracy does not necessarily mean that a corresponding degree of degeneracy of the anticodons is present on the tRNA molecules. In some cases the insertion of an amino acid into two different positions of a polypeptide is directed by different codons that specify different tRNAs for the same amino acid. However, it can also be shown that a single tRNA of known sequence can react with more than one of the codons for that amino acid. It is thought that much of this "abnormal" base pairing can be explained by the observation that the nucleotide at the 5' end of the anticodon, which pairs to the nucleotide at the 3' end of the codon, is not as restricted in its spacial conformation as the other two. It can therefore move or "wobble" to form a hydrogen bond with bases that it ordinarily would not interact with. Not all the codons for an amino acid interact with a common anticodon with the same efficiency; this might be one method of controlling the rate of peptide synthesis.

There are three codons which do not code for an amino acid, but rather for chain termination. These codons are UAA, UAG, and UGA. In some manner these codons interact with specific proteins called release factors. The details of this process are largely unknown.

When the code was completely elucidated, it was found that there was no specific codon for *N*-formyl methionine. It has now been shown that the anticodon for tRNA^{fMet} is the same as that for tRNA^{Met}. When the sequence AUG occurs at the start of a protein, fMet is incorporated, but in the interior of a chain it codes for Met. In some cases the codon GUG, which codes for valine in the interior of a chain, can code for fMet and can also be an initiation codon.

As far as is known, the code is *universal*; that is, the same nucleotide sequence codes for the same amino acid in all species that have been studied.

Mutations as Codon Alterations

Now that the genetic code is known, it is possible to look at certain mutational events in terms of what changes have actually occurred in the nucleic acids. Simple changes of one base for another can result in two different types of mutations. If the change causes the new sequence to code for a different amino acid, it is called a *missense mutation* and may or may not result in an inactive protein. A change of isoleucine, AUU, to valine, GUU, would probably have little effect on biological activity of a protein; a change from valine, GUU, to aspartic acid, GAU, would probably result in an inactive protein. If a mutation changes an amino acid codon to one of the chain termination codons, it is said to be a *nonsense mutation*. This type of mutation will result in incomplete proteins being synthesized and released from the ribosome. It has been possible in some cases to isolate these fragments and thereby to prove that this is what has happened.

In cases where the amino acid sequence of a genetically altered protein is known, it can now be shown that the amino acid substitutions (missense mutations) which have resulted are consistent with a single-letter change in a particular codon. Some examples of such changes in the protein hemoglobin are shown in Table 16-4. This type of analysis has also been used

Table 16-4 Alterations in Human Hemoglobins

| Mutant Hemoglobin | Position of Residue | | Normal Hemoglobin | | Altered Hemoglobin | | Change |
	Residue	Chain	Amino Acid	Codons	Amino Acid	Codons	
S (Sickle cell)	6	β	Glu	(GAA) (GAG)	Val	(GUA) (GUG)	A → U
M Boston	58	α	His	(CAU) (CAC)	Tyr	(UAU) (UAC)	C → U
D Punjab	121	β	Glu	(GAA) (GAG)	Gln	(CAA) (CAG)	G → C
Norfolk	57	α	Gly	(GGU) (GGC)	Asp	(GAU) (GAC)	G → A

to establish conclusively that there is a definite colinearity between the gene (DNA) and its product (the protein). One of the polypeptides which make up the enzyme *tryptophan synthetase* from *E. coli* has been sequenced. A large number of mutants having missense mutations in this gene can be obtained and these gene products have also been sequenced. It can be shown that the relative position of the various mutations, as determined by genetic mapping, correspond to the relative positions of the substituted amino acids along the polypeptide chain.

INHIBITORS OF PROTEIN AND NUCLEIC ACID SYNTHESIS

The investigation of the mechanisms of protein and nucleic acid synthesis has been aided by the availability of a number of compounds which inhibit specific reactions in these processes. Some of these are shown in Figure 16-22. The mechanism of action of puromycin is probably better understood than any of the others. The molecule is structurally very similar to the charged end of a tRNA molecule and will bind to the ribosome at the site which normally binds the incoming charged tRNA. The free amino group on the portion of the puromycin molecule, similar to a tyrosine or phenylalanine residue, can then interact to form a peptide bond with the growing peptide chain. The peptidyl puromycin derivatives which are formed will

Figure 16-22 Structures of some common inhibitors of nucleic acid and protein biosynthesis.

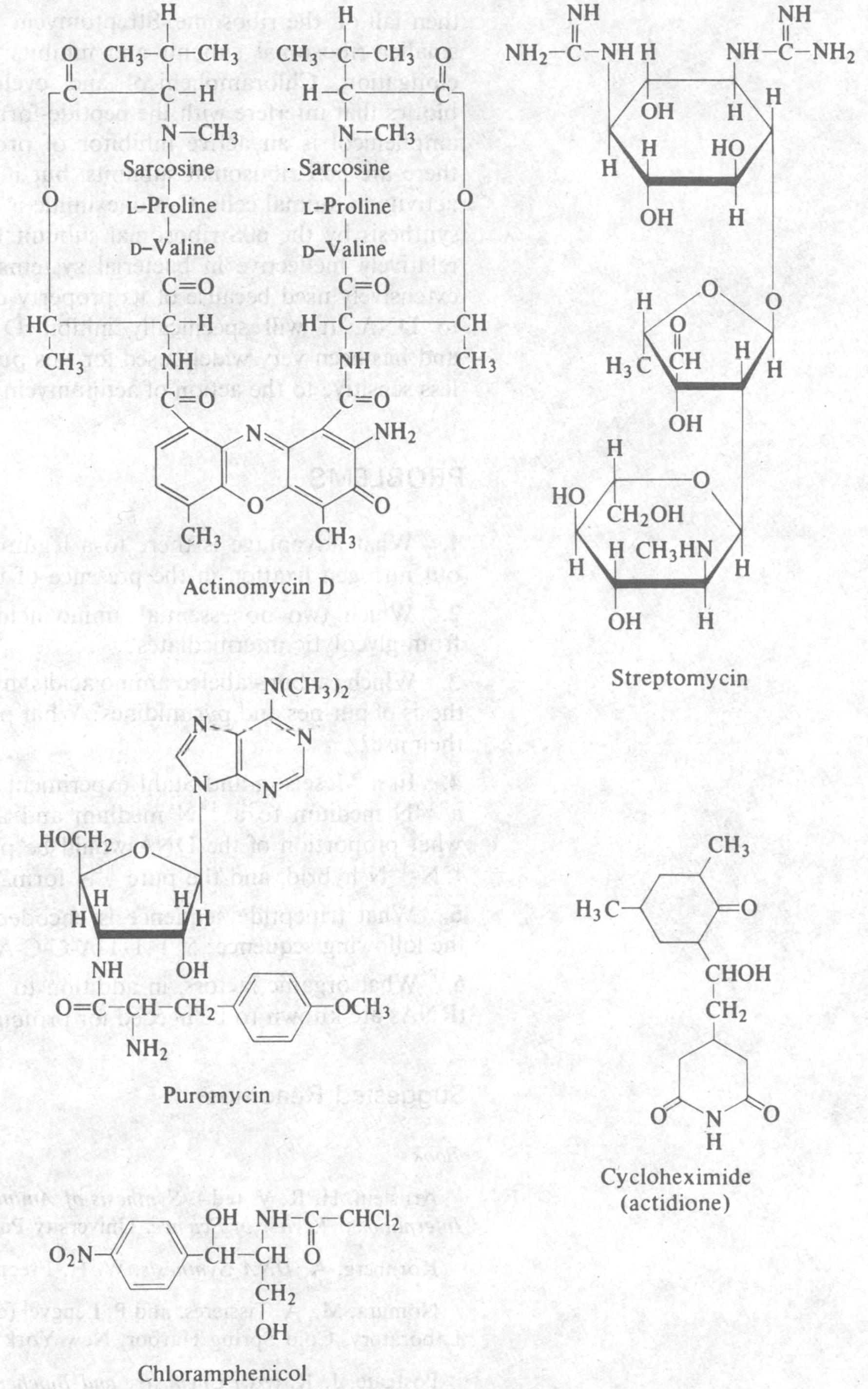

Actinomycin D

Streptomycin

Puromycin

Cycloheximide (actidione)

Chloramphenicol

then fall off the ribosome. Streptomycin is an antibiotic that binds to the smaller ribosomal subunit and inhibits both chain initiation and chain elongation. Chloramphenicol and cycloheximide (actidione) are antibiotics that interfere with the peptide-forming steps of the reaction. Chloramphenicol is an active inhibitor of protein synthesis in systems where there are 50S ribosomal subunits, but it does not have the same level of activity in animal cells. Cycloheximide is extensively used to block protein synthesis by the 60S ribosomal subunit from mammalian systems, but is relatively ineffective in bacterial systems. Actinomycin D has been very extensively used because of its property of tightly and specifically binding to DNA. It will specifically inhibit DNA-directed RNA transcription and has been very widely used for this purpose. DNA polymerase is much less sensitive to the action of actinomycin D.

PROBLEMS

1. What advantage is there to a leguminous plant if it is able to carry out nitrogen fixation in the presence of the symbiotic *Rhizobia*?

2. Which two nonessential amino acids can be formed most directly from glycolytic intermediates?

3. Which carbon-labeled amino acid(s) might be used to follow the biosynthesis of purines and pyrimidines? What problems might be encountered in their use?

4. In a Meselson-and-Stahl experiment where cells are transferred from a ^{15}N medium to a ^{14}N medium and are grown for three generations, what proportion of the DNA would be present as the pure ^{15}N form, the ^{15}N-^{14}N hybrid, and the pure ^{14}N form?

5. What tripeptide sequence is encoded by a segment of DNA having the following sequence: 5' T-T-T-A-C-C-A-G-G 3'?

6. What organic factors, in addition to ribosomes, mRNA, and charged tRNAs are known to be needed for protein synthesis?

Suggested Readings

Books

Arnstein, H. R. V. (ed.). *Synthesis of Amino Acids and Proteins,* vol. 7 of *MTP International Review of Science.* University Park Press, Baltimore (1974).

Kornberg, A. *DNA Synthesis.* W. H. Freeman, San Francisco (1974).

Nomura, M., A. Tissieres, and P. Lengyel (eds.) *Ribosomes.* Cold Spring Harbor Laboratory, Cold Spring Harbor, New York (1975).

Postgate, J. R. (ed.) *Chemistry and Biochemistry of Nitrogen Fixation.* Plenum Press, New York (1971).

Watson, J. D. *Molecular Biology of the Gene,* 3rd ed. Benjamin, Menlo Park, California (1975).

Articles and Reviews

Brenner, S., F. Jacob, and M. Meselson. An Unstable Intermediate Carrying Information from Genes to Ribosomes for Protein Synthesis. *Nature* **190**:576–581 (1961).

Crick, F. H. C. The Genetic Code III. *Sci. Amer.* **215**:55–62 (1966).

Dintzis, H. M. Assembly of the Peptide Chains of Hemoglobin. *Proc. Nat. Acad. Sci. USA* **47**:247–261 (1961).

Kornberg, A. The Synthesis of DNA. *Sci. Amer.* **219**:64–70 (1968).

Yanofsky, C. Gene Structure and Protein Structure. *Sci. Amer.* **216**:80–94 (1967).

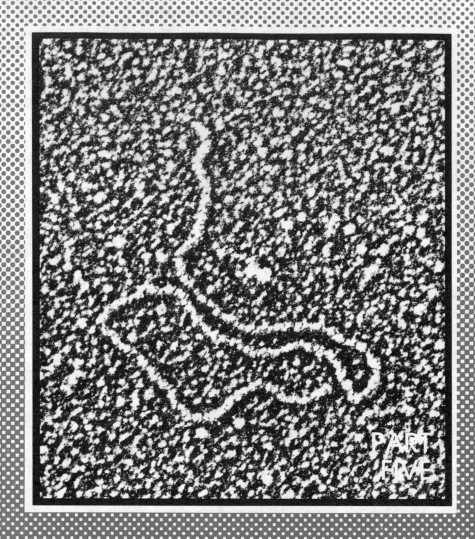

PART
FIVE

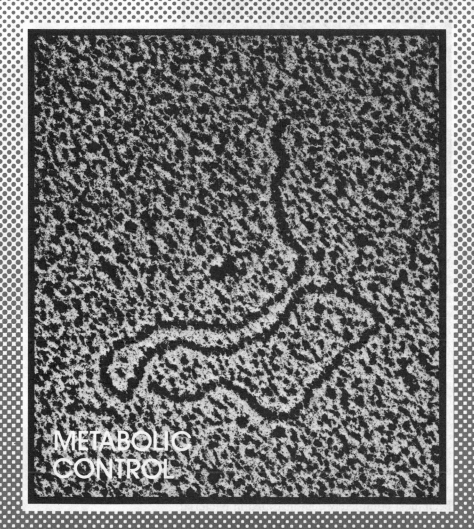

METABOLIC
CONTROL

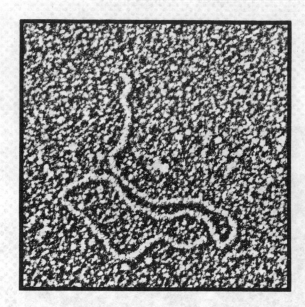

chapter seventeen

review of metabolic reactions and their control

Metabolism consists of catabolic, energy-yielding, degradative reaction pathways and anabolic, energy-requiring, synthetic reaction pathways. Many metabolic intermediates are potential substrates for a number of different enzymes, and their metabolic fate will depend on the relative activities of the various enzymes involved. Enzyme activity can be influenced by alterations in the concentrations of the substrates or cofactors for the enzyme, or by alteration in the concentration of other metabolites which serve as activators of inhibitors of the enzyme. Ultimately, enzyme activity can be regulated by altering the amount of an enzyme present in the cell. This response has been most completely demonstrated in bacterial systems through the study of inducible and repressible enzymes. The synthesis of these enzymes is controlled by regulating the amount of mRNA transcribed from the appropriate segment of DNA. Regulation of protein synthesis in mammalian systems is more difficult to study than in bacterial systems, and probably involves added complexities.

REVIEW OF METABOLISM

The various metabolic pathways discussed in the previous chapters are summarized in Figure 17-1. The reactions involved in these pathways can be divided into three general types. A series of *catabolic* reactions metabolize high molecular weight—mainly polymeric compounds—to a limited number of small molecules which are common end products of more than one pathway. These reactions are degradative and oxidative, and in these catabolic processes, high-energy phosphate compounds are synthesized and pyridine nucleotides are reduced. The second general series, the *anabolic* reactions, are synthetic and reductive. These are energy-requiring reactions, and high-energy phosphate compounds and reduced pyridine nucleotides are used to drive these reactions. As pointed out in the previous chapters, the biosynthetic pathways are in many cases completely separate from the degradative pathways. A third type of metabolic reaction series is

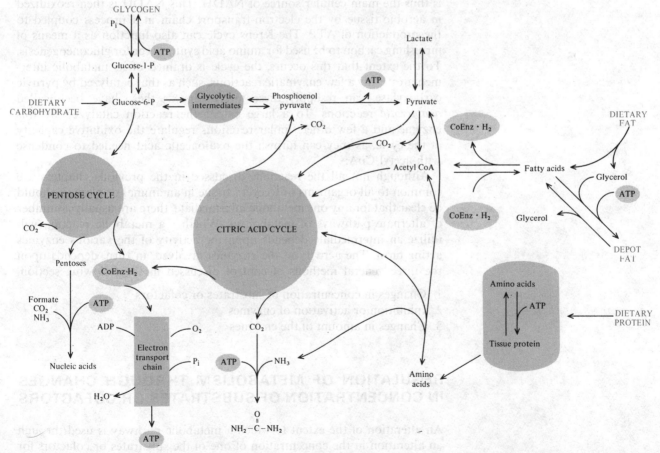

Figure 17-1 Major metabolic pathways involved in intermediary metabolism.

more properly called *amphibolic*. This reaction sequence can be used either to oxidize small molecules or to feed these small molecules into anabolic pathways. The reactions of the citric acid cycle are good examples of this function, as they both complete a catabolic sequence and feed compounds into anabolic pathways.

An inspection of Figure 17-1 indicates the key role of those metabolic pathways which have previously been discussed. The pentose cycle carries out many functions; it is a means of glucose oxidation, a means of recycling carbon in photosynthesis, a source of ribose for nucleic acid synthesis, and also a major source of reduced NADP to drive fatty acid and sterol biosynthesis. The Krebs cycle (citric acid cycle) occupies a key role in both energy production and carbon metabolism. The two major sources of energy available to animals, fatty acids and carbohydrates, feed carbon directly into it through their ultimate conversion to acetyl CoA. The degradative metabolism of amino acids results in the formation of pyruvate, acetyl CoA, or various Krebs cycle intermediates. The citric acid cycle is thus the main cellular source of NADH. This NADH is then reoxidized in aerobic tissue by the electron-transport chain in a process coupled to the production of ATP. The Krebs cycle can also function as a means of furnishing carbon to be used for amino acid synthesis or for gluconeogenesis. To the extent that this occurs, the cycle is drained of its metabolic intermediates. Only a few enzymatic reactions, such as that catalyzed by pyruvic carboxylase, can resupply these intermediates; these have been called *anaplerotic* reactions. To a large extent the reaction catalyzed by this enzyme and a few other similar reactions regulate the oxidative capacity of the cycle, as they can furnish the oxaloacetic acid needed to condense with acetyl CoA.

Although not all the reactions discussed in the previous chapters are common to all organisms or to every tissue in an animal or plant, it should be clear that for any one metabolic intermediate, there are usually a number of alternate pathways of metabolism. Whether a metabolic reaction will utilize an intermediate depends upon the activity of the various enzymes acting on it. The activity of the enzymes involved, in turn, depends upon the three general methods of control discussed in the following section.

1. Changes in concentration of substrates or cofactors
2. Inhibition or activation of enzymes
3. Changes in amount of the enzymes

REGULATION OF METABOLISM THROUGH CHANGES IN CONCENTRATION OF SUBSTRATES OR COFACTORS

An alteration of the extent to which a metabolic pathway is used through an alteration in the concentration of one of the substrates or cofactors for the reaction is one of the most direct ways of influencing metabolism.

This appears to be of particular importance in the utilization of nucleotides within cells. In a well-oxygenated tissue, the rate of respiration is regulated by the amount of ADP present to accept the high-energy phosphate formed by oxidative phosphorylation. If ATP is utilized for muscle contractions or biosynthetic reactions, the concentration of ADP is increased and the respiration rate will increase to readjust the ATP level.

The formation of muscle lactate also responds to changes in availability of reactants. When the muscle is resting, it is obtaining most of its energy from the metabolism of fatty acids or acetoacetate. If more energy is needed to support the beginning of muscular activity, glycogen is broken down to glucose and the glucose is metabolized to pyruvate. Under normal conditions the pyruvate will be oxidized to acetyl CoA, which can be further oxidized to CO_2 and water by the reactions of the citric acid cycle. The NADH formed in these reactions can be reoxidized by molecular oxygen through the electron-transport chain. During extended heavy muscle contraction, however, NADH is formed at a faster rate than it can be oxidized by the mitochondria. Under these conditions, the high concentration of cellular NADH which accumulates will begin to reduce the pyruvate rapidly being formed, and the typical high lactic acid concentration of fatigued muscle results (Figure 17-2).

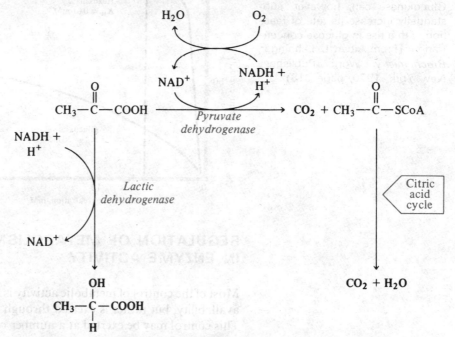

Figure 17-2 Alternate fates of muscle pyruvate. Normally, cellular NAD$^+$ is maintained in an oxidized state, and complete oxidation of pyruvate is encouraged. However, if the rate of NADH formation exceeds the capacity of the mitochondria to reduce it, pyruvate will be reduced to lactate at an increased rate.

Alterations in the cellular concentration of a cofactor or substrate will not necessarily affect the reaction rate of an enzyme. They will have an influence only if the concentrations involved are in a range that is less than, or does not greatly exceed, the K_m of the cofactor or substrate for the enzyme. A good example of this is the response of the two liver enzymes which can phosphorylate glucose: *hexokinase* which as a K_m for glucose of 0.01 mM, and *glucokinase*, which has a K_m for glucose of 20 mM. The normal concentration of glucose in the blood is about 5 mM. Hexokinase is saturated at this concentration, and an increase in the concentration of cellular glucose cannot affect the rate of glucose phosphorylation by this enzyme. On the other hand, at a cellular concentration of 5 mM, gluco-kinase would be functioning at a glucose concentration below its K_m and an increase in substrate concentration would be followed by an increase in the rate of glucose phosphorylation (Figure 17-3).

Figure 17-3 Dependence of the rate of glucose phosphorylation by the two hexose kinases on glucose concentration. Liver hexokinase is normally saturated and cannot respond to a rise in blood sugar. Glucokinase can, however, sub-stantially increase its rate of reac-tion with a rise in glucose concen-tration. (From Albert L. Lehninger, *Biochemistry*, Worth Publishers, New York, 1970, page 318)

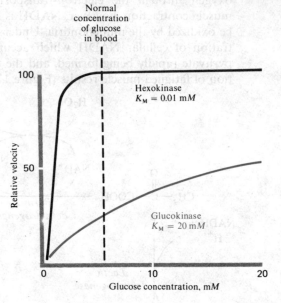

REGULATION OF METABOLISM THROUGH CHANGES IN ENZYME ACTIVITY

Most of the control of metabolic activity is not a function of simple substrate availability, but rather is exerted through an alteration of enzyme activity. This control may be exerted at a number of different levels.

Phosphorylase Activity

One example extensively studied is the regulation of phosphorylase activity in muscle (Chapter 10). Phosphorylase activity is, to a large extent, con-trolled by regulating the distribution of the total enzyme between a less active dimer (*phosphorylase b*) and a more active tetramer (*phosphorylase a*).

One of the physiological actions promoted by the release of the hormone epinephrine is an increased rate of glycogen breakdown. The hormone has no direct effect on phosphorylase itself, but, as indicated in Figure 17-4, the control proceeds through a number of steps. The less active form of

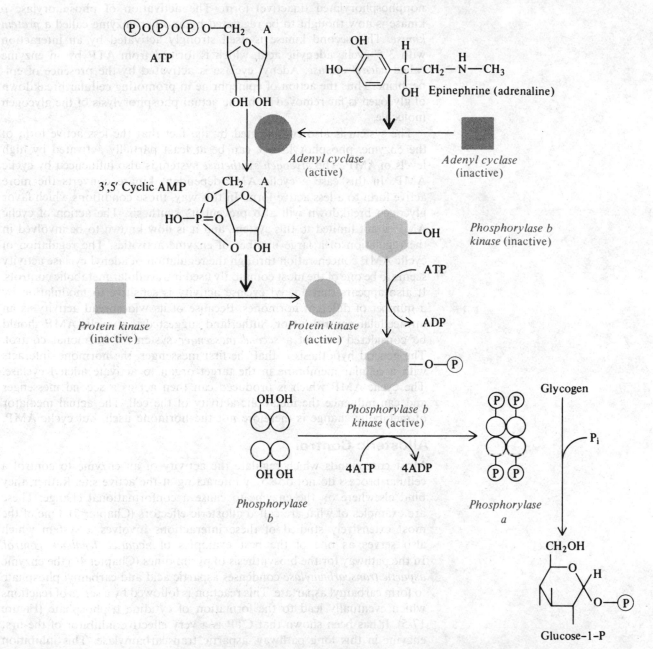

Figure 17-4 Series of enzyme activations involved in the stimulation of muscle glycogen breakdown by epinephrine.

the enzyme, phosphorylase *b*, is converted to phosphorylase *a* by a phosphorylation of hydroxyl groups of two serine residues in each of the dimers. This is an ATP-requiring reaction catalyzed by an enzyme called *phosphorylase b kinase*. This enzyme also exists in a phosphorylated (active) and nonphosphorylated (inactive) form. The activation of phosphorylase *b* kinase is now thought to be regulated by another enzyme called a *protein kinase*. This second kinase is itself strongly activated by an interaction with 3′,5′-cyclic adenylic acid, which is formed from ATP by an enzyme called *adenyl cyclase*. Adenyl cyclase is activated by the presence of epinephrine. Thus the action of epinephrine in promoting cellular breakdown of glycogen is far removed from the actual phosphorylysis of the glycogen molecule.

The system is also complicated by the fact that the less active form of the enzyme, phosphorylase *b*, can be at least partially activated by high levels of AMP. The *glycogen synthetase* system is also influenced by cyclic AMP. In this case a cyclic AMP-dependent kinase converts the more active form to a less active form. In this way, those conditions which favor glycogen breakdown will also prevent its synthesis. The action of cyclic AMP is not limited to this system, and it is now known to be involved in the regulation of a large number of enzyme activities. The regulation of cyclic AMP concentration through the regulation of adenyl cyclase activity seems to be one of the most commonly used intracellular metabolic controls. It also appears that adenyl cyclase activity is sensitive to modulation by a number of different hormones. Because of its widespread activity as an intracellular control factor, Sutherland suggested that cyclic AMP should be considered part of a *second messenger* system of hormonal control. The general hypothesis is that the first messenger, the hormone, interacts with a cellular membrane in the target organ to activate adenyl cyclase. The cyclic AMP which is produced can then act as a second messenger and can influence the metabolic activity of the cell. The actual mediator of metabolic change is therefore not the hormone itself, but cyclic AMP.

Allosteric Control

Most compounds which regulate the activity of an enzyme to control a cellular process do not do so by interacting at the active site. Rather, they bind elsewhere on the enzyme to cause a conformational change. These are examples of what are called allosteric effectors (Chapter 7). One of the most extensively studied of these interactions involves a system which also serves as one of the best examples of *negative feedback control*. In the pathway for the biosynthesis of pyrimidines (Chapter 16) the enzyme *aspartic transcarbamylase* condenses aspartic acid and carbamyl phosphate to form carbamyl aspartate. This reaction is followed by a series of reactions which eventually lead to the formation of cytidine triphosphate (Figure 17-5). It has been shown that CTP is a very effective inhibitor of the first enzyme in this long pathway, aspartic transcarbamylase. This inhibition of an enzyme that functions early in a metabolic pathway by a metabolite

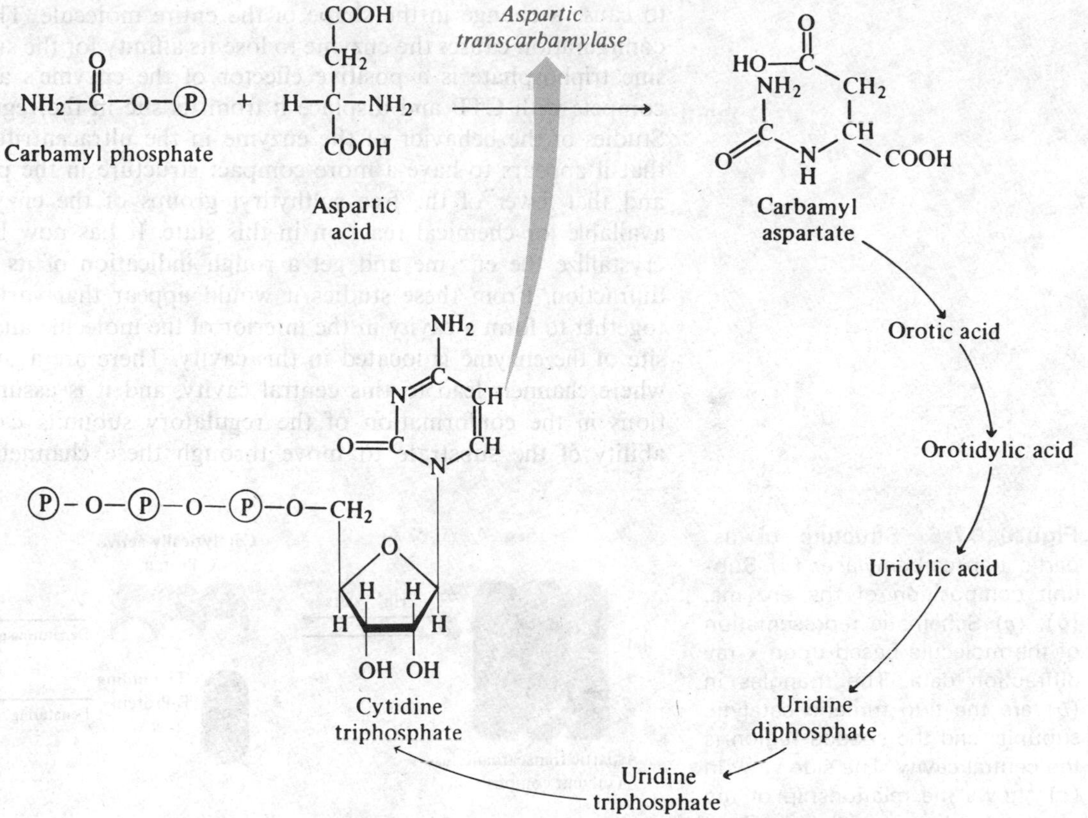

Figure 17-5 Action of cytidine triphosphate as a negative feedback inhibitor of aspartic transcarbamylase.

which is the product of one of the last reactions in the pathway is a very efficient control mechanism and has been called a *negative feedback*. Product inhibition could also decrease the rate of synthesis of carbamyl aspartate once the cell has sufficient CTP. However, since this would begin at the last enzyme in the pathway, it would mean that all the intermediates back to the first reaction in the pathway would have to build up before it becomes effective. By influencing the rate of the first reaction, a much more efficient control is achieved.

Aspartic transcarbamylase is one enzyme which can definitely be separated into a catalytic and regulatory subunit, and it has been extensively used as a model for allosteric interactions. If the native protein is treated with *p*-hydroxymercuribenzoate or other mercuricals, the enzyme can be dissociated into two types of subunits. One subunit has catalytic activity but is no longer inhibited by CTP; the other subunit, called the *regulatory subunit*, has no enzymatic activity but will bind CTP. In the intact molecule, this association of CTP with the regulatory subunit is apparently sufficient to cause a change in the shape of the entire molecule. This alteration in configuration causes the enzyme to lose its affinity for the substrate. Adenosine triphosphate is a positive effector of the enzyme's activity and will compete with CTP and displace it from its site in the regulatory subunit. Studies of the behavior of the enzyme in the ultracentrifuge have shown that it appears to have a more compact structure in the presence of CTP and that fewer of the free sulfhydryl groups of the enzyme are readily available for chemical reaction in this state. It has now been possible to crystallize the enzyme and get a rough indication of its shape by x-ray diffraction. From these studies it would appear that various subunits fit together to form a cavity in the interior of the molecule and that the active site of the enzyme is located in this cavity. There are a number of places where channels lead to this central cavity, and it is assumed that alterations in the conformation of the regulatory subunits can influence the ability of the substrate to move through these channels (Figure 17-6).

Figure 17-6 Structure of aspartic transcarbamylase. (*a*) Subunit composition of the enzyme. (*b*), (*c*) Schematic representation of the molecule based upon x-ray diffraction data. The triangles in (*b*) are the two trimeric catalytic subunits and the shaded region is the central cavity. The side view in (*c*) shows the relationship of the regulatory dimer to the catalytic subunits.

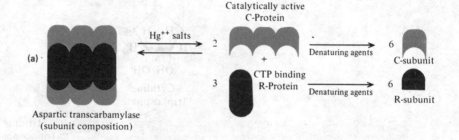

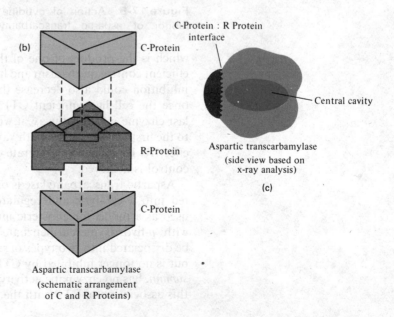

In many cases it has been observed that the change in concentration of a cellular metabolite which has a regulator function can serve a dual purpose. The same metabolite will activate a reaction proceeding in one direction and inhibit a competing reaction in the other direction. A good example of this is the metabolic regulation of a key pair of enzymes in the glycolytic pathway, *phosphofructokinase* and *fructose-1,6-diphosphatase* (Figure 17-7).

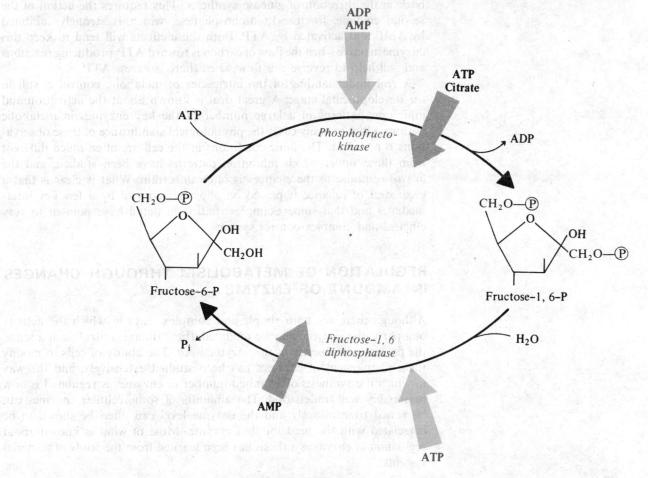

Figure 17-7   Regulation of one step in the glycolytic pathway. The direction of the flow of carbon in the glycolytic sequence can be regulated by controlling the activity of these two competing enzymes. Metabolites which cause an allosteric stimulation of the enzymes are shown as colored arrows, and those which are allosteric inhibitors of the enzymes are shown as gray arrows crossing the pathway. Note that the effects on the two enzymes reinforce one another.

Under conditions where the cell needs energy, the cellular level of ATP will be low and AMP and ADP concentrations will be increased. The latter two compounds stimulate phosphofructokinase; they therefore increase the glycolytic rate and furnish the energy needed. On the other hand, ATP, which in a sense can be considered the end product of energy metabolism, inhibits the same enzyme. High levels of citrate would be expected to accumulate when there is a rapid flow of carbon from glucose into the Krebs cycle, and citrate is also an inhibitor of phosphofructokinase. During periods when the cell has sufficient energy, it will shunt carbon back in the direction of glucose synthesis. This requires the action of the second enzyme, fructose-1,6-diphosphatase, which is strongly inhibited by AMP and activated by ATP. Both these effects will tend to keep this enzyme inactive when the flow of carbon is toward ATP-producing reactions and will help to reverse this flow when there is excess ATP.

A real understanding of the intricacies of metabolic control is still in the developmental stage. A great deal is known about the activation and inhibition patterns of a large number of the key enzymes in metabolic pathways, but in many cases the physiological significance of these observations is not clear. The ionic conditions in the cell are often much different from those under which inhibition patterns have been studied, and the *in vivo* response to the changes is often uncertain. What is clear is that a great deal of reliance is placed on allosteric control by a few key intermediates and that some examples studied in detail have pointed to very efficient and complex control systems.

REGULATION OF METABOLISM THROUGH CHANGES IN AMOUNT OF ENZYME

Although there are both simple and complex ways in which the activity of a particular enzyme may be modified, the ultimate control is, in a sense, the presence or absence of the enzyme itself. The ability of cells to modify their complement of enzymes has been studied extensively, and the way in which the synthesis of a limited number of enzymes is regulated is now reasonably well understood. The amounts of some cellular enzymes can be varied tremendously and the enzyme level can often be shown to be correlated with the need for that enzyme. Most of what is known about regulation of enzyme synthesis has been learned from the study of bacterial systems.

INDUCIBLE AND REPRESSIBLE ENZYMES

If the addition of some substance to the growth media causes an increase in the amount of an enzyme, the compound is called an *inducer* and the enzyme influenced is called an *inducible enzyme*. Inducers are usually substrates of the enzyme induced, and the enzymes induced are usually catabolic enzymes. There is, however, no correlation between how "good" an enzyme substrate a compound is, that is, a low k_M, and whether or not it is a good inducer of this enzyme. An alternate method of control is also seen. Bacteria that have the ability to synthesize all the amino acids they need from ammonia and common metabolic intermediates will often lose the enzymes of the appropriate biosynthetic pathway if one of these amino acids is added to the medium. These enzymes, which usually participate in an anabolic pathway, are called *repressible enzymes*, and the substance that turns off their synthesis is called a *corepressor*. The corepressors are often products of the enzymes whose synthesis they regulate.

This type of control of the synthesis of an enzyme could be exerted at many different points. A complex sequence of metabolic events must occur during the process by which the genetic information for the amino acid sequence of a protein is transferred into the completed enzyme (Figure 17-8). The transfer of the base sequence of a segment of DNA to a complementary sequence in a messenger RNA is called *transcription* while the large number of interactions occurring between the formation of an mRNA molecule and the final polypeptide are called *translational* events. The rate of enzyme synthesis in a cell could therefore be controlled by the rate of production of mRNA, transcriptional control, or by factors influencing the rate of translation. The latter possibility could involve a number of different events; formation of mRNA-ribosome complexes, chain initiation, rate of chain elongation, or removal of the completed peptide. Furthermore, some steps in each of these could be regulated.

The available evidence, for most of those bacterial systems that have been carefully studied, is that the production of inducible and repressible enzymes is controlled by the rate of production of the mRNA for the particular protein. It is now possible to measure the number of some specific mRNA molecules within the cell. It can be shown that the number of these molecules can be regulated by cellular *repressors*, which can combine with a site on the DNA and prevent RNA polymerase from transcribing the segment of DNA corresponding to the appropriate mRNA. The repressors are themselves proteins whose synthesis is also directed by chromosomal DNA. The genes responsible for production of repressors are called *regulatory genes*. A mutation of such a gene to cause the production of an inactive repressor is called a *constitutive mutation*. Proteins which are apparently produced continuously with no repressor control are therefore called *constitutive proteins*.

The rapidity of the change in concentration of the mRNA for the production of the enzyme β-galactosidase can be seen in Figure 17-9. When an *E. coli* cell is growing on lactose or some other β-galactoside, it has about 3000 molecules of this enzyme per cell; when it is growing on some other carbon source, it has only a few molecules. Following the addition of the inducer to the culture, the level of mRNA reaches a steady-state level following induction, and then decays rapidly as the inducer is removed. The increased rate of synthesis of the enzyme continues for only a short time after the inducer is removed, because of the rapid turnover of mRNA. However, the amount of the cellular enzyme does not drop rapidly, because

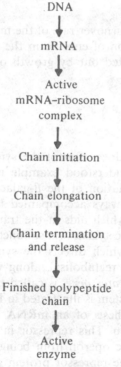

Figure 17-8 Major metabolic steps in the transfer of the genetic information for a polypeptide sequence to a completed enzyme. Control on the rate of synthesis of an enzyme could be exerted at any of these points.

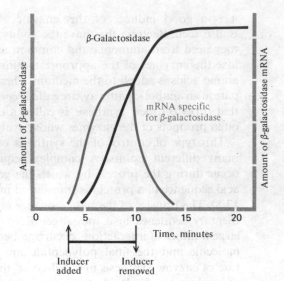

β-Galactosidase

mRNA specific
for β-galactosidase

Amount of β-galactosidase

Amount of β-galactosidase mRNA

0   5   10   15   20

Time, minutes

Inducer   Inducer
added   removed

Figure 17-9 Rapid rise and fall of β-galactosidase mRNA on the addition and removal of β-galactosidase inducers. (Reprinted by permission from James D. Watson, *Molecular Biology of the Gene*, 2nd ed., © 1970, W. A. Benjamin, Inc., Menlo Park, California)

its turnover rate is very slow compared to the turnover rate of the mRNA. In a growing bacterial culture the concentration of enzyme in the entire system will, however, fall rapidly, as it is diluted out by growth of new cells which do not contain the enzyme.

THE OPERON

One repressor can, and probably most often does, control the synthesis of more than one protein. Again, the best understood example is that elucidated by Jacob and Monod for the regulation of the β-galactoside system. The induction of β-galactosidase is always accompanied by the induction of two other proteins, a permease, which aids in the transport of the β-galactoside into the cell, and a transacetylase, whose function is not well understood. The segment of DNA which directs the synthesis of these three proteins concerned with lactose metabolism, along with a short DNA segment corresponding to the *operator gene*, is called the *lac operon*. The mechanism of regulation of this system is illustrated in Figure 17-10. The regulator gene, *i*, directs the synthesis of an mRNA which codes for the synthesis of the repressor protein. This repressor interacts with the operator gene, *o*, and prevents the lac operon from being transcribed. The interaction of the inducer with the repressor protein results in a repressor which will no longer bind to the operator gene. In this case RNA polymerase can interact with the lac operon DNA and make an mRNA from the *structural genes*; those specifying the amino acid sequence of β-galactosidase, *z*, permease, *y*, and *trans*acetylase, *a*. The efficiency of interaction of the polymerase is modified by a promotor region, *p*, on the DNA, which somehow controls binding of polymerase to DNA. In the

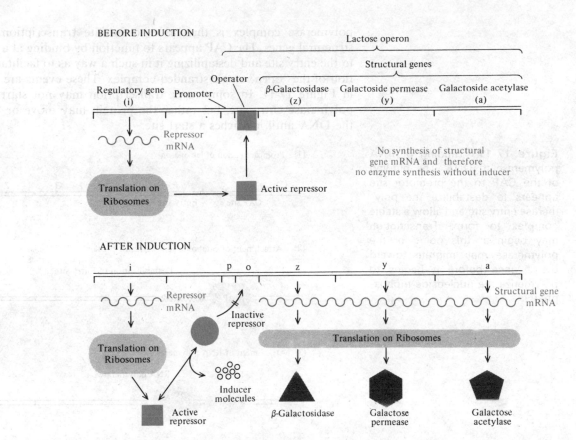

BEFORE INDUCTION

Lactose operon

Structural genes

| Regulatory gene (i) | Promoter | Operator | β-Galactosidase (z) | Galactoside permease (y) | Galactoside acetylase (a) |

Repressor mRNA

Translation on Ribosomes → Active repressor

No synthesis of structural gene mRNA and therefore no enzyme synthesis without inducer

AFTER INDUCTION

i   p o   z   y   a

Repressor mRNA

Structural gene mRNA

Translation on Ribosomes → Active repressor

Inactive repressor

Inducer molecules

Translation on Ribosomes

β-Galactosidase   Galactose permease   Galactose acetylase

Figure 17-10 Diagrammatic illustration of the effect of an inducer, in this case a β-galactoside, on the synthesis of the proteins of the lac operon.

case of the lac operon, this *p* region is prior to the *o* gene, but in other cases it could presumably be between it and the structural genes. A third protein involved in transcription of the lac operon, along with the repressor, and RNA polymerase, is a protein called the catabolic gene activator protein (CAP). It was shown that the lac operon was repressed when glucose, a more efficient source of carbon than lactose, was present. It is now known that glucose causes a decreased concentration of cyclic AMP and that cyclic AMP activates CAP, which in turn enhances the activity of RNA polymerase.

The evidence currently available allows a rather detailed interpretation of the events which must occur to initiate transcription. RNA polymerase will randomly interact with double-stranded DNA until it chances upon a site which represents a promoter region and will specifically bind to it. This is followed by an "entry" of RNA polymerase to form a complex with a short region of DNA where the helix is opened. Somehow at this stage the polymerase selects the correct strand to transcribe. This DNA-polymerase complex is then able to initiate transcription toward the structural genes. The CAP appears to function by binding at a site adjacent to the entry site and destabilizing it in such a way as to facilitate the formation of the "open" single-stranded complex. These events are diagrammed in Figure 17-11. In some cases, transcription may not start right at the polymerase entry site, but rather the protein may move or "drift" along the DNA until it reaches a start site.

Figure 17-11 Binding of RNA polymerase to DNA. The binding of the CAP to the promotor site appears to destabilize the polymerase entry site and allow a stable complex to form. Transcription may begin at this point, or the polymerase may migrate toward the *z* gene before it begins to polymerize the nucleoside triphosphates.

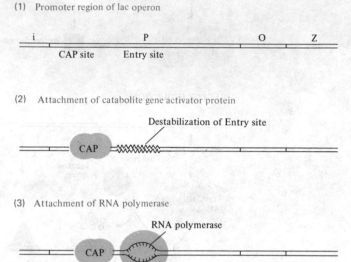

(1) Promoter region of lac operon

i   P   O   Z

CAP site   Entry site

(2) Attachment of catabolite gene activator protein

Destabilization of Entry site

CAP

(3) Attachment of RNA polymerase

RNA polymerase

CAP

Experimental Evidence for the lac Operon Concept

Much of the evidence for concept of repressor control of enzyme production has been the result of genetic investigations, and the availability of sound data is limited to a small number of systems. A great deal of validity was given to the entire concept when it was shown that it was possible actually to isolate the lac repressor protein.

Some of the best inducers of β-galactosidase are not lactose or other natural glycosides, but rather some of the sulfur analogs of these compounds. One of the best is isopropyl thiogalactoside. Gilbert and Müller-Hill obtained a mutant of *E. coli* which produced about ten times the normal amount of lac repressor, and were able to show that extracts of these cells bound radioactive isopropyl thiogalactoside. Using the binding of this compound to the protein as an assay, they were able to obtain a partially purified protein with repressor activity. They then demonstrated that this protein would bind to a DNA preparation which contained the lac operon region, and that the protein could be removed from this complex by the addition of the thiogalactoside. The repressor has been purified further and has now been shown to be a 160,000 molecular weight protein consisting of four 40,000 molecular weight subunits which have been partially sequenced. A second repressor, which is involved in phage λ infection of *E. coli*, has also been isolated and purified. These isolations have provided a real basis for the concept of the operon, but they are not the most convincing evidence.

In 1970 Beckwith and his coworkers, using a number of genetic techniques, succeeded in isolating the lac operon as a homogeneous preparation of DNA. This rather spectacular biochemical isolation could be accomplished because it was possible to incorporate the lac operon into a bacteriophage so that 5 to 10 percent of the total phage DNA consisted of lac operon. This phage DNA is normally single stranded, but again by genetic manipulation it was possible to produce a partially double-stranded phage DNA which was complementary only in the region of the lac operon. The non-base-paired region of DNA is very susceptible to nuclease action and was digested away to leave only the double-stranded region shown in Figure 17-12. The isolation of this segment of DNA by Beckwith's group represents the first isolation of a gene from chromosomal DNA. The DNA which was isolated contains the promoter region, the operator, the *z* gene, β-

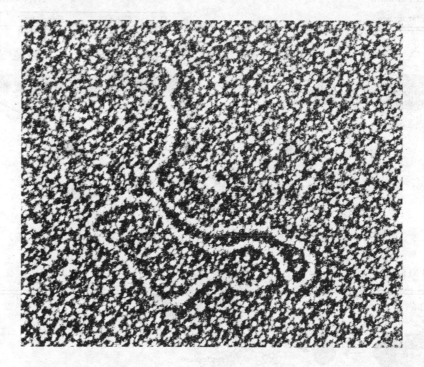

Figure 17-12 Electron micrograph of the double-stranded form of the first gene to be isolated, the majority of the lac operon. (Courtesy of J. Beckwith, *Nature*, Vol. 224, p. 798, 1969)

galactosidase, and part of the *y* gene, permease. The size of the lac operon can be calculated from a knowledge of the number of nucleotides present. Both the molecular weight of the isolated DNA, as determined by physical methods, and the length of the isolated DNA, as determined by the measurement of electron micrographs, are consistent with what the size of this segment of DNA is calculated to be.

Progress in further defining the details of the interactions involved has continued, and the base sequence in this region of the *E. coli* chromosome is now known. The sequence of about twenty-five base pairs which is protected by the repressor was first determined, and more recently the entire sequence of the *p* and *o* regions has been obtained. This was accomplished by using RNA polymerase to transcribe this region and then sequencing the RNA product. Specialized genetic techniques were used to isolate this particular segment of the total bacterial DNA. A knowledge of both the structure of the proteins interacting with these DNA segments and the sequence of bases within the segments is of course essential to a complete understanding of the interactions.

Repressible Enzyme Systems

The model outlined in Figure 17-9 is for the control of enzyme synthesis by an inducer. The situation for a repressible enzyme is outlined in Figure

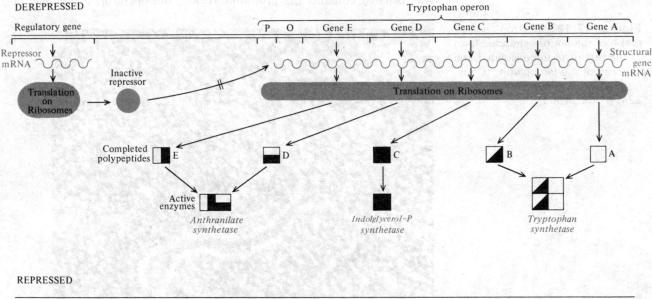

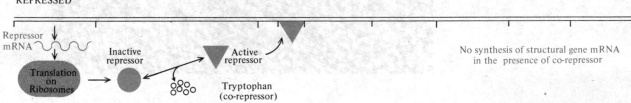

Figure 17-13 Diagrammatic illustration of the effect of a corepressor on a repressable enzyme system. The example shown is for the operon that controls the enzymes needed to synthesize tryptophan from chlorismic acid. This system is also complicated by the fact that the polypeptides produced by the structural genes are not necessarily active enzymes, but must form the appropriate complexes.

17-13. In this case it must be assumed that the form of the repressor protein which is normally produced is inactive and will not bind to the operator region unless it has interacted with its corepressor. Other than this the features of the two systems are very similar.

OTHER CONTROL FACTORS

There is no reason to believe that the synthesis of every protein is controlled by a repressor, and the synthesis of many enzymes may be merely constitutive. However, it may also be that, in many cases now thought to be examples of constitutive synthesis, the inducers or corepressors for the operon have simply not been identified. One class of proteins presumably not under repressor control are the repressors themselves. If this were the case, it would lead to the necessity for an infinite series of repressors controlling other repressors.

Assuming that there are proteins not controlled by repressors, there must still be factors which determine how fast they are produced. Some factors which could then function in regulation follow.

1. Rate of mRNA synthesis
2. Rate of ribosome attachment to mRNA
3. Rate of chain initiation and elongation
4. Rate of removal of the completed peptide from the ribosome
5. Rate of degradation of mRNA

How these various functions are controlled is largely unknown, although it does appear that the rate of mRNA synthesis is largely controlled by the efficiency with which the promoter region binds to RNA polymerase.

CONTROL OF PROTEIN SYNTHESIS
IN MAMMALIAN SYSTEMS

Almost everything that is known about control of protein synthesis through a repressor-operon mechanism has been learned from a study of a few bacterial systems. There is still some question as to how general a control mechanism this is, particularly in the case of higher animals. One difference in the protein-synthesizing system of higher animals is the relative stability of the mRNA population in differentiated cells. If a cell does not have to change its enzyme complement, it would be extremely wasteful to be able to use a messenger RNA only a few times. A good example of this situation is the immature red blood cell, or reticulocyte. This cell produces a great deal of protein, 90 percent of which is hemoglobin, after it has lost its nucleus. In this cell the mRNA is extremely stable. The same stability of mRNA is seen for those mRNAs involved in the production of plasma proteins by the liver.

411

A rather major difference in the bacterial and the mammalian systems is the type of mRNA that is formed. The mRNA in the cytoplasm seems to differ appreciably from the primary gene product. There are very high molecular weight RNA species in the nucleus, and it is usually assumed that the smaller cytoplasmic mRNAs are derived from these larger units. The mRNAs also contain a poly A region about 200 bases long at the 3' end, and this is enzymatically added in the nucleus by a process not requiring DNA. There are, therefore, more possibilities for control of gene expression than are found in bacterial cells. There would be *transcriptional control*, the rate of nuclear mRNA production; *post-transcriptional control*, RNA degradation and poly A attachment; and *translational modulation*, alterations of output of protein per mRNA used. Some of these factors are illustrated in Figure 17-14.

Figure 17-14 Possible steps in the formation of cytoplasmic mRNA molecules in mammalian cells.

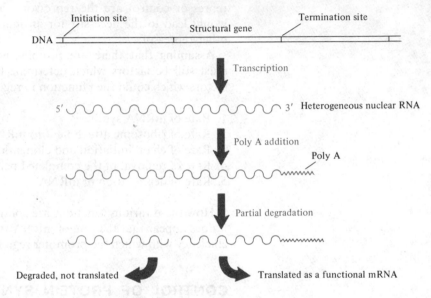

In mammalian systems it is particularly true that not all increases in the activity of an enzyme can be assumed to be "inductions" of new protein synthesis in the same sense that the term is used to apply to bacterial systems. It is possible that a high concentration of a substrate can convert an inactive form of an enzyme to an active form, or that the substrate may decrease the rate of degradation of the enzyme rather than increase its rate of synthesis. One example of such an enzyme is *tryptophan pyrrolase* whose tissue concentration is increased by hydrocortisone administration or by high levels of dietary tryptophan (Figure 17-15). It is now clear that the two agents act in completely different ways. Hydrocortisone causes an increased rate of production of an inactive apoenzyme, which is combined with its prosthetic group, hematin, in the presence of tryptophan. The holoenzyme is much more stable than the apoenzyme, so high concentrations of the amino acid lead to an apparent increase in the rate of enzyme

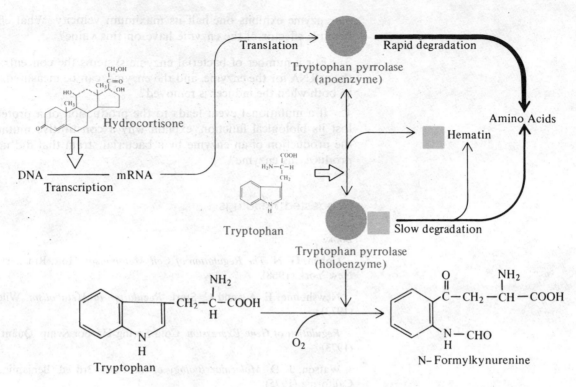

Figure 17-15 Multiple regulatory mechanisms for the control of tryptophan pyrrolase activity in rat liver. Hydrocortisone causes a stimulation of DNA transcription and tryptophan increases the activity of the enzyme by increasing the rate of production of the apoenzyme. Tryptophan causes a shift to the more stable holoenzyme and increases enzyme activity by this mechanism. The open arrows indicate that both these reactions go in the absence of the stimulating compounds, but are enhanced by their presence.

synthesis because of the decrease in degradation rate. This complex control mechanism indicates the complexity that may arise in metabolic control in higher animals.

PROBLEMS

1. A number of enzymes called transhydrogenases are able to catalyze the reaction $NADH + NADP^+ \rightarrow NAD^+ + NADPH$. If these are present and the citric acid cycle enzymes produce a large amount of NADH, why are specialized reactions which produce NADPH needed to drive biosynthetic reactions?

2. Although allosteric enzymes do not exhibit typical Michaelis-Menton kinetics, it is still possible to determine a substrate concentration at which the enzyme exhibits one-half its maximum velocity. What effect would a positive effector of the enzyme have on this value?

3. For a number of bacterial enzyme systems the concentration of both the mRNA for the enzyme, and the enzyme, can be measured. What occurs to both when the inducer is removed?

4. If a mutational event leads to the production of a protein which has lost its biological function, explain why a constitutive mutation leads to the production of an enzyme by a bacterial strain that did not previously produce this enzyme?

Suggested Readings

Books

Cohen, G. N. *The Regulation of Cell Metabolism*. Holt, Rinehart and Winston, New York (1968).

Newsholme, E. A., and C. Start. *Regulation in Metabolism*. Wiley, New York (1973).

Regulation of Gene Expression. Cold Spring Harbor Symp. Quant. Biol., vol. 38 (1973).

Watson, J. D. *Molecular Biology of the Gene*, 3rd ed. Benjamin, Menlo Park, California (1975).

Articles and Reviews

Dickson, R. C., J. Abelson, W. M. Barnes, and W. S. Reznikoff. Genetic Regulation: The *lac* Control Region. *Science* **187**:32–000 (1975).

Gilbert, W., and B. Müller-Hill. Isolation of the *lac* Repressor. *Proc. Nat. Acad. Sci. USA* **56**:1891–1898 (1966).

Jacob, F., and J. Monod. Genetic Regulatory Mechanisms in the Synthesis of Proteins. *J. Mol. Biol.* **3**:318–356 (1961).

Ptashne, M., and W. Gilbert. Genetic Repressors. *Sci. Amer.* **222**:36–44 (1970).

Tomkins, G. M. Regulation of Gene Expression in Mammalian Cells. *Harvey Lect.* **68**:37–66 (1974).

appendix

APPENDIX

answers to problems

Chapter 1

1. (a) 5.3, (b) 9.2, (c) 4.8
2. (a) 1.51×10^{-4}, (b) 2.95×10^{-7}, (c) 7.75×10^{-12} moles/1
3. 5.0, (b) 4.22, (c) 4.7, (d) 3.7
4. (a) 0.001 moles, (b) 0.01 moles
5. 8.82
6. 2.0 ml
7. 375 ml
8. Weigh out 30 mmoles of tris (hydroxymethyl) amino methane, add 20 meq of HCl from a standardized solution, and dilute to 1 liter with water.

Chapter 2

1. (a) more, (b) more, (c) less, (d) less, (e) more
2. Because the bacterial cells will have a much higher concentration of nucleic acid nitrogen which will be determined as protein in this procedure.
3. O, C, H, N
4. Analyze each for Na and Fe. The blood sample would have been about 50 percent plasma, so it would have a much higher Na content, and because it contained about 50 percent red blood cells, it would have a much higher Fe content.

Chapter 3

1.

a)

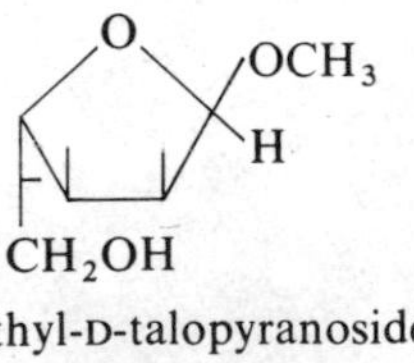

β-methyl-D-talopyranoside

b)

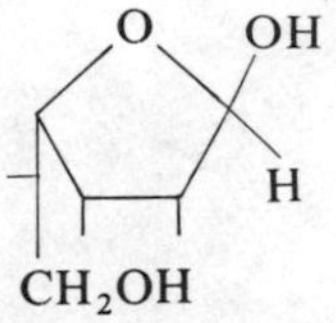

α-L-mannofuranose

c)

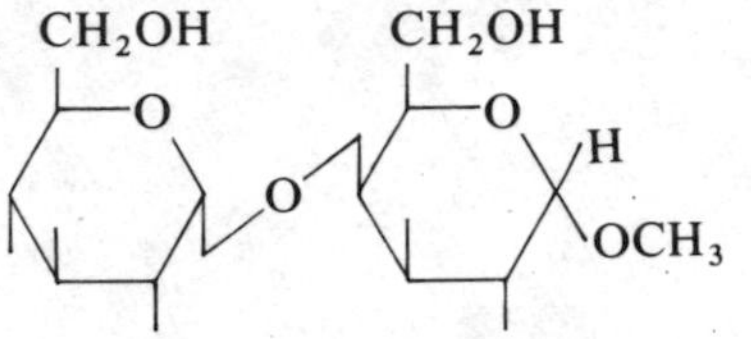

4-α-D-glucopyranosyl-α-methyl-D-galactopyranoside

d)

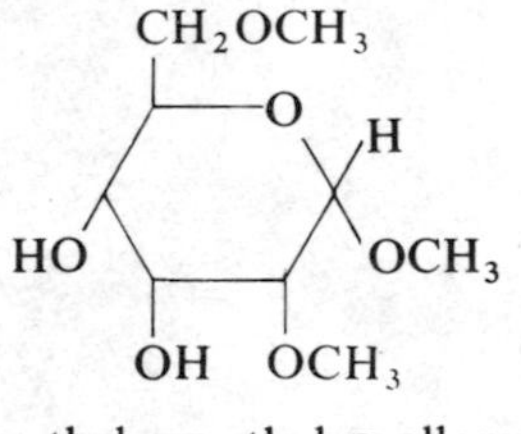

2,6-dimethyl-α-methyl- D-allopyranoside

2. D-Idose and D-gulose
3. 16.67 g/100 ml
4. The polysaccharide was composed of D-glucopyranosyl residues linked $1 \rightarrow 6$ with some $1 \rightarrow 3$ branch points.
5. 2-O-methyl-α-methyl-D-glucopyranoside

416

Chapter 4

1.

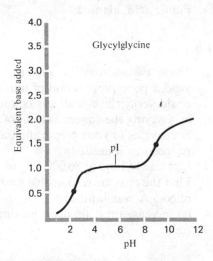

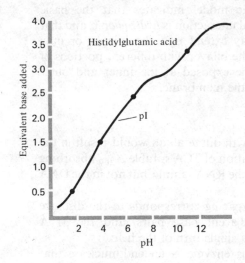

2. The peptide would split to two peptides by performic acid oxidation, and a mixture of the thiohydantoin derivatives of Ala and Gly would be found in the first cycle of an Edman degradation. It would elute from an appropriate gel filtration column after insulin.

Ala—Tyr—Cys—Phe—Glu—Asn—Arg—Gly—Ala—Lys
                    |
                    S
                    |
                    S
                    |
Gly—Ser—Cys—Trp—Met—Val—Cys—Ala—Cys—Leu
          | | |
          S   S———————S

3. Tyr-Leu-Asp-Ser-Arg
4. Glu-Leu-Asp-Ala-Ser-Thr-Arg-Arg-Lys-Asp-Glu
5. The residues Glu, Thr, Asn, and Arg would most likely be on the outside.

Chapter 5

1. Micellar particles are smaller, and their contents are in equilibrium with lipids which are in true solution.
2. Phosphatidyl glycerol, phosphatidyl inositol, cerebrosides, and gangliosides
3. Vitamin A, night blindness; vitamin D, rickets; vitamin K, hypoprothrombinemia; vitamin E, sterility and muscular dystrophy (experimental animals)
4. Proteins and phospholipids
5. Intrinsic proteins are difficult to remove from membranes without solubilizing the membrane, and when purified they have a tendency to aggregate unless they are in the presence of a detergent. Extrinsic proteins can be more easily removed and can usually be purified like soluble cellular proteins.
6. The Davson-Danielli-Robertson model as-

417

sumed that proteins were held on each surface of the lipid bilayer by *ionic* interaction with the polar head groups of the phospholipids. The fluid mosaic model indicates that the basic protein-lipid interaction is *hydrophobic*, and that proteins may extend into the bilayer or may even span the bilayer with different portions of the molecule exposed at the inner and outer surfaces of the membrane.

Chapter 6

1. Digestion with dilute alkali would result in the rapid formation of TCA soluble A_{260} absorbing material in the RNA sample, but not in the DNA sample.
2. The 3.4 Å spacing corresponds to the distance between adjacent base pairs, and the 34 Å spacing to a single turn of the helix.
3. Because this enzyme is an endonuclease that cleaves at cytosine or uracil residues to yield a pyrimidine-3-P.
4. pTpCpCpApGpC
5. The hyperchromic effect is the result of a separation of the two strands of DNA. If the solution is cooled too rapidly, complementary strands do not have time to form the correct base pairs.
6. No, the structure could be (p)ApUpGpCpA, or (p)ApCpGpUpA.

Chapter 7

1. 0.42 units/min
2. $k_M = 5 \times 10^{-4}$ and $V_{max} = 0.5$ units/min
3. (a) 26.6 μmole/liter-min, (b) 37.7
4. The initial velocities will be increased five times in each case.
5. (a) 5.4 μmoles/liter, (b) 16.2, (c) .081 percent
6. The inhibition is competitive.

Chapter 8

1. $\Delta G^0 = +58$ cal/mole, and prod/React = .18
2. -1363 cal/mole
3. $K_{eq} = 1.8 \times 10^{-7}$, $\Delta E'_0 = -0.2$ v, and $E^{0\prime}$ for X-CHO/X-COOH $= -0.12\,v$

4. (a) $= +5,540$ cal/mole, (b) $+7,380$ cal/mole
5. C, B, A, D, O_2
6. 1-propanol, 1,3-propandiol, propionamide, propionic acid, alanine

Chapter 9

1. These tissues may be sufficiently different to yield a poor preparation of mitochondria. You could verify the cleanliness of your preparation by assaying the concentrations of various marker enzymes in your preparation and modify the procedures if needed.
2. Increase to time at $80,000 \times g$ to about 100 min
3. That the enzyme responsible for the conversion of $S \rightarrow A$ was inhibited
4. (a) increase time, (b) decrease time, (c) decrease time

Chapter 10

1. salivary amylase, pancreatic amylase, maltase and isomaltase
2. Gal-1-P $\rightarrow$ Glu-1-P $\rightarrow$ Glu-6-P $\rightarrow$ F-6-P $\rightarrow$ F-1,6-P
3. ATP and $NADP^+$
4. glycolysis: glyceraldehyde-3-P-dehydrogenase; pentose phosphate pathway: glucose-6-P dehydrogenase, and 6-phosphogluconate dehydrogenase
5. glucose-1-^{14}C $\rightarrow$ $^{14}CH_3$-$CH_2OH + CO_2$
   glucose-3-^{14}C $\rightarrow$ CH_3-$CH_2OH + ^{14}CO_2$
   glucose-5-^{14}C $\rightarrow$ CH_3-$^{14}CH_2OH + CO_2$
6. Fructose-2-^{14}C and erythrose-1-^{14}C

Chapter 11

1. The butyric acid would be hydrolyzed, pass through an intestinal cell into the portal blood, and diffuse from there into a liver cell. The oleic acid would be hydrolyzed, pass into the intestinal cell, be resynthesized to a triglyceride which would pass into the lymphatic system as a chylomicron, pass into the veinous system through the thoracic duct, be hydrolyzed at the surface of a liver cell, and diffuse in.

2. In C-2 of the acetyl CoA, and C-3 of succinyl-CoA
3. Octanoate $+ ATP + 4CoASH + 3NAD^+ + 3FAD \rightarrow 4$ acetyl $CoA + 3NADH + 3H^+ + 3FADH_2 + AMP + PP$
4. Pantothenic acid, nicotinic acid, riboflavin, biotin, and vitamin B_{12}
5. In the carbonyl carbon

Chapter 12

1. Because of the high α-keto glutarate transaminase activity in many tissues, the amino group of other amino acids can be transferred to α-ketoglutarate to form glutamic acid.
2. Carbamyl phosphate and aspartic acid
3. Leucine is ketogenic and is converted to acetyl CoA and acetoacetate. Isoleucine is partially glucogenic through the production of propionyl CoA, and this requires biotin for its further metabolism.
4. Pyridoxamine phosphate, and pyruvate
5. No, because the end product of purine excretion in the human is uric acid, rather than its further metabolites.

Chapter 13

1. The lipoic acid is able to swing between the different subunits of the complex and can be alternatively reduced when it picks up the 2-carbon unit, and reoxidized by the flavoprotein containing dihydrolipoyl dehydrogenase.
2. The glutamate would be labeled in the ω-carboxyl group.
3. In both carboxyl groups. As succinate is a symmetrical compound, the label will get randomized at this step.
4. $C_{16}H_{32}O_2 + 23O_2 \rightarrow 16CO_2 + 16H_2O$
   $RQ = 0.69$
5. The glucose will be labeled because it is derived from the oxaloacetate pool which becomes labeled by the metabolism of acetyl CoA through the citric acid cycle enzymes. However, to the extent that oxaloacetate is shunted toward glucose production, it is replenished from gluconeogenic precursors, and two moles of CO_2 are still produced from the cycle for each mole of acetyl CoA metabolized.

Chapter 14

1. The label will be in carbons 3 and 4 of glucose. Even at early periods there will be some label in other carbons, and eventually the entire molecule will become uniformly labeled as part of the fixed carbon passes into the pool of intermediates needed to reform ribulose-1,5-diphosphate.
2. NADPH and O_2
3. As NADPH was a known product of the reaction, it could be postulated that some compound with a more negative redox potential than the $NADP^+/NADPH$ pair $(-0.32\ v)$ must be present to drive the formation of NADPH.
4. The equilibrium for pyruvate kinase is far in the direction of pyruvate formation, whereas the enzyme in plant tissue must function to form PEP.

Chapter 15

1. Predominantly in carbons 1 and 6; however, as the oxaloacetate, which is an intermediate in the pathway, equilibrates with the symmetrical fumarate pool, there will also be a substantial amount of label in carbons 2 and 5.
2. The synthetase system provides the possibility for a much more elaborate control of its activity than would be possible if the same enzyme had both to synthesize and to degrade glycogen.
3. Because the CO_2 incorporated into malonyl CoA is lost when it is condensed to form the acetoacetyl derivative.
4. Because it can form this compound by the reduction of dihydroxyacetone phosphate.
5.
```
      M
       \
        C—M—C—M
       /           \
      M             C—M—C—OH
                   /
                  M
```
6. Biotin, H_4folate, and S-adenosyl-methionine

Chapter 16

1. The limiting nutrient in many soils is nitrogen, and these plants could not grow at their maximum rate without the ammonia provided from this source.
2. Alanine by direct transamination of pyruvate, and serine from 3-phosphoglyceric acid.
3. Aspartic acid for pyrimidines and glycine for purines. Both of these amino acids are actively metabolized, and the label would rapidly be diluted in many systems.
4. There would be no pure ^{15}N DNA left, and would be $\frac{1}{4}$ of the ^{15}N-^{14}N hybrid, and $\frac{3}{4}$ of the pure ^{14}N DNA.
5. Pro-Gly-Lys
6. GTP; initiation factors F-1, F-2, F-3; and transfer factors I and II. It is probable that other factors will be characterized as the system is more clearly understood.

Chapter 17

1. Because the majority of the NADH generated in the cell is intramitochondrial, and the NADPH is needed for extramitochondrial biosynthetic reactions.
2. A positive effector would shift the entire v versus [S] curve so that a lower concentration of substrate would be needed to observe half maximum velocity.
3. The mRNA has a much higher rate of turnover, and its concentration rapidly decreases. The rate of turnover of the protein is probably one to two orders of magnitude slower and its concentration will fall more slowly as it is diluted out by new cellular growth.
4. Because the mutation is not in the structural gene which produces the enzyme, but is rather in a gene which now produces an inactive repressor. This allows continual production of the enzyme.

index